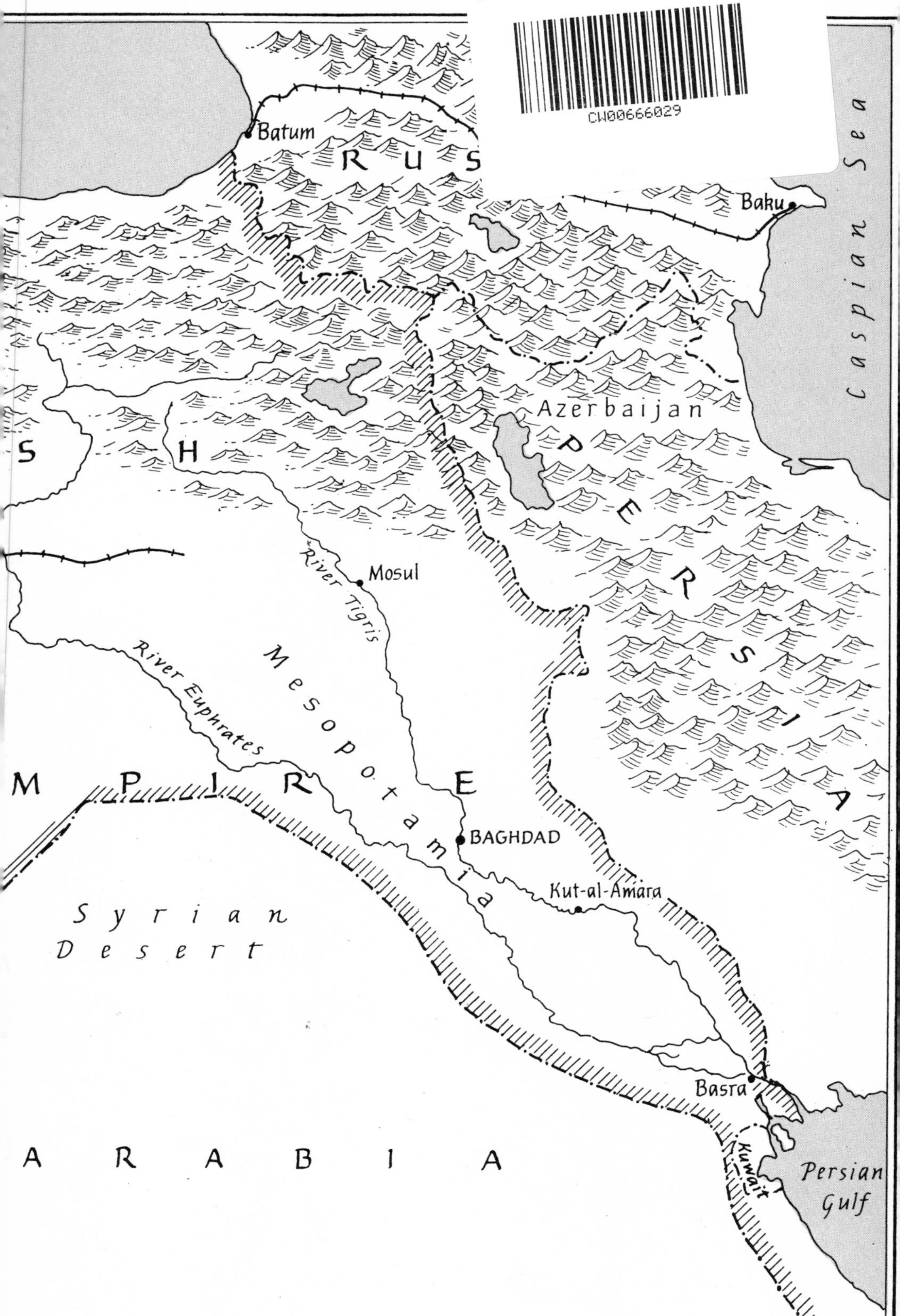
Batum
R U S
Baku
Caspian Sea
Azerbaijan
P E R S I A
S
H
M P I R E
River Tigris
Mosul
Mesopotamia
River Euphrates
BAGHDAD
Kut-al-Amara
Syrian Desert
Basra
Kuwait
Persian Gulf
A R A B I A

A HISTORY OF THE
BRITISH CAVALRY
1816 to 1919
VOLUME 5
EGYPT, PALESTINE and SYRIA
1914 to 1919

By the same author

THE CAPEL LETTERS, 1814–1817 (CAPE, 1955)
ONE-LEG (CAPE, 1961)
SERGEANT PEARMAN'S MEMOIRS (CAPE, 1968)
LITTLE HODGE (LEO COOPER, 1971)
A HISTORY OF THE BRITISH CAVALRY, 1816–1919
  VOLUME I: 1816–1850 (LEO COOPER, 1973)
  VOLUME II: 1851–1871 (LEO COOPER, 1975)
  VOLUME III: 1872–1898 (LEO COOPER, 1982)
  VOLUME IV: 1899–1913 (LEO COOPER, 1986)

A HISTORY OF THE BRITISH CAVALRY
1816 to 1919

by

THE MARQUESS OF ANGLESEY
F.S.A., F.R. HIST S.

VOLUME 5
EGYPT, PALESTINE and SYRIA
1914 to 1919

LEO COOPER
LONDON

First published in Great Britain in 1994 by
LEO COOPER
an imprint of
Pen & Sword Books Ltd,
47 Church Street, Barnsley, South Yorks S70 2AS

© The Marquess of Anglesey, 1994

A CIP record for this book is
available from the British Library
ISBN 0 85052 395 8

DEDICATED TO THE MEMORY OF THE VALIANT SOLDIERS
OF THE TURKISH ARMY WHO PUT UP SO STOUT A FIGHT
AGAINST THE HEROIC MEN OF THE AUSTRALASIAN LIGHT HORSE,
OF THE YEOMANRY AND OF THE INDIAN CAVALRY.

CONTENTS

Acknowledgements

Preface

1	Turkey enters war – first yeomen, Indians, Australians and New Zealanders arrive in Egypt – Suez Canal defences – birth of Australian Imperial Force – officers and men of Australian light horse – Australian equipment – the Imperial Camel Corps	1
2	The Turkish Army – German influence on it – types of officers and men	21
3	The Senussites – the Western Frontier Force – Wadi Senab – Wadi Majid – Halazin – Agagia	27
4	The Jifjaffa and other raids – *fanatis* for water – the Egyptian Camel Transport Corps	33
5	Sir A. Murray forms EEF – railway and pipe line begun	37
6	The return from Gallipoli – Kress advances on Katia – the yeomanry disasters at Oghratina and Katia	39
7	Anzac Mounted Division formed – extreme heat in desert – Egyptian Labour and Transport Corps – the *khamsin* – treatment of drinking water – spearpoint pumps	45
8	Food and drink – lice – flies – *gunyahs* – health: septic sores, 'Barcoo Spew', malaria, cholera – Australians abandon formal clothing – transport of sick and wounded: camel cacolets, sand-sleighs, stretchers – desert monotony	50
9	Chauvel – Cox – Ryrie – Antill – Royston – Chaytor – Romani – the pursuit to Bir el Abd	55
10	Cabinet's policy changes – Mazar – Maghara – arrival of Chetwode – Magdhaba – Rafah	74
11	The first and second battles of Gaza	90

12 Six months' stalemate starts – Chetwode succeeds Dobell – Chauvel succeeds Chetwode – Chaytor succeeds Chauvel – re-organization of mounted divisions – arrival of Barrow – expedition to destroy Asluj–Auja railway 108

13 Murray replaced by Allenby – effect of Allenby's arrival – Allenby's first review 119

14 Chetwode's plan for Third Gaza accepted by Allenby – Eastern Force abolished – Chauvel succeeds Chetwode – Chetwode commands XX Corps – Bulfin commands XXI Corps – reconnaissances, transport, water and deception preparations for Third Gaza 125

15 The enemy situation before Third Gaza – preparations and preliminaries: Third Gaza 136

16 Third Gaza: XX Corps attack – actions of Descorps – charge of 4th Australian Light Horse Brigade – capture of Beersheba 142

17 XXI Corps attacks Gaza – Newcombe's detachment surrenders – XX Corps attacks Hareira 163

18 Barrow's 'special detachment' – XX Corps takes Hareira – XXI Corps enters Gaza – action at Kh. Buteihah – Ameidat taken – pursuit starts – Jemmame taken 167

19 The charge of the Warwick and Worcester Yeomanry at Huj 173

20 Pursuit continues – Huleikat taken – enemy panics at Et Tine – action at Balin 183

21 El Mughar – Junction Station entered – Abu Shushe – Jaffa entered – end of pursuit over Philistian plain – Turkish VII and VIII Armies divided 191

22 The battles for and entry into Jerusalem 205

23 Westerners v Easterners – Supreme War Council, January, 1918 – Smuts's mission to Near East – Ghoraniye bridgehead across Jordan established – Jericho entered – the Amman Raid – re-organization of EEF on withdrawal of troops to France – Second trans-Jordan operation 213

24 Summer stalemate – Descorps stationed in Jordan Valley – horrors of life there – Abu Tullul – El Hinu 229

25 Railway improvements – Turkish Armies' situation – Allenby decides to attack in late summer – his strategy – EEF's orders and objectives – supply and communications arrangements 242

26 Elaborate means of deceiving the enemy 253

27 Battle of Sharon – arrangements for cavalry to enter gap – barrage – infantry assault 258

28 Battle of Sharon: 5th Cavalry Brigade advances – 4th and 5th Cavalry Divisions go through gap made by XXI Corps – Nahr Falik 262

29 5th Cavalry Division ordered to Nazareth – 13th Cavalry Brigade fails to capture the town or Liman von Sanders – Kelly sacked – 4th Cavalry Division crosses Musmus Pass – Howard-Vyse, 10th Cavalry Brigade, sacked 268

30 Birket el Fuleh – El Afule taken – Beisan taken – Nablus taken – Jenin taken 279

31 The 'Haifa Annexation Expedition' – Acre taken – Haifa taken 295

32 4th Cavalry Division: Makhadet Abu Naj – Masudi – Ausdiv: Semakh – Tiberias occupied – Chaytor's Force: Jisr el Damieh, Es Salt and Amman captured 301

33 Irbid – Kuneitra – Sasa – Kaukab – Kadem – Kiswe – El Mezze – Barada Gorge – El Hayira – Damascus entered – Khan Ayash 315

34 Advance on Aleppo – malaria – influenza – Haritan – end of war with Turkey 332

Epilogue 341

Appendix 1: Desert Mounted Corps Operation Order 345

Appendix 2: A note on the photograph of the charge at Beersheba 349

Abbreviations used in the footnotes and source notes 353

Source notes 360

Index 379

ILLUSTRATIONS

		Facing page
1.	The battle of Agagia, 26 February, 1916. Painting by Lady Butler (Dorchester Town Museum)	38
2.	Loaded camel transport ready to move from the rail-head. Hill, A.J. *Chauvel of the Light Horse,* 1978, 77	39
3.	Brigade headquarters office ready for the road. Powles, Lt-Col C. Guy, *The New Zealanders in Sinai and Palestine,* 1921, 11	39
4.	Starting out on a desert journey. *The Times History of the War*, X, 389	70
5.	Camels bearing water in the desert. *The Times History of the War*, XV, 165	70
6.	Two Cacolet Camels. Sitting-up cases from the battle of Romani. Powles, Lt-Col C. Guy, *The New Zealanders in Sinai and Palestine,* 1921, 40.	71
7.	'Our Water Supply.' Gullett, H.S. and Barrett, C. (eds.), *Australia in Palestine*, 1919, 127.	71
8.	Mysore and Bengal Lancers with the Bikanir Camel Corps in the Sinai Desert, 1915. (Imperial War Museum)	102
9.	Men of the 10th Australian Light Horse leaving the Suez Canal defensive line on 10 June, 1916, to destroy the supplies of water in the Wadi um Muksheib. (Imperial War Museum)	102
10.	'We capture and question a lone prisoner as to his intentions of why he is alone and get some useful information; needless to say he is a deserter.' Bostock, Henry P. *The Great Ride*, 1982, 170.	103
11.	When in action, cover for horses had to be found. The usual practice was for one man to look after four horses. Bostock, Henry P., *The Great Ride*, 1982, 340.	103
12.	A photograph purporting to be the charge at Beersheba, 31 October, 1917. (Australian War Memorial)	134
13.	The charge at El Mughar, 13 November, 1917. By J.P. Beadle, *Cavalry Journal* XXII, July. 1922.	135
14.	This artist's impression, after the capture of the Turkish guns at Huj, illustrates the tasks confronting the victors after an engagement. (Royal Gloucester Hussars Museum)	135

15. Brigadier-General Granville Ryrie. 166
Hill, A.J., *Chauvel of the Light Horse*, 1978, 61.
16. Brigadier-General William Grant. 166
Hill, A.J., *Chauvel of the Light Horse*, 1978, 61.
17. Major-General Edward Chaytor. 166
Hill, A.J., *Chauvel of the Light Horse*, 1978, 61.
18. Major-General Henry Hodgson. 166
Hill, A.J., *Chauvel of the Light Horse*, 1978, 61.
19. Chauvel a few days after Romani, August, 1916. 167
Hill, A.J., *Chauvel of the Light Horse*, 1978, 61.
20. Lieutenant-General Sir Philip Chetwode. 167
The Times History of the War, XIV, 300
21. General Liman von Sanders. 167
The Times History of the War, XVIII, 231.
22. General von Falkenhayn. 167
The Times History of the War, XV, 169.
23. General Sir Archibald Murray. 198
The Times History of the War, X, 367.
24. General Sir Edmund Allenby. 198
By David Barker. Gillett, H.S. and Barrett, C. (eds.) *Australia in Palestine*, 1919, frontispiece.
25. Djemal Pasha. 199
Livesey, A. *Great Battles of World War One*, 1989, 176.
26. Colonel Kress von Kressenstein. 199
Wavell, General Sir A., *Allenby: a Study in Greatness*, 1940, 207
27. Major-General Chaytor with Ali Bey Wahaby, Commander II Corps, 29 September, 1918. 199
(Australian War Memorial)

between pages 214 and 215

28. A regiment of Australian light horse on the march up to Jerusalem.
Hill, A.J., *Chauvel of the Light Horse*, 1978, 41.
29. A wire-netting road across the desert.
The Times History of the War, XVII, 226.
30. Water express on desert railway drawn by London and South-Western Railway engines.
Massey, W.T., *Allenby's Final Triumph*, 1920, 14.
31. Dummy horses in the Jordan Valley.
Nicol, Sgt C.G. *The Story of Two Campaigns*, 1921, 216.
32. Watering horses in the Jordan Valley.
Berrie, Lt George L. *Under Furred Hats (6th Light Horse Regiment)*, 1919, 140.
33. The camp of 'A' Squadron, 9th Australian Light Horse, near Jericho, August, 1918.
Hammerton, Sir J., *The Great War, I Was There*, XLIII, 1939.

34. Sand sleighs with wounded. 230
Gullett, H.S., *The Australian Official History*, 1923, 105.

35. Machine-gunners of the Australian light horse. 230
Gullet, H.S., *The Australian Official History*, 1923, 160.

36. Megiddo. Transport moving up across the Turkish Lines. On right, wounded coming back on camels, 12:30 pm, 19 September, 1918. 231
(Imperial War Museum)

37. Megiddo. 5th Cavalry Divisional transport crossing one of the two pontoon bridges over the River Auja, 19 September, 1918. 231
(Imperial War Museum)

38. The Inverness Battery, RHA, attached to the 3rd Brigade, going into action. 262
Bostock, Henry P., *The Great Ride*, 1987, 130.

39. A section of the Auckland Mounted Rifles Regiment. The leather buckets hanging round the horses' necks had to be put on their noses to prevent them sucking the sand for its salt. 262
Nicol, Sgt C.G., *The Story of Two Campaigns*, 1921, 88.

40. Royal New South Wales Lancers watering their horses at Edsud, 1918. 263
Vernon, P.V. (ed.) *The Royal New South Wales Lancers 1886–1960*, 1961, 101.

41. Indian lancers at Jisr Benat Yakub on the advance to Damascus. 263
(Imperial War Museum)

42. General Chauvel, escorted by a squadron of the 2nd Light Horse Regiment, as bodyguard, riding through Damascus, 2 October, 1918, followed by units representing his three cavalry divisions including men from the United Kingdom, Australia, New Zealand, India and France. 294
(Imperial War Museum)

43. Allenby, 'saluting some civilians on a balcony. . . . Mysore Lancers lining the street', probably in Damascus. 294
Scrapbook of Captain A.C. Alan Williams (Imperial War Museum).

44. The Es Salt raid, 1918. 295
By George Lambert (Australian War Memorial).

45. Charge of the 2nd Lancers at Birket el Fuleh, 20 September, 1918. 295
From the painting by T.C. Dugdale (Imperial War Museum)

46. Looking westward from Damascus can be seen the narrow Barada Gorge through which ran the Abana River, a metalled road, a single-line railway and a telegraph line. Through this gorge six railway trains and a column of 14,000 Turks tried to escape. The head of the column was caught by machine-gun fire from the hill on the left of the photograph. The Gorge was turned into a shambles. 326
(Australian War Memorial)

47. 'A' Squadron, 10th Light Horse Regiment, Tripoli, 1919. 327
Bostock, Henry P., *The Great Ride* 1982, 210.

TEXT ILLUSTRATIONS

Australian Light Horse types. 7
Gullett, H.S. and Barrett, C. (eds.) *Australia in Palestine*, 1919

An Australian Light Horseman. 10
Gullett, H.S. and Barrett, C. (eds.) *Australia in Palestine*, 1919

An Australian Light Horseman. 13
Gullett, H.S. and Barrett, C. (eds.) *Australia in Palestine*, 1919

'A Waler's Story' 20
Gullett, H.S. and Barrett, C. (eds.) *Australia in Palestine*, 1919

Charge of the 4th Australian Light Horse Brigade at Beersheba (12th ALH Regiment pictured) 157
(Australian War Memorial)

The Qatra-Maghar Position. 192
The Official History of the War, Egypt and Palestine, 1930

'COO-EE'. 212
Bostock, Henry P. *The Great Ride*, 1982

'Where did you get that bunch?' 227
Punch, 7 November, 1917

'Off again are you? . . .' 241
Punch, 28 August, 1918

'Aleppo, entrance to the Citadel'. 338
by H.M. Tulloch, Wylly, Col. H.C. *The Poona Horse II*, 1933

'The Horses Stay Behind' 343
Gullett, H.S. and Barrett, C. (eds.) *Australia in Palestine*, 1919

MAPS

	Middle East	end-papers
1.	Western Desert	29
2.	Action at Agagia, 26 February, 1916	29
3.	Eastern Desert: Port Said to Gaza	34
4.	Katia and Oghratina, 23 April, 1916	40
5.	Battle of Romani, 4–5 August, 1916	61
6.	Action at Magdhaba, 23 December, 1916	79
7.	Gaza I and II, March, April, 1917	95
8.	Palestine: Rafah to Jaffa	109
9.	Gaza III	131
10.	Gaza III, October, 1917	137
11.	Gaza III, Turkish entenchments	152
12.	Gaza III, November, 1917	169
13.	Action at Huj, 8 November, 1917	174
14.	Pursuit after Gaza III, November, 1917	186
15.	Action at El Mughar, 13 November, 1917	193
16.	Action at Abu Shushe, 15 November, 1917	201
17.	Jordan Valley, January to May, 1918	216
18.	Jordan Valley, 1918. Action at Abu Tellul, 14 July, 1918	235
19.	Action at El Hinu, 14 July, 1918	238
20.	Palestine: Jaffa to Haifa	247
21.	Haifa to Tyre and Damascus	280
22.	Action at Birket el Fuleh, 20 September, 1918	281
23.	'Haifa Annexation Expedition', 23 September, 1918	296
24.	Advance to Lake Tiberias, September, 1918. Action at Semakh, 25 September, 1918	309
25.	Advance into Syria. Action at Irbid, 26 September, 1918	318
26.	Advance to Damascus. Action at the Barada Gorge, 30 September, 1918	324
27.	Advance to Aleppo. Action at Haritan, 26 October, 1918	336

ACKNOWLEDGEMENTS

Without the wholehearted aid of successive Chief Librarians of the Ministry of Defence Whitehall Library and of their staffs, especially Miss Judithe M. Blacklaw, none of the volumes of this work could have been written. Their forbearance in the face of innumerable importunities and their swift and accurate replies to them are beyond praise. My gratitude knows no bounds. The same applies to the Imperial War Museum, particularly to Mr Roderick Suddaby, the Keeper of Documents, who has without stint given me helpful answers and made available to me numbers of unpublished documents. Other institutions which have been of the greatest assistance are the National Army Museum, the India Office Library, the Australian War Memorial and indirectly the Liddell Hart Centre.

Of the numerous individuals who have been kind enough to give me of their wisdom or to lend me documents and illustrations or both, the following stand out: Dr S.D. Badsey, Professor Brian Bond, Peter Chapman, Dr John Godwich, Dr Prys Morgan, I.D. Leask and Captain C.H. Perkins, MC. To them and others too numerous to name I offer my most sincere thanks.

The extraordinary exactness and thoroughness which characterize the editorial skills of Tom Hartman have once again ensured that my confidence in them has not been misplaced.

The persistent patience of my publisher, Leo Cooper, and his constant comforting during the very lengthy period of gestation of this and all the other volumes have been well in excess of what this always behindhand author deserves.

The dexterity of Neil Hyslop in translating my rudimentary instructions into maps of the highest clarity cannot be too warmly commended.

I find it not in the least invidious to say that the warmest of all my acknowledgements goes to Mrs Pat Brayne. The extraordinary expertise which she has invariably devoted over numbers of years to making sense of my intolerably unclear manuscripts has been equalled only by her marvellous expedition in producing immaculate typescripts from them. I am eternally beholden to her.

Finally I must confess to a guilty feeling at having exacted from my wife lengthy periods of patience. Those and her continual reassurances warrant my profoundest appreciation.

'History and more especially military history is dry, misleading stuff without a clear understanding of the character and motives of the chief actors. It is like tinned food: it lacks the vitamins necessary for health.'

WAVELL
Allenby: A Study in Greatness

'If every man of both armies wrote his experiences of one day's fighting only, it would take a great library to house the books. And yet we hear a rumour that the Government is sending out a journalist to write a "history of war". What tommy rot!'

* * *

'What with Samson and the Israelites and Philistines, Abraham and Moses and Isaac and the Aussies and New Zealanders, not to mention the Yeomanry and Coeur de Lion, a man doesn't know in whose dust he is riding.'

TROOPER ION L. IDRIESS
5th Australian Light Horse Regiment

'We've been to Pompey's pillar,
We've fished in the Canal,
If we haven't got a sunstroke,
No doubt in time we shall!
They've placed us East of Suez
Our heads are fit to burst,
And we quite agree with Kipling
That "A man can raise a thirst."

'We've felt those gentle showers,
Whose very rain is sand,
We've seen, like Joseph's brethren
The bareness of the land.
We've tried the plagues of Egypt,
We know the flies and lice,
We sympathise with Pharoah,
Who hadn't any ice.'

CAPTAIN C.G.B. MARSHAM
West Kent Yeomanry, 1915

'Altogether 80,000 British officers and troops and 210,000 Indian officers were sent from India overseas during the first six months of the war. I would here remark that the largest Indian expeditionary forces ever previously sent from India overseas amounted to 18,000 men.... Of nine British cavalry regiments seven were sent overseas.... Twenty out of thirty-nine Indian cavalry regiments ... were also sent overseas.... India supplied England in her need within the first few weeks of the war with 560 British officers of the Indian Army who could ill be spared, 70 million rounds of small-arm ammunition, 60,000 rifles, more than 550 guns of the latest pattern, together with enormous quantities of material such as tents, boots, clothing, saddlery, etc.'

LORD HARDINGE OF PENSHURST in
My Indian Years 1910–1916

PREFACE

In the Preface of Volume 3 of this work I wrote:

> 'I have decided to make the projected final volume into two. The penultimate one [Vol 4] will embrace only thirteen years, 1899 to 1913, as opposed to the present one which covers twenty-seven. The fifth and concluding volume will be devoted to the First World War and its aftermath. It will also comprise an epilogue in the shape of an extended *envoi* or denouement and a summing up of the last hundred years of cavalry and cavalrymen in Britain.'

To that programme I've been unable to adhere. The 'fifth and concluding volume' has become three: the present one and volumes 6 and 7.

Until I had begun to gauge the size of the task, I had believed that the part played by the mounted arm on the Western Front in the First World War was so small that it could easily be accommodated in a single volume together with the chief 'side shows': that, in short, it would take up only a slight proportion of the cavalry history of the whole war. I assured myself that, compared with the more obviously spectacular part played by the mounted troops in Egypt, Palestine and Syria, there was little to say. Led on by the miniscule portion allotted to it in France and Flanders by the leading historians, I supposed that it was hardly worth paying much attention to. Before long I realized that I had been misled, chiefly by Liddell Hart, but hardly less so by Edmonds, Falls, Cruttwell and to a lesser degree by Terraine and others. In what I confidently expect will be the final volume, the seventh (which will be longer than its two predecessors), I hope to show that the much despised mounted troops played, if not a vital part in the western campaigns, an important, extremely interesting and much neglected part.

When it came to considering the campaign in Mesopotamia, it became clear that the rôle played by the cavalry had also been largely overlooked by virtually all those who have written about it. The whole of volume 6, (the shortest of the three), therefore, will be entirely given over to it (with a note on the other 'side shows' in which mounted men operated, namely Italy and Salonika). To have tacked the campaign on to the present volume would have been too unwieldy. There are certain chronological disadvantages in splitting one volume into three. The war against Turkey, dealt with in this volume and the next, started after and ended before that in the west, which is treated in the final volume. Nevertheless it seemed logical to end the whole work with the main Armistice in November, 1918, though this has meant that the story of the Curragh incident (which of course preceded the opening of hostilities on the Western Front) has had to be told at the beginning of the last volume, followed by the early stages of the war in France and Flanders.

* * *

As in earlier volumes I have furnished much in the way of details of mounted actions, sometimes quite limited ones, and I have stressed officers' and men's social life, though not to the same degree as in volume 4. To politics and strategy I have devoted only what seemed essential to a proper perception of the part played by the soldiers on horseback. To the questions of horsemanship, horsemastership, and, indeed, horses themselves I have dedicated rather less space than in earlier volumes, since these have already been fully discussed, particularly in volume 4.

* * *

The present volume, though recounting the well known exploits of the yeomanry, the Australasian and the Indian mounted forces in Murray's and Allenby's campaigns, will be found to throw a few revisionary, perhaps controversial, lights on a number of accepted interpretations of the rôles played by commanders, other officers and men.

* * *

It is easy to dismiss as largely unnecessary the Near and Middle Eastern 'side shows', or rather the extent to which they escalated. Certainly the only real interests that Britain had were, in Egypt, the security of the Suez Canal and, in Mesopotamia (now Iraq), the protection of the oil fields of Persia (now Iran).

There will for ever be a keen debate as to whether, and if so, how, when and where in both theatres a halt in a secure position could have been called, thus achieving these two objectives without embarking on lengthy, expensive campaigns.* All sorts of political and strategic considerations which cannot be entered into here have been put forward against such a policy, but the most telling reason was probably the perennial difficulty in justifying a military force sitting in a purely defensive position over a number of years. Further there is a great deal to be said for the maxim that the best form of defence is attack. This does not, of course, rule out the eschewing of territorial gains, yet continual 'returning to base' is neither easy nor good for morale.

What is certain is that in both theatres the Turks rendered a signal service to Germany by containing a far greater number of allied troops than they themselves placed in the field, to say nothing of the vast quantity of matériel employed. That the Western Front would have benefited hugely had these been available in France cannot be doubted.

* * *

From August, 1914, onwards for the first time in history, systematic censorship was increasingly imposed on letters home. As an officer of the Dorset Yeomanry wrote in 1917, he was 'obliged by the censor to write the dullest possible letters, when there is so much one could say and would like to'.[1] This inhibits everyone who writes about the war in a way which did not apply in earlier conflicts. In the archives of many institutions and in the collections of private individuals are numerous letters and diaries from the various fronts. More often than not these prove, alas, to be, as a result of censorship, of 'the dullest possible' description. Consequently, readers of the earlier volumes of this work will notice how comparatively bereft of first-hand comments and anecdotes these last three are.

Equally, with a few sparkling exceptions, the regimental histories which proliferated after the war make generally dull reading. I have spent much time in trawling through these books, so often

*Wavell discusses why Britain's overwhelming naval superiority was not employed to hasten, even replace, the long 500-mile haul from the Suez Canal to Aleppo. (Wavell: *Pal.*, 15–16).

overlooked, even despised, by most historians. A thorough study of them can be rewarding. Embedded in a mass of ill-written verbiage there sometimes appear nuggets of real value to an understanding of the day-to-day life of regiments both on and off the field of battle.

* * *

As in the first four volumes I have dealt with the army's other arms only to the degree necessary for an appreciation of the cavalry's actions. There is something to be said, but not much, for Enoch Powell's idea that to try to disentangle 'the cavalry from the rest of the British and Indian armies of which it was an integral part' is 'hopeless. The history of a regiment there could be – just', he says in a review of volume 3. 'The history of an army – perhaps. But the history of an arm . . . I think not.' He does not vouchsafe a reason for this perverse opinion; indeed if there is any logic in that brilliant man's view, might it not be extended to include histories of political parties, or, say, the House of Lords, even perhaps biographies of individuals?

One of the motives that has driven me on during the compiling of this chronicle has been the desire, almost I felt the self-imposed duty, to do homage to the memory of those stalwart members of the mounted arm who throughout the war in all theatres have been so largely disregarded, not to say slighted, by those who have entered the minefield which is the history of the First World War.

To those who may doubt the justification for so extensive a study of an arm which was virtually on its death bed, I would only say that wherever an undoubted gap appears, it seems to me to be important that it should be filled for the edification of posterity. Especially is this so when the material for filling it is of intrinsic military and social interest. In the words of a German dragoon colonel, writing in 1934:

> 'Libraries are glutted with war books, and the military student can learn all that he wants, and more, of campaigns in Europe, Asia, Africa, and away in the far corners of the Wide World, to where the reverberations of the great storm extended. Yet, among all this mass of professional literature, no befitting monument to the cavalry – as well deserved as that of any other arm – appears to be forthcoming. Is it

> possible that posterity perhaps after all will come to believe that cavalry did no more in the Great War than: "Lend a tone to what otherwise would have been a vulgar brawl"?'[2]

It is my intention and my hope that future generations will be at least a little enlightened as to the part played in 'The Great War' by the army's mounted branch by reading these last three volumes of this history.

The Adjutant of the 7th Hariana Lancers wrote home from Mesopotamia in December, 1915: 'The charm of war is less apparent to anyone involved and is probably only appreciated by a historian.'[3] Before 1914 it was just possible for a historian to detect an element of charm in war; after that, in the First World War especially, he would have to be excessively insensitive to discover more than minute traces of it. The present writer has at times been overwhelmed by the deepest depression whilst researching for these volumes.

* * *

In spelling Arabic place names I have not taken the trouble to be scholarly. Since, I hope without exception, those mentioned in the text and notes appear on the maps, their identities ought to be clear enough.

* * *

Mr Ian J. Crow has kindly pointed out to me that there are a number of inaccuracies in my table of regiments' stations from 1872–1898 on pp. 430–1 of volume 3. The most important of these concern three regiments which were not at home as shown but in India or South Africa during the following periods:

3 DG India 1884–92; South Africa 1892–95
7 DG India 1884–94
12 L India 1877–87

Lieutenant-Colonel E.T. Lummis has obliged me by showing that on p. 197 of volume 4 I have inadvertently stated that the Suffolk Regiment was at Zilikat's Nek, 11 July, 1900. In fact it was the 2nd Lincolns.

Major J.D. Harris has been good enough to point out convincingly that on p. 266 of volume 4 I wrongly stated that Lieutenant W.J. English won his VC at Vlakfontein, near Naauwport on 29 May, 1901, whereas in fact it was on 3 July at Vlakfontein, near Machadodorp.

1

'All roads led to Victoria Barracks. From city, village, farm and farthest-out station they came during the dying months of 1914; a never-ending stream of men borne thither by the variable tide of their own reasons . . . . The "old soldier" came back. There was the man who sensed change, excitement and travel; the gambler who decided on the spin of a coin; and there was he who felt instinctively that man's greatest job was at hand.'

The Historian of the 6th Australian
Light Horse Regiment

'The Banners of England unfurled across the sea,
Floating out upon the wind, were beckoning to me.
Storm-rent and battle-torn, smoke-stained and grey:
The Banners of England – and how could I stay!'

CORPORAL J.D. BURNS (An Australian light
horseman killed at Gallipoli)

[The light horseman] 'combines with a splendid physique a restless activity of mind. This mental quality renders him somewhat impatient of rigid and formal discipline, but it confers upon him the gift of adaptability and this is the secret of much of his success, mounted or on foot.'

ALLENBY in 1918

'They are not soldiers at all; they are madmen.'

A German officer prisoner on the Australian
light horsemen in Palestine, 1917

'Fine fellows these Aussies. They made light of everything, and grander troops to fight with could not be found the world over.'

S.F. HATTON, Middlesex Yeomanry

'Sentry: "Halt! Who goes there?" Voice from the dark: "****!! ****!!" Sentry: "Pass Australian." '

H. ESSAME in *The Battle for Europe, 1918*[1]

Turkey enters war – first yeomen, Indians, Australians and New Zealanders arrive in Egypt – Suez Canal defences –

Britain declares war on Turkey, 5 November, 1914

birth of Australian Imperial Force – officers and men of Australian light horse – Australasian equipment – the Imperial Camel Corps

With the intrigues and the inducements with which over a lengthy period Germany persuaded the Turks to enter the conflict on her side, completed before the first shot was fired in Europe, fascinating though they are, this work is not concerned. The initial German setbacks in Belgium and France and the need to prepare for the coming conflict delayed the implementation of the Turkish commitment so that it was not until 1 November that Sir Louis Mallet's forlorn but skilful and patient exertions at Constantinople came to an end. Five days later Britain declared war on Turkey. Among the immediate menaces which had now to be faced were the cutting of the shortest route to Britain's Russian ally through the Dardanelles and the loss of the Anglo-Persian oil fields. Longer-term and pre-eminently vital was the threat to the connection with India and Australasia through the Suez Canal, in German eyes Britain's 'jugular vein'[2] and in Sir Henry Wilson's 'the Clapham Junction of Imperial Communications'.[3]

The garrison of Egypt,* which included the 3rd Dragoon Guards, its sole regular cavalry regiment, and a battery of Royal Horse Artillery, was sent home in late September, destined for France. Despatched to take its place were two yeomanry regiments, accompanied by their horses, the Herts Yeomanry and the 2nd County of London Yeomanry (which on 19 January, 1915, were formed into the Yeomanry Mounted Brigade) and a Territorial infantry division with 'A' Squadron of the Duke of Lancaster's Own Yeomanry acting as Divisional Cavalry. From India between October and December came two Indian infantry divisions, the Bikanir Camel Corps and the Imperial Service Cavalry Brigade made up of the Hyderabad, Mysore and Patiala Lancers, together with small detachments from several other Indian States.† In the

*In December the country was officially brought under British protection as it had been unofficially since 1882.

†The Imperial Service Troops evolved in the 1880s. They consisted of units of the Princes' small armies which were earmarked for the defence of the Indian Empire. They were trained to a standard which would make them efficient to stand in the battle-line with regular Indian troops. There were units of infantry, engineers and transport as well as of cavalry and the Bikanir Camel Corps. All were officered by Indian 'gentlemen' and financed by the

(continued over)

hands of these troops for some months the defence of the Canal chiefly rested. The infantry dug and manned entrenched posts on both sides of the waterway, while horse and camel patrols frequently rode some miles into the unoccupied Sinai Desert to the east of the Canal. However, the likelihood of a Turkish invasion was not at first considered very highly by the British command.

* * *

By the end of the year the War Office had decreed that Egypt should be both a training centre for reserves from most parts of the Empire and a base for the various theatres of war in the Near and Middle East. On 23 November Kitchener decided to bring the Australian and New Zealand contingents to Egypt for training, intending them for France.* Between December, 1914, and April, 1915, the 2nd Mounted Division (arrived from home), and before it the Australian Mounted Division and the Australian and New Zealand Mounted Division together with the Australian Infantry Division.

* * *

The alacrity with which the Governments of the Dominions offered their aid to the Home Government is well known. Less well known are the nature and extent of their armed forces in 1914. When, for example, New Zealand, with a population of 1,100,000, offered 8,000 soldiers as well as her small naval force, it was decided that to constitute her expeditionary force each of the regiments and battalions of her small volunteer army should provide a squadron or a company.

The Australian situation was more complex. There was a small 'citizen army', about a brigade in size, for home service. Beyond that there were various militia units. In 1910, however, Kitchener had advised the Government to adopt, which they promptly did, a

---

States. Training was supervised by officers from the regular Indian Army. On the declaration of war the Princes offered the whole of their resources to the King Emperor. Some 18,000 out of the 22,000 of all ranks served overseas. The Jodhpore Lancers which had served in France was brought to Palestine for the 1918 operations.

*They would have gone straight to England had camp accommodation been available. (Bean, 111–2; see also *Chauvel*, 45–6).

totally new army system. The old militia arrangements were given up and instead the country was divided into 224 training areas. In each of these under the Compulsory Service Act of 1911 boys from the age of twelve were compulsorily trained as cadets. At eighteen they passed into the 'active' units of what were now called the Commonwealth Military Forces for short annual periods of training up to the age of twenty-five. By 1914 the number had risen from the 23,000 of militia days to 45,000 under the compulsory scheme. But in 1914 that scheme was still five years from its intended full development and since from 1911 onwards the only new blood allowed had been the drafts of eighteen-year-old trainees, no one from the old militia being permitted to join except officers and non-commissioned officers, the other ranks of the army consisted almost entirely of men from nineteen to twenty-one. It would hardly do to send away an army of boys. Consequently it was determined to raise an altogether separate force. It was christened the Australian Imperial Force (AIF) and as such it fought throughout the war.

The initial offer promised 20,000 men embarking within four to six weeks. Beside an infantry division, there was to be raised a light horse brigade of 2,226 men and 2,315 horses.[4] In fact so great was the rush to the colours, especially for the light horse, that a second mounted brigade was offered on 3 September and, since many country-bred men fit for mounted work were enlisting in the other arms because the first two brigades were full, a third was formed a month later. These three brigades, each consisting of three regiments, lacked any horse artillery or field engineer support, but had attached to them signal troops – without, surprisingly, any field telephones – field ambulances and brigade trains.[5] One regiment, the 4th Light Horse, formed at first the infantry divisional cavalry and in due course four further regiments came into being. These were often employed as detached units, although later in the campaign they formed part of the 4th and 5th Light Horse Brigades.

* * *

Something like half the Australian light horsemen were aged between twenty and twenty-five. About a quarter were civilians who had no previous military experience and of the older men most had served in the South African War, including every one of the original commanding officers, as had also numbers of the junior officers. Eleven of the 5th Regiment's, for instance, had done so.* Most of the other older officers had served in the militia.[6] The regiments were recruited pretty strictly on a local and territorial basis. Throughout the war reinforcements almost always came from each regiment's home States. Queensland furnished a considerable proportion of the whole, the men of the cattle-stations being peculiarly suited for mounted work. Two of the regiments were exclusively Queenslanders. The two largest States, New South Wales and Victoria, were the recruiting areas of five of the regiments. South Australia and Tasmania provided the men of one, South Australia and Victoria another and West Australia another. Only a small number of men came from the big towns compared with those who joined the infantry. Squatters, stockmen, shearers, dairymen, small cultivators, timber-getters, prospectors, farmers and labourers were the chief types represented. In New South Wales alone 164 students of the State Agricultural College and 140 policemen enlisted. Bank clerks and more surprisingly clergymen (not as chaplains) also joined. Some eleven officers of the Commonwealth Military Forces for whom no officer vacancies were available in the AIF threw up their commissions to join as privates.[7]

There were extraordinary instances of determination to enlist. One man rode 460 miles and then took a train which covered an even greater distance so as to join in Adelaide. Finding there that the ranks were full he sailed to Hobart, finally enlisting in Sydney. Another who was accepted there had been refused four times in Melbourne. Many who were refused at home took ship for England to enlist in British regiments.[8] There were numbers living on remote outback stations who were exempt from compulsory military training and who had never seen a military uniform in their lives. 97% were of pure British stock and they were 'the children of the most restless, adventurous and virile individuals of that stock'.[9] The men who composed the later contingents were in many cases those who had some business or farm to dispose of before they were able to enlist. They were naturally less bold and reckless than

*The 5th Regiment was raised from men of the Upper Clarence Light Horse (later the Northern River Lancers) and from the Queensland Mounted Infantry which had served in South Africa. (Chauvel's Foreword to Wilson, 9). (See Vol. 4, 174 and 202).

their comrades of the first contingent who for the most part were 'the romantic, quixotic, adventurous flotsam that eddied on the surface of the Australian people'.[10]

In view of the very large number of applicants, the medical inspection could afford to be, and for the first contingent always was, exceedingly severe. One medical officer reported that many men had 'thrown up good jobs and travelled hundreds of miles. They have been feted as heroes before leaving, and would rather die than go back rejected. Some I have to refuse and they plead with me and almost break down – in fact some do go away, poor chaps, gulping down their feelings and with tears of disappointment in their eyes.' By early 1918 conscription in Australia was failing to provide the necessary flow of reinforcements. By then medical standards fell of necessity.[11]

* * *

The Australian light horsemen differed enormously from their British equivalents. There was hardly one of them who was not a horseman of some degree of excellence. From boyhood onwards 'the breaking and backing of bush-bred colts and the riding of any horse that came their way'[12] were generally matters of necessity. There were exceptions, though. One youth failed thrice to mount his horse when being tested as a recruit for the 10th Regiment. 'My boy,' asked his officer kindly, 'what made you think you could ride?' 'I don't, sir,' he replied. 'Have you ever been on a horse in your life?' 'No, sir.' 'Well, what the hell did you come here for?' 'I didn't say I could ride, but I don't mind having a try and I want to join the regiment.' 'That's the way to talk,' said the officer. 'Go away for a week and practice and I'll give you another chance.' In a week's time he passed![13]

The young Australian countryman's sense of direction and distance and his expertise in judging country and generally as an observer came from constantly riding great distances along country tracks by day and by night. Most of them were good shots with breech-loading shotguns, though the rifle was virtually unknown to them, there being no big game in Australia. It is probably true that the light horseman was at least as good a shot as any other combatant in the world. Wild, unaimed shots were not for him. 'To waste his effort and ammunition in a fight was to his eyes an offence against his personal intelligence.'[14] Very high summer

Australian light horsemen types

temperatures, dust and meagre water supplies such as he was to meet in Sinai and Palestine were not new to him. They were an integral part of his home life.

He was not in the conventional military sense a good horsemaster. At home he would hardly ever walk when a horse was available. If he overworked one so that it lost condition he could always bring another into use. Grazing areas were usually unlimited, cold weather was a rarity. Cherishing and fostering of his horses were therefore far less important than in the colder countries of the north. The light horseman had to learn that there was value in constant grooming, in regular feeding and watering and the balanced packing of saddles. It did not take him long. Many recruits brought their own horses with them. Most of them were fine specimens which were bought by the remount department and then issued to their former owners. As one commanding officer put it, there were a few men, like the 10th Regiment's non-riding recruit, whose inadequacies as horsemen were made manifest at the first mounted parades. Some of these had 'joined the light horse because they were "too tired" to enlist in the infantry'.[15]

When General Sir Ian Hamilton as Inspector-General of Overseas Forces visited Australia early in 1914 he reported the light horsemen as being 'the pick of the bunch .... They are real thrusters who would be held up by no obstacle of ground, timber or water, from getting in at the enemy.' He thought them less disciplined than the British Yeomanry but showing more go and initiative. 'Take them as they are,' he wrote, 'they are a most formidable body of troops who would shape very rapidly under service conditions.'[16] Two years later General Sir Archibald Murray, Commander-in-Chief in Egypt, wrote to Sir William Robertson, then CIGS: 'The Australians are the finest body of men I have ever seen.'[17] Self-discipline as opposed to corporate discipline was a strong element in the light horseman's makeup, but under his own officers he very speedily learned those special disciplines necessary for efficiency in the field.

> 'When away from his own officers, he was somewhat indifferent to the rigid rules of saluting; and this attitude [as the Australian Official Historian puts it] together with the disdain with which he regarded all army formality and etiquette which did not, to his rational mind, have some direct bearing on his work as a fighting soldier, produced much embarrassment, and at times even strained relations between the light horse commands and the British General Staff in Egypt and Palestine.'*

Typical of his lack of conventional respect for his seniors was a story current in August, 1917. Allenby, during an inspection of light horsemen, asked one of them, 'Well, are we going to give these Turks a hiding?' To which the reply was, 'I'm — if I know; that's your — business!' The Commander-in-Chief is said to have merely smiled.

The light horseman's capacity for looking after 'number one' was infinitely greater than that of the average Tommy. In Sinai he

*The senior Australian chaplain in Palestine described one aspect of the light horseman when he called him, with, perhaps, some exaggeration, 'a man who, while denying he is a Christian, practises all the Christian virtues.' (Gullett, 552).

never lost an opportunity of acquiring extra water bottles and small sacks for carrying extra horsefeed. It was said that the average light horseman 'rode out upon all serious missions with double the normal supplies for himself and his horse'. On at least one occasion a brigade spent over £1,000 at a single field canteen in the few hours before a prolonged operation. General Chauvel (see p. 56) who commanded the light horse throughout most of the campaign found that the British infantry had 'the greatest admiration for our men, but seem to think they are more or less savages. The latest rumour in their lines is that I have issued a Divisional Order to the effect that "all prisoners are to be brought in alive"! As a matter of fact they have the time of their lives when they get into our hands, the men crowd round them and give them their cigarettes and share their water and rations with them, and put them up behind them on their horses to bring them in because they are footsore.' No wonder the Turks called them the 'mad bushmans'! All accounts stress that the Australian other ranks had deep respect for the Turks. This did not stop them, when they saw that rigor mortis had caused the arms of buried Turks to stick up out of the sand, placing pieces of cardboard in the protruding fingers, marked 'Baksheesh'! His *dry* sense of humour led a man of the 10th Light Horse to chalk on a well:

'Good morning John
Your water has gone
Without the slightest doubt,
And before we go
We should like you to know
It was the 10th who bailed it out.'

Yet another example of their humour is given by Barrow (see p. 113): Ryrie (see p. 58) once asked a trooper why he was wearing his shorts inside out. 'Well it's like this, Brig.,' he replied, 'when the little chats [lice] gets too troublesome we turn our shorts inside out; they crawl inside and we turn them round again; they crawl inside again and we turn the shorts around again; they crawl inside again and we turn the shorts round again; the chats crawl out once more and when we turn our shorts again that breaks their bloody little hearts.'[18]

The officers of the Indian Army when they arrived in Palestine in 1918 took some time to get used to the Australians. They 'don't seem too popular with others' discovered Lieutenant Roland Dening of the 18th Lancers,* 'though in official circles they are IT. They have no discipline – very badly turned out and don't seem to fight any better than anyone else.' The first view that Captain E.B. Maunsell of Jacob's Horse had of the light horsemen was near Jericho.

'The stench of some unburied horses, to leeward of them but to windward of us,' he wrote, 'proclaimed their notions on disposal of carcasses differed from ours . . . .

'Their appearance was impressive and distinctly original, reminiscent more of the South African War than of the Great War . . . . In small parties at all events, if not in large formations as is quite understandable, they were the real stuff, men you would like to go into a dirty show with.

'Their "uniforms" consisted, in many cases, of a pair of breeches and boots only. Gaiters were fairly common. Shirts of the British greyback pattern were not in fashion, the sleeveless, collarless singlet being the garb affected. Many did not even affect this, but were stripped to the waist, a white band of skin marking where the bandolier went.

'The men seemed of a very good class, better than the

*Dening became DAAG 4th Cavalry Division in 1918 and, as a major-general, commanded the 1st Abbotabad Infantry Brigade in 1940. He died in 1978 aged ninety.

> [Australian] infantry in France, and one heard far fewer tales of ruffianism among them. As one got to know them better, a very genuine liking arose between us. [A visit to a bivouac] reminded one rather of a visit to Buffalo Bill's Wild West show, minus the squaws, wigwams and feathers.
>
> 'Considerable difficulty was experienced in "spotting the orficer". His relations towards the "Boys" seemed rather on the lines of "Now, boys, don't spit, and for God's sake don't call me Alf." Socially one found, not infrequently, better men in the ranks than among the officers, and men, moreover, who seemed to be more natural leaders. "Imperial officers", as they termed British Regular officers, were popular among them, but there seemed an antipathy to Yeomanry.'[19]

Of all the troops engaged in the war, few endured separation from their homes so protracted and complete as the Australasians in Sinai and Palestine. Many were stuck there without leave to Europe, let alone their own countries, for over four years. Civilized white society, except for a few of the Jewish villagers in 1918, was unknown to them. It was a scandal that they were not allowed into the better hotels in Cairo which was the only sophisticated city which they ever saw. Consequently, as the Official Historian puts it, 'they got so little pleasure out of their occasional leave from the front line that men frequently applied for permission to return to their regiments before the completion of their holiday.' Much was done to succour them by such philanthropic organizations as the Australian Comforts Fund, Mrs (later Dame) Alice Chisholm's Soldier's Club and Rest Camp at Kantara railway station, the YMCA and the Australian Imperial Force Canteens. These last, from a modest beginning, soon expanded until there was a mobile canteen with each brigade, employing a number of motor cars. During the summer of 1918 they opened an iced-soda fountain near Jericho. This reminded the Official Historian that 'in Antony's sensuous days the imperial Roman toasted Cleopatra at Jericho in bumpers iced by snow carried all the way from Hermon.'![20]

* * *

The Australian trooper's 'active' pay was 5s a day plus another shilling of deferred pay which he did not receive till he was discharged.* In practice, though, it seems that 3s were 'banked' for him regimentally. One trooper of the 5th Regiment, whilst complaining that he was drawing only 2s a day, compared that to the 4s 2d which the Scottish infantry were allowed per *month*. 'We are millionaires to them,' he said.[21] The Australians' was the highest pay given to privates in any army. On top of that each man had 7½d over and above the British scale of rations.[22] The New Zealander's total pay was 5s, and the American's was 4s 7d a day. The British private was still being paid his 1s, though cavalrymen got more. (See Vol. 4, 480-1.) This wide difference caused surprisingly little envy and embarrassment, but it meant that on leave, especially in Cairo, the Australian was much pursued by prostitutes and every type of charlatan. An officer of the South Notts Yeomanry discovered that the Egyptians 'hated the Australians and small wonder; many of them never lived in a civilized place before coming here . . . . It is quite usual for them to get helplessly drunk and spend the night in the gutter unless fetched in by the police.'[23]† The Military Police did indeed have a great deal to cope with when the men were on leave, but in the field there was remarkably little indiscipline. The commanding officers' summary powers were quite sufficient to deal with minor irregularities. For instance in the 5th Regiment there were only two courts martial in three years.[24]

The men of the New Zealand Mounted Rifles Brigade, made up of the Auckland, Canterbury and Wellington Regiments, like the Australians, were nearly all pioneers or children of pioneers, practised horsemen from country districts and fine shots. They were perhaps more 'colonial' and less 'national' in their outlook.

*There was no separation allowance for Australian married men, as in Britain until 1915, when only soldiers receiving less than eight (later ten) shillings a day received it.

†Major-General James Doiran Lunt, serving with the Arab Legion at the time of the Anglo-Egyptian dispute over the Suez Canal zone in 1952, was asked 'in all seriousness' by an Arab 'why the British did not send a few Australian troops to Port Said. "The Cairo shopkeepers would at once persuade their government to allow the British to remain, providing the Australians went home," he said.' Lunt also tells a story of a monocled British officer sent temporarily to command a light horse regiment. His first parade revealed to him every man wearing a monocle. He at once tossed his in the air and caught it in his eye. 'Bet you can't do that,' he said. 'From then on he could do anything with them. But he was a good soldier as well as a good monocle-juggler.' (Lunt, 200)

Australian light horsemen types

Their fighting qualities were at least as admirable, though they were a little less aggressively independent. 'I don't think they grumble as much as we do' was the view of one Australian trooper. 'They shave oftener, anyway.'[25]

* * *

Australian junior officers' pay was greater than in the British army. A lieutenant received £1 1s a day while overseas and a captain £1 6s, but brigadier-generals at £2 12 6d were less well paid than their British equivalents. In the higher ranks the difference increased. As was to be expected the Australasian officer was very unlike his British counterpart. There was virtually no 'officer class'. Indeed the ranker would have laughed and indeed did laugh at the typical British officer, probably thinking that he was guilty of affectation or, in 'strine', of 'putting on dog'.* 'At first,' as one of the Australian official historians wrote, 'there undoubtedly existed a sort of suppressed resentfulness, never very serious, but yet noticeable, of the whole system of "officers".'[26] The question of selection and training of officers was therefore the most vital of all the questions facing Brigadier-General William Throsby Bridges, Inspector-General of Commonwealth Forces. There were very few regular officers available. The two chief sources were the officers of the militia and any past or present officers of the British army who happened to be in Australia. There were also the cadets of the senior year of the Royal Military College at Duntroon. Throughout the war and particularly in its earlier stages officers who failed were relieved with a firmness 'scarcely to be expected in a young citizen army . . . . Neither social nor political influence, nor considerations of personal friendship, played a sinister part in the selection or advancement of leaders. The result was a degree of efficiency not excelled in any force engaged in the war.'[27]† So long as they gave themselves no

*In late 1916, the commander of the Desert Corps [Chetwode (see p. 77)] wrote to the commander of the Anzac Mounted Division (Chauvel) complaining that 'not only do your men fail to salute me when I ride through your camps, but they laugh at my orderlies.' (Gullett, 207)

†About one third of the light horse officers were married. This was a higher proportion than in the infantry because they came from country districts where Australians married earlier than in the towns. (Bean, 60)

airs and were truly efficient a close personal association with their men developed. In Sinai and Palestine this was fostered by the relatively low casualties there as compared with those in France. The degree to which promotions from the ranks took place was high. Up to the end of 1917 nearly all these were made in the field, causing certain unavoidable strains and embarrassments. Early in 1918, therefore, all candidates who had proved themselves in action were withdrawn for a three months' course at the officer training school which had been established at Zeitoun near Cairo. The material differences between officers and men in camp and field were not very great. In theory every officer had a batman, but 'as good personal servants were scarce among the independent light horsemen, officers seldom fared better than the troopers in the lines – often not as well'.[28] They had a little more space in their tents than the other ranks had in their 'bivvies' and they could buy spirits which the men were not allowed to do.*

* * *

There was no proper cavalry in the Australasian armies. All the mounted units were armed with rifle and bayonet, none, until 1918, with swords. Though later on in the campaign under certain circumstances the light horse regiments came a little to resemble true cavalry, they were always in fact if not in name mounted rifles and their training was as such. Each regiment consisted of headquarters and three squadrons, each of four troops. A troop was divided into eight four-man sections. The establishment was twenty-five officers and 497 other ranks.

Each regiment left Australia with two Maxim machine guns. In July, 1916, these were taken to form brigade machine-gun squadrons each with eight officers and 221 other ranks. At first each squadron possessed twelve Maxims which were later replaced by Vickers. At the same time (August, 1916) three Lewis guns – one per squadron – were issued to each regiment. In April, 1917, these were withdrawn and twelve Hotchkiss guns, or automatic rifles,

*During the cold winter of 1916 when the 1st Regiment was guarding the Freshwater Canal the want of fuel was keenly felt. The officers shared their issue with their men buying kerosene and Primus stoves for cooking. (Vernon, 102; Gullett, 57)

took their place – one per troop.* These became very popular and substantially increased the firepower of the light horse. All machine guns were usually carried into action on pack-horses and they were frequently employed as advance guards on flanks. A fully trained team could open fire in as little as twenty seconds after receiving the order to halt their galloping horses. To help in achieving this remarkable feat the 5th Regiment, with the aid of its Canteen Fund, 'improved on the MG equipment issued'.[29] Where possible a reserve team was trained for each gun.[30]

Rifles were of the latest short-barrelled type. The 'citizen forces' in Australia held over 87,000 of them and these were at once handed over to the AIF. Others were manufactured by a Government factory in New South Wales. Two other factories which had

*The Maxim was a revolutionary water-cooled, recoil type, automatic gun, which, unlike its hand-actuated Gardner, Gatling and Nordenfeldt predecessors, harnessed the force of the explosion to increase its rate of fire. Its prototype had first been used by the British army in 1889. Fed by belts of 250 cartridges, it could fire some 450 rounds of .303 ammunition per minute. With its water-jacket full it weighed 70 lbs and its tripod weighed 48 lbs. Sighted up to 2,900 yards and operated by two men it could deliver as great a volume of fire as thirty riflemen who required at least fifteen times as much frontage. It was very noisy and when its boiling water gave off steam was apt to betray its position. To avoid excessive waste of ammunition it required very careful laying. The Vickers gun was much the same as the Maxim, but at 38½ lbs was considerably lighter. A major disadvantage of both was the need for cooling water. Especially was this so in such places as the Sinai Desert. It took but a minute and a half of consecutive firing (about 600 rounds) for the gallon of water in a Maxim's water-jacket to come to the boil. However, the provision of a condenser in the form of a canvas bag connected to the jacket by a flexible metal tube meant that a good deal of water could be recovered for re-use.

The Lewis and Hotchkiss guns were air-cooled, their barrels being fitted with fins to act as radiators. They were only 26 and 27 lbs in weight respectively. The Lewis was fed by fan-shaped or circular hopper magazines each containing forty-seven cartridges. It took only four seconds to replace an empty one with a loaded one. The cartridges of the Hotchkiss were fed through it from left to right by a belt holding fifty cartridges. Both Lewis and Hotchkiss were of the 'gas-engine' type, in which a minute part of the explosion gases is diverted from driving forward the cartridge and applied to operating the gun's mechanism. Both types were liable to overheating, but the changing of barrels took but a few seconds. The Hotchkiss was fitted with a single-shot mechanism. The Maxim, Vickers and Lewis were not.

There were a few complaints that the Hotchkiss suffered from jamming due to percussion caps becoming separated from the cartridges in extraction and 'getting into the runners and other mechanisms of the gun'. (Fox, 167, 180).

been set up over the previous three years provided ammunition, clothing and saddlery, of which to start with there was a serious shortage, especially of stirrup irons.[31] Private firms produced at great speed every sort of equipment including wagons, carts and boots of an exceptionally stout and pliable sort – 60,000 in the first month. Uniforms were of a pea-soup khaki colour, brass buttons were oxidized to a dull black. Collar badges exhibited the rising sun and on each shoulder-strap appeared the word 'Australia'. The Norfolk type jacket had a cloth belt and an oxidized buckle. The sleeves were very loose and buttoned at the waist and the light horsemen wore leather leggings. On their heads they wore the typically Australian wide-brimmed hat, looped up on the left side. The New Zealanders wore their hat turned down with the colour of their arm of the service streaked through the puggaree wound round its crown. The officers' uniforms differed little from their men's.[32]

* * *

One of Murray's earliest decisions in January, 1916, was to form four companies of infantrymen on camels chiefly at first for patrolling the Western Desert against incursions by the Senussites (see p. 27). From the Australian infantry divisions recently returned from Gallipoli he called for volunteers. These, like their numerous successors, were given elementary training first at the Camel Training School at Abbassia Barracks in Cairo and later at the Imperial School of Instruction at Zeitoun, a suburb of Cairo, where before long every sort of officer and man was trained in all branches of warfare. These camel companies were soon expanded into the Imperial Camel Corps Brigade (ICC) under Brigadier General Clement Leslie Smith, VC, who had special knowledge of camel work from pre-war service in Egypt and the Sudan. The brigade consisted of four battalions made up of ten Australian, six British* and two New Zealand companies. Each company numbered six officers and 169 other ranks and roughly corresponded to a cavalry squadron. Each man carried on his camel 240 rounds of rifle ammunition. On one side of the wooden saddle, which had a hole for the hump and provided a comfortable seat when padded with blankets, was a five-gallon fanatis (see p. 35), balanced on the other side by a canvas bag which held up to fifty pounds of durra.† An apron of sheepskin tanned with bark covered the upper part of the neck to prevent friction by the rider's boots which rested on it, there being of course no stirrups. The reins, tied to the nose, were a simple rope; the headstall and girth were made of leather. Thrown across the saddle, covering the fanatis and durra, were two canvas bags which held the ammunition, extra clothing and food, as well as everything the ingenuity of the men could conjure up. Loads of over 450 lbs, including the rider, were not uncommon.

The brigade's artillery was supplied by the six mountain guns, firing nine-pounder shells, of the Hong Kong and Singapore Battery (known for some reason as 'the Bing Boys'[33]). This was manned by 240 Sikhs and Mohammedans, sturdy and well-trained ex-Indian Army regulars recruited in those two places, under six British and four native officers. The guns were drawn by mules. British personnel manned the attached machine-gun squadron of

*One Australian battalion was made up entirely of light horsemen. The British at first came mostly from yeomanry regiments, but later on from the regular infantry.

†A kind of millet, common all over Asia, a variety of *Andropogon sorghum*.

eight Vickers guns, while each of the eighteen companies had its section of three Lewis guns. There was also a mobile veterinary section of 142 Egyptians under a British officer. With all details, the ICC had over 2,900 men. For the firing line it had about 1,650 rifles, equal approximately to the fighting strength, dismounted, of two light horse brigades.

The men of the battalions, which more often than not were employed split up, included a sprinkling of volunteers from the Rhodesian Mounted Police, a South African mining prospector who had fought with the Boers in the South African War, a fruit grower from the Canadian Rockies, a noted polo player from Argentina and an American.[34] Some of the later recruits came direct from Australia where a certain Abdul Wade, 'the "camel king" of Central Australia, had supplied the future Cameliers with four of these animals, and . . . they served to accustom the men to the smell, bad temper and peculiar habits of the tricky brutes.'[35]

All the brigade's camels were single-humped Arabians, known as dromedaries. Most and the best came from the Sudan, others from Somaliland, India and, of course, Egypt. A good specimen can keep going for up to thirty miles a day for weeks on end at a speed of between four and ten miles an hour with a full load.* Normal trotting speed, the only one comfortable for the rider, is six miles an hour. An especially well broken in animal could go for up to six days without water but three to four days was the period usually counted upon. At the end of such a march it can drink some thirty gallons at a time. Camels varied in character from extreme aggressiveness to almost excessive docility, the latter it is said being 'more the result of habitual nonchalance than any outcome of intelligent subservience'.[36] Under fire they were quite unmoved.

Because of the camel's inability to move at speed the ICC could only ever be used as mobile infantry; once brought by their steeds to the scene of action, cameliers were committed to act on foot. Their mounts (though on occasions they could be made, most uncomfortably for the rider, to gallop) could not generally be whistled up to carry them from position to position as in the case of the light horsemen. The brigade proved its worth from Katia

*The standard laid down for the Transport Corps by the Director of Veterinary Services, included the 'capacity to carry 300 lbs at continuous daily work for three months'. 'Heavy burden' camels were used to 350 lbs as well as the saddle. (Blenkinsop, 154, *et al.*)

(see p. 42) onwards, even after the desert had been left behind, its superior strength providing the light horse brigades with a body of infantry which, unlike the slow-moving regular infantry and 'unlike the slender light horse line, could attack a sector firmly in depth [as it did at Magdhaba and Rafah, for example (see p. 79)], while the horsemen could be moved about as the circumstances of the field dictated.'[37] One of the more salient disadvantages of camel units was the awful noise the animals emitted. 'Owing to the loud roars and grunts they made when barracking or getting up they had to be kept well in the rear on night marches.' The mounting of the ICC was accompanied it was said by 'a noise which could be heard all over Asia'.[38]

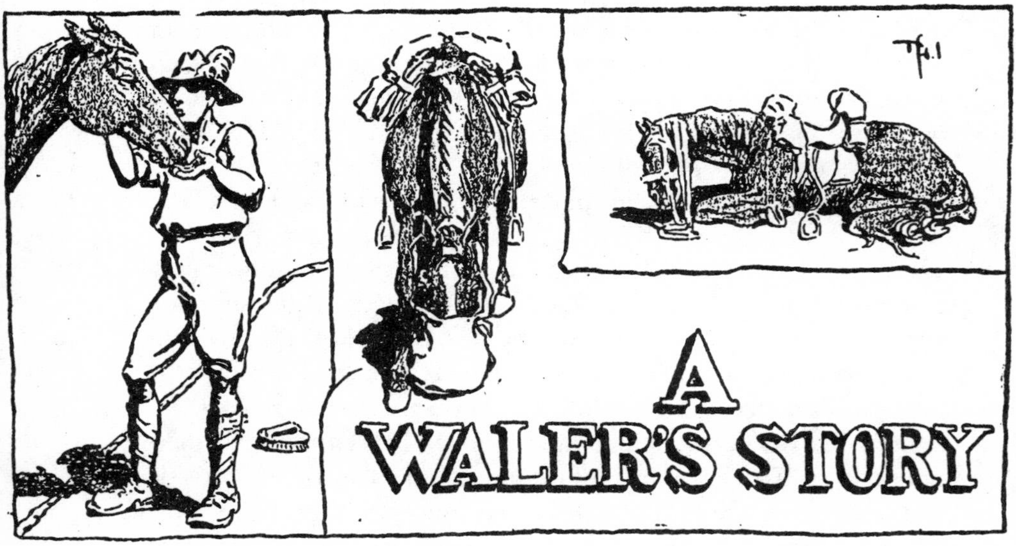

2

'The British Army certainly cannot belittle the men who forced them to abandon the Gallipoli enterprise, who captured a complete division at Kut, who crossed the Sinai Desert to the Suez Canal and who checked the first two onslaughts at Gaza.'

WAVELL in *The Palestine Campaigns*

'The seed of von der Goltz and Liman von Sanders was sown in barren soil. The consequence was a machine without precision and without motive power.'

BUCHAN in *A History of the Great War*

'The Turk, once settled down, is a dangerous man, and the best method of dealing with him is to rush him and keep him moving.'

COLONEL E.B. MAUNSELL, historian of Jacob's Horse

'It is never forgotten by the Turk that his enemy is not merely a foreigner but also an infidel.'

Field Notes: Mesopotamia, 1917[1]

The Turkish Army – German influence on it – types of officers and men

The Balkan War of 1912 against Bulgaria and Serbia had produced savage casualties in the Turkish army, many of the best of her regular army officers and men being lost. In 1914 there were thought to be about a million men of military age in the Ottoman Empire not including Arabia. All Turkish statistics are peculiarly suspect, but it is just possible that some 2,700,000 were called up in the course of the four years of the war.[2] What is certain is that a very large proportion of the available military population was called to the colours. Yet the maximum strength that Turkey mustered in the field is not likely to have exceeded 650,000.* For

*The total effectives in 1914, including 17,000 officers, numbered about 232,000. After something like 335,000 casualties, the 1916 numbers were calculated at 520,000 of which 362,000 were first line troops. The reserves

(continued over)

this heavy losses and large numbers of desertions were partly responsible.

From as long ago as 1883 the Turkish army had been advised by German officers, at that date under Feldmarschall Baron Kolmar von der Goltz Pasha, but in 1913, after the Balkan War, a seventy-strong military mission was sent to Constantinople under General Otto Liman von Sanders who became Inspector General of the Army.* He had little enough time to apply to his task of regeneration: the weaknesses and deficiencies which confronted him were chronic. However much he managed to improve organization at the centre, the further they were from the capital, the worse armed, equipped and trained were the troops. Heavy artillery (which had suffered severely in the Balkan War (see below), technical units and supply services were virtually non-existent. Of mechanical transport there was none. Horse transport included every possible type. Indeed, it was laid down that the first priority on mobilization was to be the commandeering of all the local transport.[3] As for medical services they were either deplorable or totally lacking. In August, 1914, there were in existence thirty-six Turkish divisions, very few of them up to strength. In the course of the war a further thirty-four were raised, but at no time were there more than forty-three in the field. Each division comprised ten battalions with a machine gun section and from twenty-four to thirty-six field guns. An Army Corps contained either two or three divisions.

Before long, in spite of the difficulties of communications between the Central Powers and Turkey,† it became true to say that Germany and Austria supplied both the mechanism and most

---

numbered 163,000, 'of which A. Actually under training (half unfit for various reasons) = 68,000. B. Available could Government equip and maintain them = 45,000. C. Could be made available if Lebanon Syrians were enrolled and capitation tax cancelled = 50,000.' (*Field Notes*, 86–7).

*This mission had been increased nearly sevenfold by December, 1914. (*Field Notes*, 85).

†These became easier in the autumn of 1915 when Bulgaria joined the Central Powers and Serbia was overrun. The direct rail link from Germany then reached as far as Mesopotamia and south Palestine. Communications within the Turkish Empire were far from speedy. To convey troops from Constantinople to Aleppo for instance took twelve to sixteen days and from Aleppo to south Palestine ten to fourteen.

of the brains of the Turkish Army, especially in Sinai and Palestine. The artillery,* the machine guns, the flying corps, the signals, engineering and supply services as well as the motor transport were either all manned or closely controlled by Germans or Austrians. A few infantry battalions were also sent from Germany. As the war went on all the principal staff appointments as well as the working of the railways were taken over. Wavell thought that, with their arrogant tactlessness, the Germans 'put almost as much grit as oil into the military machine'. They failed to appreciate 'the idiosyncrasies of the fighting methods of their allies. They frequently ordered counter-attacks or movements requiring a promptness of action and a precision of manoeuvre unknown to the Turks, with consequent failure and mutual recrimination.'[4] The United States Ambassador writing from Constantinople found that the army 'had been completely Prussianised. What in January [1913] had been an undisciplined ragged rabble was now parading with the goose-step; the men were clad in German field-grey† and they even wore a casque-shaped head-covering which slightly suggested the German *pickelhaube*.'[5]

The Turkish infantry soldier was a very variable commodity. He came from a number of different races. In the Balkan War the Slavs and Greeks were, not surprisingly, unreliable, while the Armenians fought well.‡

The true Anatolian Turk was the best of all. He was (and perhaps still is) extraordinary in that he could endure conditions of appalling wretchedness, semi-starvation, hopelessly defective equipment and the depression of constant retreats and still keep his morale. 'He will continue,' in the words of one of the Australian official historians, 'month after month and year after year, a

*In the Balkan War the entire artillery of three army corps and three independent divisions, as well as more than half of the other four army corps' guns, had been lost. Since the beginning of 1913 these losses had only been very partially made up, chiefly from German and Austrian sources. (*Field Notes*, 88–9).

†In Sinai and Palestine the uniform was generally of a yellowish hue, and a red sash was not uncommon. (Idriess, 80).

‡In 1908, when the Young Turks came to power, liability for compulsory service was extended for the first time to the empire's whole population. In 1914 the majority of the Christians and Jews thus conscripted were enrolled in unarmed labour battalions. (*Field Notes*, 85).

dangerous foe to troops of a higher civilisation fighting under the happiest conditions.'[6] His capacity for surviving on really brackish water and his resistance to the diseases which normally arise from living in a state of constant filth were a perpetual cause of wonder to the British and Australasians. He was a splendidly dogged marcher, often bare-footed in the desert. His understanding of his 1903 pattern 7.65 mm Mauser rifle and his precision and steadiness in its use, particularly when in a defensive position, were probably unsurpassed by any other troops in the war. 'On the defensive,' wrote Wavell, 'his eye for ground, his skill in planning and entrenching a position and his stubbornness in holding it made him a really formidable adversary.'[7] Many of the best of their officers had been lost in 1912 and more of the cream were to become casualties at Gallipoli. Few of the remainder were much more literate than the peasants whom they led.

It is usual to decry the Turkish cavalrymen as of practically no account, but in spite of their very small numbers compared with those of their enemies, there is considerable evidence that they were not inefficient in reconnaissance. They were mostly equipped with curved, broad-blade swords attached to the saddle as well as with 7.65 mm Mauser carbines or rifles 'in some cases slung over the left shoulder, in others in a leather bucket, projecting above the right thigh, attached to the rear of the saddle on the off side'. Their Arab ponies were usually quite small but exceptionally sturdy. The units very seldom showed fight, although many efforts were made to get at close quarters with them. This was surely a wise use of very limited mounted forces. A four-man patrol of the 5th Australian Light Horse Regiment tried on one occasion to induce a troop of Turkish cavalry to attack it. 'The Turks lowered their lances and charged, but turned back after going a few yards when their bluff was called;'[8] very sensible, too. Only on two or three occasions are they known to have attempted a proper mounted action. One was just before the Third Battle of Gaza, when several attempts at charges were beaten back by the quick action of the Hotchkiss guns of the 5th Australian Mounted Brigade.[9] Another was four days later, on 27 October, 1917, when two dismounted posts of the Middlesex Yeomanry were attacked. It was said that as many as seventy saddles were emptied (which is almost certainly an exaggeration), 'but the

Turks rode on like men and galloped right over the post' which the yeomanry were defending.[10*]

The historian of the 10th Light Horse Regiment tells of what was practically that regiment's first experience of close contact with the Turkish horsemen just before the Second Battle of Gaza. They found them 'nippy' and 'extremely smart and bright. Their Arab horses stepped jauntily in the sunlight and were very sure-footed over rough or broken ground. Whilst our troops remained in observation, the Turks kept well out of range, but as soon as the regiment began to withdraw – the reconnaissance being now completed – they galloped smartly forward and opened a brisk fire, following from ridge to ridge.'[11] There are numerous instances of similar tactics being employed. The mounted Turks on every single occasion avoided being ambushed in spite of many attempts to trap them.

* * *

In late January, 1915, Ahmed Djemal Pasha, ardent Pan-Islamist and Britain-hater, who had been instrumental in influencing Turkey to join the Central Powers, led between 10,000 and 20,000 men, many of them in Arab units of the Turkish army, with ten artillery batteries and 'floating bridges', from Beersheba towards the Suez Canal. For ten days they marched mostly by night over the largely waterless Sinai Desert, a most brilliant feat which had never been attempted before. Apparently not a single man or beast was lost. On 2 February attacks were made on the Canal's defenders, but according to the Bavarian Colonel Baron Kress von Kressenstein,

*'Several squadrons' of the Turkish cavalry were engaged. The post held out for just under seven hours. There were only three survivors. The last message received from Major A.M. Lafone, in command of the post, said, 'I shall hold on to the last.' He 'sprang out into the open to meet the last charge and was ridden down.' He received a posthumous Victoria Cross. The total losses of 8 Mounted Brigade in this action were ten officers and sixty-nine other ranks, mostly from the Middlesex Yeomanry. This regiment's gallant resistance gave time for the 3rd Light Horse Brigade and some infantry to force the enemy to retire. Had this not been possible, work on the railway, which was vital to the coming battle (Gaza III), could not have been carried out according to the precise timetable. (Falls, 38–9).

A very full and highly interesting account of this desperate action is given in Osborne: I, 339–45. This includes two official reports made by officers who took part.

who was at Djemal's elbow, the Arabs 'went over in large numbers to the enemy'.[12] The Indians of the Canal defence force repulsed the remainder. Little damage was done to the Canal which was only closed for one day and no bridgehead across it was established. Djemal withdrew, taking most of his men, animals and guns with him. His casualties are said to have amounted to 2,000. The British lost 163. It is supposed that the Turks believed that once the Egyptians learned that the Canal had been crossed they would rise against their British 'oppressors'. This did not happen.[13] There was no pursuit because the Indian mounted force was far too small. Further, to advance any distance into the desert required much preparation particularly with respect to water supplies.

This attempt to block the Canal illustrated the ease with which a defeated force in the desert could break off an engagement and escape over its lines of communication, 'retreating upon sharply defined wells or oases which its rearguard can deny to the pursuers long enough to exhaust their offensive capacity'. This explains in part the Turks' audacity in daring to cross so barren a region to attack a superior force based upon rail and water communications. The lesson was increasingly taken to heart in the years to follow. Over the next nine months numerous small-scale raids were made up to the Canal. They ended when the Indian cavalry killed the Bedouin Rozkalla Salim who had led most of them. During this period the numbers of Indians and others defending the Canal were steadily eroded. Many were sent off to Gallipoli, Mesopotamia and France. The remainder were put strictly on the defensive ranging no further than fifteen miles into the desert on reconnaissance.[14]

3

'The whole thing was a marvellous instance of the awful terror inspired by galloping horses and steel.'

2ND LIEUTENANT BLAKSLEY, Dorset Yeomanry, after the charge at Agagia, 26 February, 1916[1]

The Senussites – the Western Frontier Force – Wadi Senab – Wadi Majid – Halazin – Agagia

In mid-winter, 1914, the Turks under Enver Pasha invaded the Caucasus. The Russians appealed to Britain for a diversion. The Gallipoli campaign was the result, though, as is well known, there was more to its inception than the Russian appeal. The landing at Helles took place on 25 April, 1915, the naval attempt to force the Dardanelles, which had started on 18 March, having failed. The Turks were thus so fully occupied as to be unable in 1915 to attempt a second attack on the Canal.* In Egypt, too, the numbers of British fighting troops were drastically reduced, even on the Canal, as men were fed into the trenches of Gallipoli.

General Sir John Grenfell ('Conky') Maxwell, a soldier with a long experience of wars in the Nile Valley, (see Vol. 3, 369), who commanded the forces in Egypt, was faced at this critical time with a new danger. In the Western Desert, where, a quarter of a century later, desert warfare reached its apogee, there lived the Senussites, an Islamic religious sect. Its leader at this time was the highly respected Sayyid Ahmed, known as the Grand Senussi. In 1911 he had helped Turkey to fight the Italians in Cyrenaica (unsuccessfully) and was perpetually quarrelling and battling with them, chiefly over commercial matters. The Turks, egged on by the Germans, urged the Senussites to attack Egypt, sending them military missions, munitions, money – and medals. Sayyid needed much persuading and bribing, for he was on good terms with

*Kress, however, who had remained in the Sinai with a small force, launched a number of small-scale raids with the intention of sowing mines and wrecking the railway on the western bank. He was persistent but so were the small number of defending troops. No serious damage was done to the Canal or the ships in it.

Maxwell and had nothing particular against the British. Skilful propaganda and the entry of Italy into the war on the Allies' side helped him to be swayed into declaring a holy war against the infidel, especially as he was assured that the British were about to be thrown into the sea at Gallipoli. As Wavell put it, 'The danger to the British lay not so much in the size of the military force at the command of Sayyid Ahmed as in the spiritual authority he exercised. With memories of the Mahdi (see Vol. 3, 308 *et seq.*) in his mind, Maxwell feared the possibility of serious religious and internal disorders should the Senussites once gain any striking success.'[2]

The Senussian force consisted of about 5,000 so-called 'regulars' well armed with modern rifles, varying numbers of irregulars and some mountain and machine guns. The Western Frontier Force which Maxwell hastily threw together included a Composite Mounted Brigade made up chiefly of yeomanry and Australian light horse convalescents and horseholders numbering in all about 1,000 men, who had been left behind when their regiments, some twenty in number, had gone off to Gallipoli, a composite infantry brigade and the Territorial Notts Battery, Royal Horse Artillery. Other rather less makeshift formations were added after the Gallipoli evacuation in late December. There were also a section of the Rolls-Royces of the Royal Naval Armoured Car Division under Major the Duke of Westminster and four Royal Flying Corps aircraft. The section acquired so many cavalry officers that it was nicknamed 'the petrol hussars'.

The first action took place on 13 December, 1915, to the west of Mersa Matruh at Wadi Senab. The scouts of the 2nd Composite Yeomanry Regiment were not sufficiently far ahead to save the right flank guard from being suddenly fired upon at short range. Though the yeomanry charged ineffectually[3] and the armoured cars tried to retrieve the situation, the attack had to be recalled. The guns now came into action and a squadron of Australian light horse helped to drive the enemy from his position. There were thirty-three British casualties, including sixteen dead. Twelve days later, on Christmas Day, at Wadi Majid, another sharp but inconclusive check was administered to the Senussites, in which the mounted troops played an undistinguished part. At Halazin on 23 January, 1916, yet another engagement proved quite indecisive. This was largely because for lack of water the horses were unable to pursue, as also were the armoured cars because of the soft going

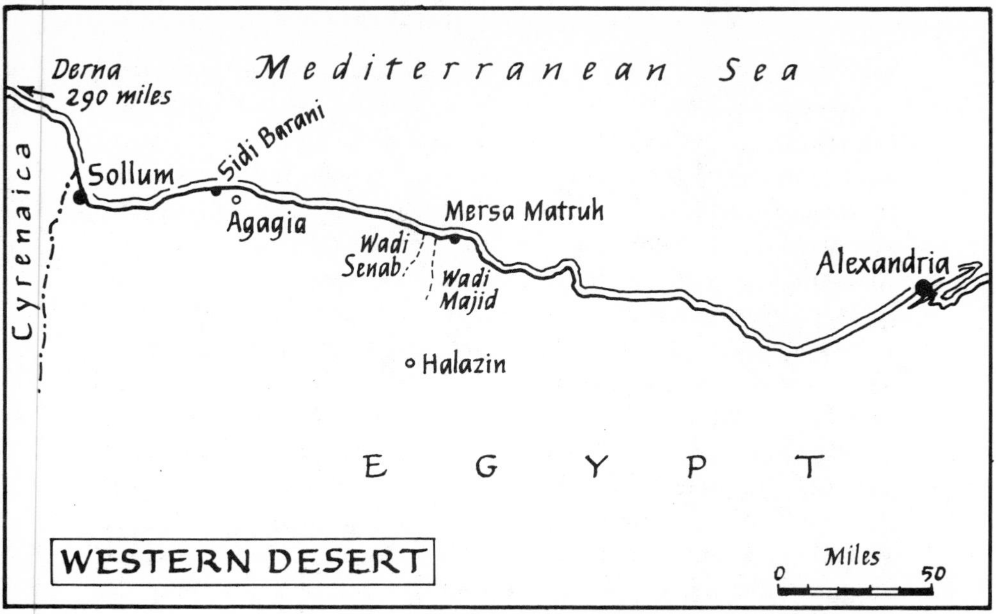
Mediterranean Sea
Derna
290 miles
Sidi Barani
Sollum
Cyrenaica
Agagia
Mersa Matruh
Wadi
Senab
Wadi
Majid
Alexandria
Halazin
EGYPT
WESTERN DESERT
Miles
0
50

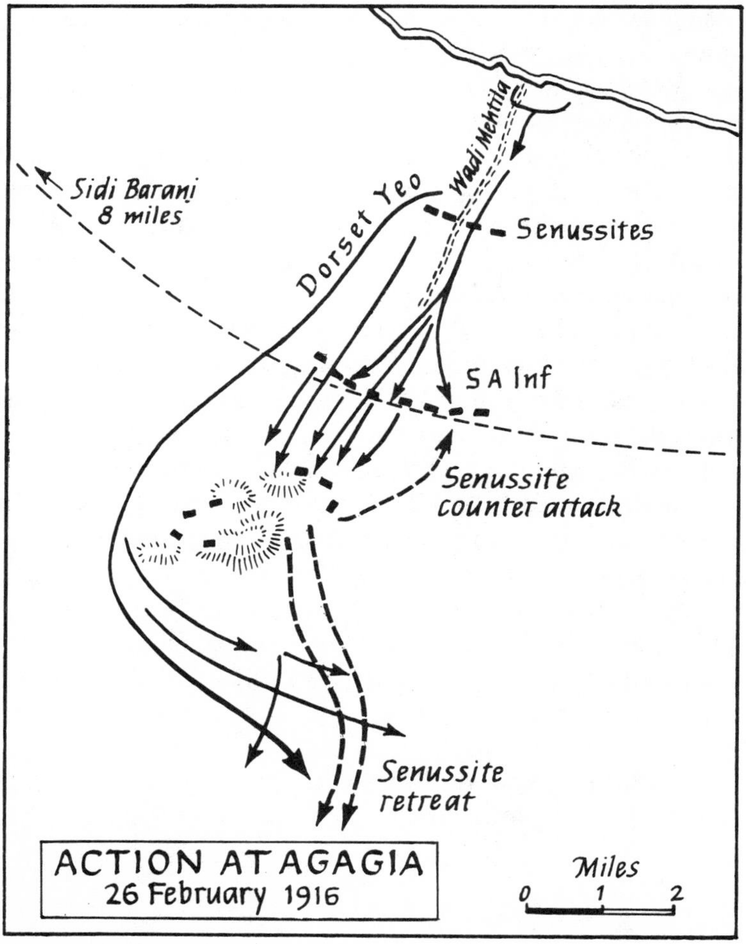
Wadi Mehtila
Sidi Barani
8 miles
Dorset Yeo
Senussites
SA Inf
Senussite
counter attack
Senussite
retreat
ACTION AT AGAGIA
26 February 1916
Miles
0
1
2

after heavy rains. The British casualties amounting to 312 were chiefly among the infantry and were probably not much fewer than the enemy's. In February 2,000 camels were assigned to the force, making it able at last to advance into and remain in the desert for prolonged periods. The heavy Rolls-Royces were replaced by seventeen light armoured cars and the Cavalry Corps Motor Machine Gun Battery was added. The Sikh regular infantry was replaced by three battalions of regular South Africans.

It was these which on 26 February, acting on information from air reconnaissance, attacked a well entrenched position at Agagia and drove some 1,500 of the enemy from it. This time the mounted troops were well placed and prepared for the pursuit. Brigadier-General (Sir) Henry Timson Lukin, commanding the South Africans, had made sure that Lieutenant-Colonel Hugh Maurice Wellesley Souter, commanding the Dorset (Queen's Own) Yeomanry, had time to get into a good position on the flank of the enemy so as to be able to pin him to his ground while the infantry column advanced. The Dorset Yeomanry, accompanied by two armoured cars, also helped to keep down the enemy's machine-gun fire. 'We would fire at them from one position,' wrote 2nd Lieutenant J.H. Blaksley of the Dorsets, in a letter home a week later, 'then get on our horses and ride quickly to another and then fire at them again.' The range was about 950 yards for most of the time. After a three-hour running fight through seven miles of sand dunes the South Africans lost touch, the armoured cars became bogged down in soft sand and the force's single squadron of the Buckinghamshire (Royal Bucks Hussars) Yeomanry was pursuing a camel convoy which had broken away from the enemy's column. Only some 180 men of the Dorsets – three squadrons less one troop (including the regimental cooks who rode swordless, brandishing their cleavers) – and a few stray men of the Bucks, had the main enemy in sight.

Souter now had in front of him the Senussite rearguard. It consisted of about 150 to 200 *muhafizia* (regulars) with three or four Maxim machine guns. Most were officered by Turks. He waited until the mass of the retreating Senussites – a column of camels a mile long and 300 yards broad protected by the rearguard in flank and rear – were clear of the trenches and wire of their old camping ground. While waiting,

> 'the led horses were whistled up, we were ordered to "mount" and "form line". Then and not till then,' wrote

> Blaksley, 'we knew what was coming. Imagine a perfectly flat plain of firm sand without a vestige of cover and 1,200 yards in front of us a slight ridge; behind this and facing us were the four machine guns and at least 500 men with rifles. You might well think it madness to send 180 yeomen riding at this. The Senussi, too, are full of pluck and handy with their machine guns and rifles, but they are not what we should call first-class shots; otherwise I do not see how we could have done it. We were spread out in two ranks, eight yards roughly between each man of the front rank and four yards in the second. This was how we galloped for well over half-a-mile straight into their fire. The amazing thing is that when we reached them not one man in ten was down. At first they fired very fast and you saw the bullets knocking up the sand in front of you, as the machine guns pumped them out. But as we kept getting nearer, they began to lose their nerve and (I expect) forgot to lower their sights. Anyhow the bullets began going over us and we saw them firing wildly and beginning to run; but some of them – I expect the Turkish officers – kept the machine guns playing on us.'

At this point Blaksley's horse was shot dead. He grabbed a riderless replacement, but 100 yards on that too fell dead. 'There was no other horse to get and I was alone.' At this moment six or seven of the Senussites rushed up to him, recognizing him as an officer, and begged for mercy. Among them was Ja'far Pasha, the force commander, who was wounded. Colonel Souter, who had also been dismounted, now joined Blaksley and, getting hold of horses, placed Ja'far and his staff upon them, sending them to the rear under escort.* Meanwhile the rest of the Senussites were 'running in all directions, shrieking and yelling and throwing away their arms and belongings; the yeomen after them, sticking them through the backs and slashing right and left with their swords . . . . Some stood their ground, and by dodging the swords and shooting at two or three yards' range first our horses and then our men, accounted for most of our casualties.'[4] At this time the British machine-gun section arrived on the scene and finally decided the

*Ja'far Pasha said of the charge: '*C'est magnifique, mais ce n'est pas selon les règles*', adding: 'No one but British cavalry would have done it!' (Parham, A.G., 9 Nov. 1936, letter to *The Times*)

issue.[5] The enemy machine guns and their crews escaped. It is tantalizing to be unable to say how many Senussites were killed and wounded. The Dorsets lost four officers and twenty-seven men killed and two officers and twenty-four men wounded. One other officer was killed, 2nd Lieutenant J.C. Bengough of the Royal Gloucestershire Hussars Yeomanry, acting as aide-de-camp to Souter, who wrote of him: 'As we galloped to the charge he said, "Isn't this splendid", and I said, "Yes, I have waited twenty-four years for this" and he replied, "I always was a lucky fellow" . . . . He must have gone on for nearly a mile . . . . He ran two men completely through . . ., when a third shot his horse and then shot him on the ground.'[6]

The Agagia charge has been dealt with in some detail because it was one of the last against 'uncivilized' troops in the annals of the mounted arm, while being also the very first where armoured cars and machine guns cooperated with the mounted troops to an appreciable degree. The pattern, as shall be shown, was often to be repeated in the Sinai and Palestine campaigns against the Turks and Germans. It was perhaps a risky undertaking and the casualties were high, but it is certain that the news of it had a very salutary effect among the Egyptians.

The victory of Agagia was more or less decisive. There was little further serious fighting, though another Senussite force under Sayyid himself had to be dealt with in the oases further south. This occupied a whole infantry division and more than one brigade of dismounted yeomanry during the summer of 1916.[7] In the autumn a camel corps and light car patrols reoccupied the oases. Sayyid himself was defeated in February, 1917, by a force of armoured cars, but certain camel and light armoured car units remained on watch along the 700-mile frontier till well into 1918. The Turks could congratulate themselves on having invested in the Senussites, for Sayyid had achieved for them a satisfactory dispersal of British forces which would have been very useful elsewhere.[8]

4

'A bit of a jaunt after Turks . . . . They ran for their lives but didn't have a chance for our fellows ran them down with their horses.'

One who took part in the Jifjaffa raid.[1]

The Jifjaffa and other raids – fanatis for water – the Egyptian Camel Transport Corps

Early in 1915 an aerodrome was constructed at Ismailia. From it the small number of planes available managed over the following months to make limited reconnaissances over the Sinai Desert in conjunction with the light horse. On one occasion, so as to extend the range, which was at most sixty miles, an Australian mounted detachment deposited a store of petrol in the desert. This made possible two-stage flights of up to eighty miles from base.[2]

During this period and well into 1916 the Australian light horse launched a number of raids designed in part to break up the storage tanks and cisterns on the central route by which Djemal had come in January, but also to get information about the country ahead, the habits of the enemy and to stop him gaining information. Between early April and mid-July there were sixteen of these expeditions, varying in size from two brigades to one squadron. The most spectacular of them – to Jifjaffa, fifty-two miles from the Canal – took place between 10 and 13 April, 1916. In less than three and a half days 160 miles were covered by a squadron made up of men from the 8th and 9th Australian Light Horse Regiments, accompanied by a Royal Flying Corps ground signal section, Royal Engineers, with a wireless section and a field ambulance. They came up against considerable opposition. After making a stout resistance eleven of the enemy protecting the tank system became casualties and an Austrian engineer and thirty-four Turks were captured. One Australian trooper was killed.[3]

The raid is of special interest as it was the first demonstration of a swift and successful blow by mounted troops across many miles of waterless desert. It is, too, an early instance of the successful use of wireless in air cooperation. Above all it well illustrates how large a supply train was always essential where water had to be carried. For 320 of all ranks, 436 animals were required: 175 horses and

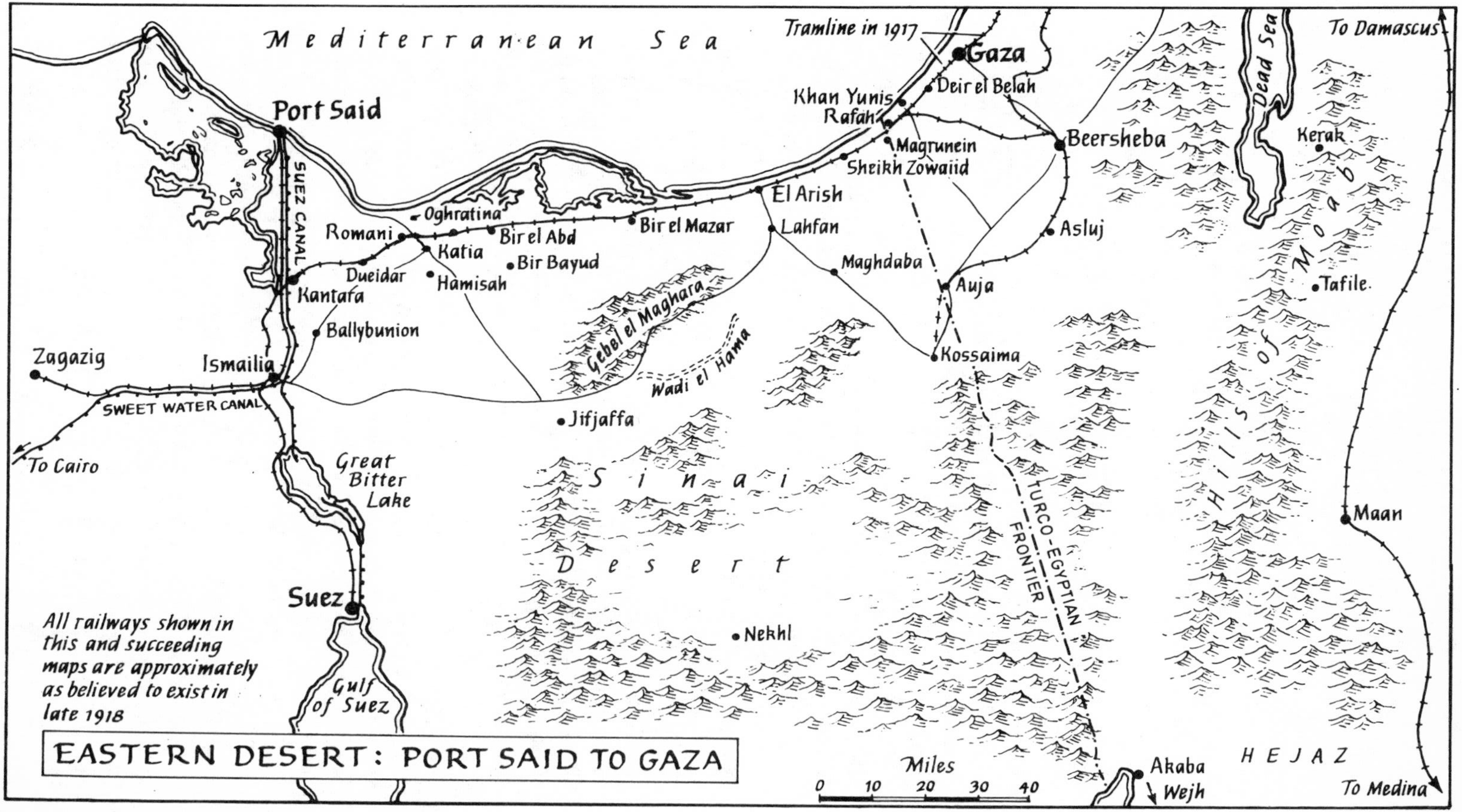

EASTERN DESERT: PORT SAID TO GAZA

261 camels. Each of eighty-eight of the camels carried thirty gallons of water. These were held in *fanatis*, galvanized-iron or copper tanks, one slung on either side of the saddle, each having a hole with a screw plug near the top for filling and a tap near the bottom for emptying. These were to become a very familiar and often desperately longed-for part of the desert scene. Where there were standing patrols each camel brought out two *fanatis* every other day to each post, one being buried in the sand and used the following day.[4]

In no other theatre were the fighting troops so utterly dependent for their supplies upon four-legged beasts, and without the services of the Egyptian Camel Transport Corps neither Murray's nor, later, Allenby's men could have carried on the struggle. During the whole campaign from December, 1915, to demobilization in 1919, some 72,500 camels were employed in it. About 170,000 Egyptian drivers passed through the Corps. Of these 222 were killed in action and 1,458 were wounded while 4,010 died of disease, seventy-eight went missing and sixty-six were taken prisoner.

For most of the time that the operations were taking place in the desert the corps consisted of ten companies, each with ten British or English-speaking Egyptian officers and ten senior non-commissioned officers. A company consisted of 1,168 Egyptians, 2,020 camels and twenty horses. The corps was an accepted unit of the Egyptian Expeditionary Force (EEF). As with the labour units, the transport corps had difficulty in obtaining sufficient British officers who could speak Arabic. In 1916 classes of instruction in colloquial Arabic were set up for both officers and non-commissioned officers. The earliest companies were officered by Egyptian Government officials and 'gentlemen established in Egypt in business'. Later on volunteers from the Australasian forces proved, as in so much else, quick learners and hard workers. A total of 419 officers served in the corps during the campaign, of whom 111 were attached from the Army Service Corps* and 205 transferred from units. Numbers of these were Australian and yeomanry troopers who had already seen service in Egypt. Of the 527 who joined the corps, ninety were later commissioned.

There was difficulty at times in recruiting the fellaheen, for the Labour Corps was a counter-attraction. The earliest recruits were

*The ASC had the distinction Royal conferred in November, 1918.

all 'volunteers': of every three, one enlisted to avoid the police, one was sent by the police and one was 'a respectable wage-earner'! Each man received seven piastres a day, while *reises* and *bash reises*, the equivalents of non-commissioned officers, received nine and twelve piastres respectively. Each member of the corps was allowed, at first, one blanket and later two. Service was usually for six months.

Eight different types of camel were employed, originating from the Delta, Middle and Upper Egypt, Somalia, Algeria, Syria, the Sudan and India. Some were particularly subject to diseases such as mange, diarrhoea and arthritis; some moved faster than others, the slowest being unable to achieve much more than two miles an hour and the fastest going along at a steady three. The heavy-burden camel carried up to 350 lbs, as well as its saddle, its own forage and its driver's kit.[5]* Light-burden camels carried up to 250 lbs. Examples of 'freak' loads were 200-gallon fanatis, four and a half feet square, for conveying water from pipehead to reserve water dumps, the large wine barrels which were used for the storage of water in the first line,[6] telegraph poles and five-seat latrine tops.

*Typical loads were:
'6 cases preserved meat 4 cases jam (@ 90 lbs) 4 chests tea (@ 90 lbs) 2 cases bacon 4 cases tobacco or cigarettes 4 bags rice (@ 80 lbs) 2 large or 4 small sacks grain 6 cases biscuits 4 cheese 6 cases milk 4 cases candles 4 bags sugar (@ 80 lbs) 2 bales compressed forage 2 fanatis water.' (Badcock, 81)

See p. 19 for dromedaries in the Imperial Camel Corps.

5

'I desire to express my indebtedness to my predecessor . . . . I reaped the fruits of his foresight and strategical imagination.'

ALLENBY'S despatch, 28 June, 1919[1]

Sir A. Murray forms EEF – railway and pipe line begun

In the spring of 1916 the EEF was formed under Lieutenant-General Sir Archibald ('Archie') Murray who had been French's Chief of Staff during the Mons retreat and later. Attached to his orders was a memorandum which said: 'The force under your command is in a sense a general strategic reserve for the Empire . . . . The War Committee has decided that for us France is the main theatre of the war. It is therefore important, as soon as the situation in the East is clearer, that no more troops than are absolutely necessary should be maintained there.'[2] He at once decided that the passive defence of the Canal was wasteful of men and material and that a mobile force to be based eventually near El Arish was the best way to do the job. He estimated that if that place were to be occupied he would only need five infantry divisions and four mounted brigades. To occupy Katia which is about twenty-five miles from the Canal was the first step. For this, though, major railway construction was essential. The first priority was to construct swing bridges across the Ship Canal to connect the Egyptian State Railway lines to the depots on the east bank. It was also necessary in view of the military traffic envisaged to double the seventy-nine kilometre line from Zagazig to Ismailia. For this and for the large numbers of extra stations and sidings required there were available 150 kilometres of new track which fortunately had been bought just before the war for routine renewals up to the end of 1917. The line reached a point forty-seven kilometres from the Canal in mid-July, 1916. Many kilometres of small-gauge lines had also to be laid. During 1916 2,144,411 officers and men, 274,065 animals, including 56,198 camels, 28,043 guns and vehicles and 1,126,970 tons of supplies were carried on trains of which 2,714 were 'specials', 577 of them for mounted troops.[3]

Beside the railway line was laid a pipeline (see p. 45) for carrying drinkable water from the Sweetwater Canal. To get it across the Ship Canal and to ensure by filtering that it was free of the deadly bilharzia worms (known to the troops as 'Billy Harrises') with which the Sweetwater Canal was infested, were major problems. At the crossing points filtering plants were installed and from these the purified water was carried by syphon-pipes laid across the bed of the Ship Canal into reinforced concrete reservoirs on the east bank. Pumping machinery then drove the water along the pipelines.[4] About every twenty miles beside the railway were more pumps and reservoirs.[5] For water transport and the carrying of other supplies beyond railhead and to camps either side of the line large numbers of camels were employed. By June there were some 50,000 in constant use.

1. The battle of Agagia, 26 February, 1916 (see p.30)

2. Loaded camel transport ready to move from the railhead (see p.35)

3. A brigade headquarters office ready for the road

6

'At Oghratina the Yeomanry were surprised and overwhelmed after a gallant resistance of three hours. The Turks then pressed on to Katia, and at 8 a.m. attacked the squadron of Gloucesters, which was presently joined by the squadron of Worcesters from Bir el Hamisah. The two squadrons held out till 3 p.m., when they were overborne by weight of numbers.'

WAVELL in *The Palestine Campaigns*[1]

The return from Gallipoli – Kress advances on Katia – the yeomanry disasters at Oghratina and Katia

The evacuation of Gallipoli was successfully effected in late December, 1915, and the shattered regiments of the yeomanry and the Australasians began the task of rebuilding their strength. There were emotional scenes as the survivors were reunited with their horses and it took some time for the men to abandon the 'Gallipoli Stoop' which they had acquired in the trenches of the peninsula. There was the problem, too, of reverting from infantry to cavalry both as regarded equipment and training. After seven months out of the saddle, the 'first few rides could not be classed as the ultimate in pleasure; meals in consequence,' wrote one of the 3rd Light Horse, 'were taken with more comfort in a standing position!' Where insufficient care had been taken to guard equipment left in depots, units came back to find much of it stolen by the ubiquitous Egyptian syces and other natives. One of the Sherwood Rangers Yeomanry wrote that 'it would be no exaggeration to say that there was not a curb chain or cloak strap left in the regiment and scarcely a bridle had more than one rein.'[2]* Reinforcements arrived in trickles from home, their passage made dangerous from the increasing submarine menace in the Atlantic and the Mediterranean.

*The Australian horses had been looked after partly by unfit men left behind, by the farriers and by a body of public-spirited Australians, most well advanced in middle age who had been refused as too old for active service but who had enlisted and gone to Egypt as grooms in the light horsemen's absence. Many of them found employment later on in the remount depots. (Darley, 7; Gullett, 54)

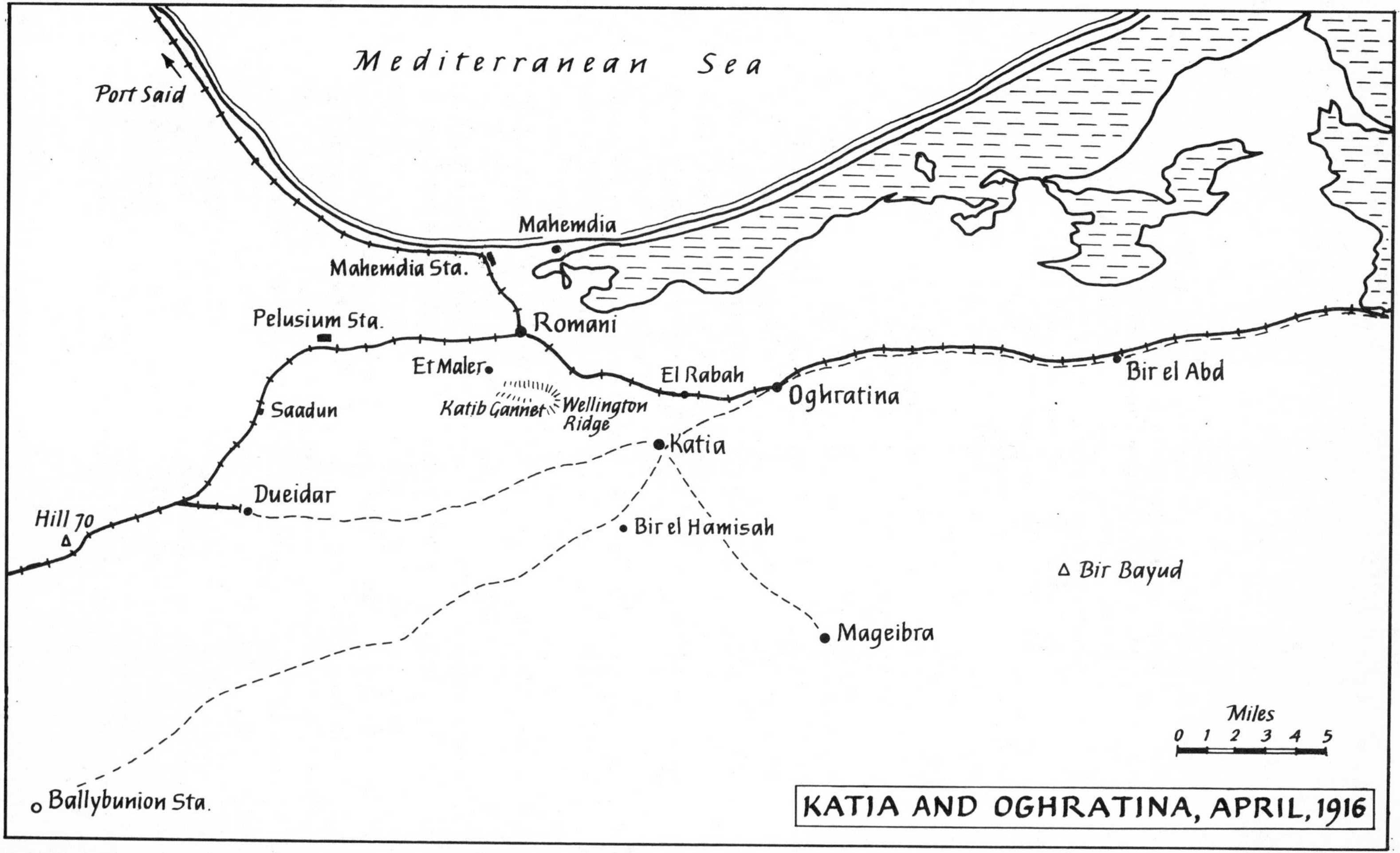
Mediterranean Sea
Port Said
Mahemdia
Mahemdia Sta.
Pelusium Sta.
Romani
Et Maler
El Rabah
Oghratina
Bir el Abd
Saadun
Katib Gannet
Wellington Ridge
Katia
Dueidar
Hill 70
Bir el Hamisah
Bir Bayud
Mageibra
Miles
0 1 2 3 4 5
Ballybunion Sta.
KATIA AND OGHRATINA, APRIL, 1916

Fortunately there followed a three-month respite, for though the Turks intended in the early spring of 1916 to launch a second expedition against the Canal, due partly to the losses incurred at Gallipoli, the necessary reinforcements were tardy in arriving. Kress, therefore, decided upon a comparatively minor advance with the object of alarming the British and thereby preventing the departure of further troops for the Western Front. This 'reconnaissance in force' was carried out by two and a half battalions, an Arab camel corps, six guns, at least four machine guns and a small force of cavalry: in all about 3,600 men.[3] Covering the construction of the railway which was approaching Romani was the 5th Mounted Brigade consisting of the Warwickshire and Worcestershire Yeomanry and the Gloucestershire Hussars. Because of the soft sand it was totally without artillery.

On 21 and 22 April two squadrons, less one troop, of the Worcester Yeomanry moved out to the oasis of Oghratina. The rest of the brigade was at Katia and Hamisah with the Gloucesters, less one squadron and its machine-gun subsection which were at Katia, in reserve at Romani. It was thus perilously split up should the enemy attack. On 22 April Brigadier-General Edgar Askin Wiggin decided to raid a small force which had been reported at Mageibra. Finding only six Turks there, he returned to Hamisah by 9 a.m. on 23 April, Easter Sunday. There he heard of an attack upon Oghratina. He at once ordered 'C' Squadron of the Worcesters to collect 'A' Squadron of the Gloucesters from Katia and to push on to Oghratina. At that isolated post, rising sixty feet out of the desert, 'A' and 'D' Squadrons of the Worcesters had recently arrived to protect a party of fifty Royal Engineers which had been sent to bore for water. They had not been there long enough to complete entrenching the post.* Major Frank Williams-Thomas, in command of the post, had asked for the engineers to be mounted so that the cavalry's horsemen should not be compromised. 'He was told that there was no likelihood of an attack and that his

*To make a proper trench three feet wide in soft sand invariably meant opening 'some fifteen feet of ground, putting in battens with canvas backs and anchoring them and then refilling the spaces behind with the excavated soil, and when that was done a tiny rent in the canvas allowed sand to filter through at such a rate that a portion of the trench would be filled in in twenty-four hours.' (Massey, 26). Wire entanglements were often buried two or three feet under the sand during dust storms. (Foster, 7). Fire trenches for temporary occupation were merely wide shallow ditches giving little protection.

trouble would be to get near any Turks.'[4] On the morning of 23 April in a thick fog, which did not disperse till 9 a.m., three mounted patrols had been sent out before daybreak. Two returned with nothing to report; the third disappeared never to be seen again. At this moment a small number of enemy Arabs on camels materialized out of the night and were fired on at close range. On hearing this a large body of Turkish infantry some 2,000 strong, which was spear-heading Kress's advance, realized that the post was occupied. They therefore began to develop an assault upon it. The first attack was repulsed, for the Turks found it difficult to climb the hill in the soft sand. Their casualties were numerous. A second attack from a side easier of approach was accompanied by accurate machine-gun fire. Unable to use his horses for a fighting withdrawal without abandoning the dismounted engineers, Major Williams-Thomas felt that he had no alternative but to make a last stand. He was forced before long to contract the circle of defence as more and more Turks pressed forward. His one machine gun and its crew were speedily put out of action. After two hours there were not more than thirty rifles remaining in action. One of the squadron leaders and all four troop leaders were killed. The total casualties were eleven officers and 135 other ranks. The remnant, including Williams-Thomas, surrendered.

The enemy now pressed on to attack Katia. Using his mountain guns, he managed to inflict heavy casualties on the horses of both 'A' Squadron of the Gloucesters under Captain M.G. Lloyd-Baker and 'C' Squadron of the Worcesters under Lieutenant-Colonel the Hon. Charles John Coventry. Dismounted, the two squadrons manned the entrenchments and fought on gallantly from 8 a.m. to 3 p.m. The reinforcements which Wiggin was pushing forward, parts of two squadrons of the Warwicks under Lieutenant-Colonel Hugh Annesley Gray-Cheape and others under Lieutenant-Colonel Ralph Maximilian Yorke, consisting of five troops and a machine-gun subsection of the Gloucesters from Romani, failed to coordinate their movements. Had they done so, it is possible that they would have been able to save the defenders of Katia. In fact both men, perhaps wisely, decided to retire when they saw how hopelessly outnumbered they were. Indeed Gray-Cheape's and Yorke's men were subjected to intense artillery and rifle fire before they withdrew, suffering numerous casualties. The Gloucesters lost four officers and sixteen other ranks killed, fifteen other ranks wounded and had sixty-four taken prisoner. All except sixty-one of the

Warwicks' squadrons also fell into the enemy's hands. At Oghratina and Katia the Worcesters lost in all nine officers and 101 other ranks killed. Most of the remainder were taken prisoner.* The loss in horses was 425.[5]

* * *

A minor action was fought on the same day at Dueidar but the infantry there held the enemy off. Had they not done so, the line of retreat of the remnants of the yeomanry must have been entirely cut. The 5th Australian Light Horse of the 2nd Australian Light Horse Brigade which had been preparing to move to Katia arrived on the scene in time to cover the retreat of the surviving yeomen. The Turks, having inflicted a spectacular defeat on their enemies, withdrew to Bir el Abd.

These two engagements were the worst defeats inflicted on the British in the course of the whole campaign. The yeomen owed their catastrophic reverse to failure to send their probing patrols far enough forward, to chronically defective dispositions and to poor communication. Their gallantry was of the highest order but availed them nothing. They were raw troops who had a vast amount to learn. It was to be some time before they did so. Never again, though, throughout a campaign of lengthy fronts and widely dispersed units was a British force surprised and overwhelmed.

Kress had achieved a remarkable tactical success which much boosted Turkish morale, but it did not interfere with the programme of withdrawal of British troops to the Western Front. The blow he struck was skilfully planned and boldly executed. The destruction of three and a half squadrons was no mean feat. Good luck certainly attended him. The fog could not have been foreseen and the timing of the attack just before Katia was to have been reinforced cannot have been planned, though the local bedouin may well have afforded him some intelligence of the yeomanry's movements.[6]

* * *

*The Worcesters' machine gun was buried in the sand by Private J.C. Ratcliffe on his own initiative. It was recovered intact some months later. (Cobham, 54–55)

Two months later Murray showed how low was his opinion of the yeomanry at this time. 'The bulk of the yeomanry officers,' he advised the War Office in June, 'are ignorant of the rudiments of mounted work . . . In any serious mounted work I rely entirely on my Anzac Mounted Division, who are excellent under hard conditions.' He asked for three good regular cavalry officers to be sent out for each yeomanry regiment, so that he could make drastic changes among their higher ranks. These arrived in due course. The Worcestershire Yeomanry for instance by mid-September found itself commanded by an officer of the 1st Dragoon Guards and two of its squadrons by officers from the 9th and the 21st Lancers.[7]

7

'[The Anzac mounted troops] have a genius for this desert life.'

SIR ARCHIBALD MURRAY to the CIGS,
10 May, 1916

'Fortunate are those commanders and troops who are granted a period of training and minor operations on the ground where they are to fight their first battles.'

A.J. HILL in *Chauvel*[1]

Anzac Mounted Division formed – extreme heat in desert – Egyptian Labour and Transport Corps – the khamsin *– treatment of drinking water – spearpoint pumps*

After Katia there was a prolonged period of comparative inaction, one of a number in the course of the campaign. For nearly three months Kress made no further move. This gave the British time to consolidate in the area, to form static defences and to train intensively. The railway from Kantara, which had now become a very substantial supply base, reached Romani in mid-May, but the pipeline had not kept pace. In consequence, the water supply for Romani had to come chiefly by lorries from Kantara or by camels from the pipehead at Pelusium Station. Further, the six-inch pipes were proving inadequate and had to be replaced by ones of twelve inches and these took months to arrive. There was increased activity in the air on both sides, but bombs dropped on Port Said by the very superior Fokker and Taube aeroplanes were answered by bombs dropped on the Turkish airfield at El Arish and no great damage was inflicted on either.

The yeomanry were now replaced by the Anzac Mounted Division which had been formed on 16 March under Major-General Chauvel. It consisted of the 1st, 2nd and 3rd Australian Light Horse Brigades and the New Zealand Mounted Rifles Brigades, the 2nd and the New Zealand Brigades taking over the forward patrolling and railway construction duties.* As there were

*The 52nd (Lowland) Division moved to Romani in May, to be followed later by a brigade of the 53rd. This division and the 42nd were in the rear and echeloned along the railway.

(continued over)

no Dominion horse artillery units available, four Territorial batteries, the Ayrshire, Inverness, Leicester and Somerset were attached to the division and remained with it throughout the campaign. Relations between these batteries and the Australasians became exceptionally close. The gunners even applied (unsuccessfully) for permission to wear emu plumes in their headdress.[2]

Numbers of reconnaissances were undertaken at this time in conjunction with the Royal Flying Corps and on 9 June the cisterns on the central Sinai route containing five million gallons of water were blown up and sealed, thus preventing the enemy from using that route again. Five days earlier there began the revolt of the Arabs in the Hejaz, in which Lawrence of Arabia was to play a prominent part. One result of this was that the idea of a forward movement to El Arish was now seriously considered both in London and Cairo. October was thought to be the earliest possible date for this forward movement.

* * *

As the hottest period of the year approached the conditions under which the troops lived became progressively more uncomfortable. 115°F and up to 126°F in the shade were not uncommon. By 10 a.m. such tasks as plate-laying had to be suspended even by the tens of thousands of devoted men of the Egyptian Labour and Transport Corps. The historian of the transport services in the campaign wrote thus of the long-suffering fellaheen who manned the corps: 'Tried by extremes of heat and cold, always his own worst enemy, not accounting bravery a virtue or cowardice a crime, scourged by fever, cheerful, miserable, quarrelsome, useless, wonderful men,

---

The 1st ALH Brigade consisted of the 1st, 2nd and 3rd Regiments. The 4th was attached to it. It was commanded by Lieutenant-Colonel Charles Frederick Cox (see p. 58). The 2nd Brigade was commanded by Colonel (temporary Brigadier-General) Granville de Laune Ryrie (see p. 58) and consisted of the 5th, 6th and 7th Regiments with the 12th attached to it. The 3rd Brigade was commanded by Brevet Lieutenant-Colonel (temporary Brigadier-General) John Macquarie Antill (see p. 58) and consisted of the 8th, 9th and 10th Regiments and had the 11th attached to it. Brigadier-General Edward Walter Clervaux Chaytor (see p. 60) commanded the New Zealand Brigade, which consisted of the Auckland, Canterbury and Wellington Mounted Rifles Regiments with the Otago Mounted Rifles Regiment (less one squadron) attached.

their name is for ever written *honoris causa* in the records of war.'[3]

Although the heat was severe by day, in the early morning 'men rose from sleep stiff with cold'. Very often a man's only covering at night when units were travelling light on patrol was 'the saddle blanket, sopping wet and malodorous from the perspiring horse'.[4] When in camp single bell tents were found to provide insufficient covering. By early June double (lined) bell tents had been issued to all units. These were usually erected over holes about three feet deep dug in the sand.[5] The *khamsin*, a searing south-westerly wind from the Sahara, made things even less tolerable. When it blew, as it did for short periods every now and again, the atmosphere was satiated by a haze of floating sand particles. It made one yeoman want to take off his flesh and sit in his bones 'to let a cool draught blow right through'.[6] Lips cracked and the small amount of water allowed for drinking scarcely relieved parched and swollen throats. An officer of yeomanry wrote during a mid-May heat wave: 'One does not want to eat anything, but just to pour liquid down one's throat; very bad for one's "little Mary" I feel sure and it seems to visibly swell!!!!!'[7] The daily water ration per man was never more than one gallon of which one pint was intended for washing, while two were for drinking and five for cooking and tea.[8] At times the heat was so intense that packets of candles melted leaving only the wicks.[9]

Many men returning from reconnaissances became unconscious from heat stroke and had to be evacuated to field hospitals. 'We had to wear spine pads to protect our backs from the heat of the sun,' wrote one yeomanry private. The Australians, not surprisingly, coming from the bush, suffered a good deal less than the men from home. What local well water was available dried up in the extreme summer heat and when it could be had it took some time before men became accustomed to it. When boiled it was apt to curdle, which diminished its use for tea making. Nor, because of its salinity, could it be used in the boilers of locomotives. Springs existed in the date-palm groves, or hods, but where they were touched by palm tree roots the water proved particularly bitter.[10] In case he had to resort to well water which he was forbidden to do except in emergency each man was issued with sulphate chlorinating tablets, two of which were to be added each time he filled his water bottle, and the water was not to be drunk for half an hour afterwards.[11] These water purification tablets were not in fact always used for their proper purpose because they made the water very offensive to the taste, according to one yeomanry officer, 'like water someone else had drunk before!' They were much esteemed, however, for their rust-removing qualities! Many a stirrup-iron was thus rendered rust-free.[12]

When on patrol the water-duty man in each squadron carried out sterilizing procedures and mapped new wells. He was allotted a strip of the area covered by his advancing unit and maintained touch with the men on either side of him. He was meant to note the positions of all wells treated and the time at which the treatment took place. The water near the coast had a very low degree of salinity but inland it was always brackish. Men could just drink water of a salinity of 150 parts to 100,000, but it ought normally to be a long way under 100 parts. The Turks seem to have thrived on water which the British could seldom stomach. Horses had to be excessively thirsty to touch water that contained more than 150 parts. When they first drank any brackish water they lost condition until they became used to it. Camels suffered from its effects far less.[13] Medical officers erected notices over wells proclaiming: 'Drinking', 'Horse' or 'Not fit for horse'. These were not always paid attention to. Trooper Idriess of the 5th Light Horse wrote in his diary: 'We have been slipping away to drink from the horses' well. It is forbidden. The water is awful, but a man must drink something.' Later on wherever possible guards were posted over water-tanks to prevent this. 'Few men,' wrote an officer of the 6th Regiment, 'returned from a watering parade without a bucket on either foot or their section's water-bottles round their necks.'[14]

Lieutenant-Colonel Lachlan Chisholm Wilson, commanding the 5th Light Horse (who from October, 1917, was to command the 3rd Light Horse Brigade) had seen what were known as 'spear-point' pumps working in Queensland. In fact they were also well known in Egypt where they were called 'Abyssinia Wells' or 'Norton Tubes'. Wilson had several of them made in Cairo, each squadron soon carrying a set on pack horses. It consisted of a two and a half inch tube with a solid point. The tube sides were perforated and covered with a sheet of strong brass gauze. Wherever there was water within ten feet of the surface, these tubes were driven down by means of a small pulley bar and monkey or by a sledge hammer and when they reached the water-bearing strata, an ordinary General Service 'Lift and Force Pump' was attached to the tube. It took two men at most half an hour to hammer the tube in, whereas it took them half a day to dig a well to the same depth. The tubes seldom choked and the water brought up was, of course, free from surface pollution. The immense value of these pumps was soon recognized and several hundreds were soon being manufactured by the Australian and the Royal Engineers. It would be hard to exaggerate their importance to the mounted troops, for horses, especially, require a vast quantity of water to keep them going, each requiring five gallons daily to remain in condition for hard work. When, later on, the fertile regions of the Philistian plain were reached, the task of watering was, surprisingly, sometimes more difficult than it had been in the desert using spearhead pumps.[15]

8

'League after league of blazing silver sand;
Mile after mile of camel-weed and scrub;
Black marshes with their brine-encrusted ground,
Where sandhills roll out and spread around,
Ridge upon ridge of wind-swept shifting sand;
Deep sudden hollows where an army hides.'

'R.T.T.' in 'The Desert of Sinai.'

'How we wish we were going to France.... Numbers of our men have volunteered for the infantry, odd ones have cleared out to stow away with them. Anything to escape from this flying sand.'

TROOPER ION L. IDRIESS[1]

Food and drink – lice – flies – gunyahs *– health: septic sores, 'Barcoo Spew', malaria, cholera – Australians abandon formal clothing – transport of sick and wounded: camel cacolets, sand-sleighs, stretchers – desert monotony*

By the standards of past wars all ranks were well fed when in camp, especially of course the officers. 'Our appetites are splendid,' wrote Second-Lieutenant Blagg of the South Notts Hussars in May, 1915. 'I can manage tea and biscuits at 4.30 a.m. (reveille is now 4.15 a.m.) – porridge, eggs and bacon and marmalade at 7.30 a.m. and lemonade and bun at 11 or so – bread and cheese at 1 p.m. (& perhaps a slice of cold meat). Dinner – joint and two vegetables and pudding at 6 p.m. and *sometimes* I have a plate of rice and figs (very good) at 8 or so from the dry canteen.' Blagg's men got '1 lb meat, 1 lb bread, ½ lb mixed vegetables, ¼ lb potatoes, tea, sugar, pepper, etc. and 8½d a day for messing. The money does not go nearly as far as you would expect: bacon 1/10½d a pound, milk 2½d a pint . . . eggs cheap but very small.'[2] The Australians were if anything better fed still. Trooper Idriess wrote in August, 1916:

'We are being issued with good food, bread, bully-beef, occasionally Maconochie rations, potatoes and onions, jam, tea and sugar, occasionally coffee and bacon – far and away the best rations we have yet had. The Heads [familiar

> Australian term for generals] must have dug up all the camels in Egypt to supply us. Each section does its own cooking and these palms smell some tasty dishes. We cook in empty meat and biscuit tins and the ever handy quart-pots [known as "jackshays"]. Those of us not lucky enough to own knives use sticks.'

Though usually very few useful rations were recovered from captured Turkish positions, on one occasion the 6th Regiment acquired from a deserted camp 'tea, dates, meal, olives and several hundredweights of dried apricots pressed into rolls like brown paper. Under the comprehensive name of "mungaree" we learnt to envy the Turks its possession as a ration issue.'[3] 'Jackshays' had been issued to all ranks before leaving Australia and they were found 'much more suitable than the Army Canteen issue'. Later on, in Palestine, cooking in standing camps was done regimentally or by squadrons because of a shortage of fuel, all wood having to be brought from Egypt.[4] The Australians received very considerable quantities of goodies from home. Trooper Idriess often exults on the generosity of people and organizations. On one occasion 'the regiment received £100 worth of bucksheesh tinned fruit. Great! Good luck to whoever sent it.' On another 'the Victoria Racing Club sent us a splendid gift of lollies, 1st-class tobacco and some shirts marked "from the Mayoress of Melbourne" ', while the Sydney War Chest Fund sent 'a bottle of pickles, a tin of peaches and a tin of Bonza golden syrup' for each four-man section. When a consignment of tinned milk and fruit arrived from Brisbane, Idriess was 'sorry the ladies of the Cooee Cafe cannot realize how much we appreciate their gift.'[5]

Lice were a constant problem. 'A man's daily tally is about thirty,' wrote Trooper Idriess, 'except when he's right off duty and gets a chance to take his pants down, and then God knows how many he gets.' Private H.C. Brittain of the West Somerset Yeomanry dipped his shirt 'in a creosote bucket and dried it in the sun.'[6] An easier way of getting rid of what Idriess called those 'wretched damn things' was to enlist the help of colonies of ants. Any articles of clothing or bedding exposed on sand hills inhabited by them were speedily cleansed. The first 'delousing train' seen on the Sinai front arrived in January, 1917.

The swarms of flies always to be found in the desert were sometimes effectively reduced by the 'numerous tame chameleons of variegated hues' which lived on the tent-poles of standing mess tents. In the Worcestershire Yeomanry there was one nicknamed Cuthbert who 'held the record for eating fifty flies before breakfast!' The problems of refuse disposal in large standing camps had to be solved before the danger to health from the clouds of flies that bred in it could be banished. Since sand is weak in bacteriological activity the burying of refuse was no answer and shortage of firewood prevented simple burning. Improvised incinerators made of thousands of empty bully beef tins cunningly arranged to ensure plenty of draught proved to be the answer; but sieves had to be used to clear all rubbish of sand before it could be burned in them.[7]

Throughout much of the campaign there was a shortage of tents, which were not anyway required in the hot weather. A great deal of ingenuity was shown in creating 'bivvys' when units were stationary in hods. One such was made by Lieutenant McGrigor of the Gloucesters with 'a piece of matting, a waterproof sheet, two sacks and ten sandbags, four thin palm leaves upright and some string'. Less elaborate ones were made by the Australians. These were modelled on the aboriginal huts known as *gunyahs*. They were 'simply palm branches thrown together under the palms. We crawl inside,' wrote Trooper Idriess, 'exactly as the blacks do in their gunyahs.'[8]

Health in the desert was on the whole good, the worst affliction after heat stroke* and dysentery seems to have been septic sores known by the Australians as 'Barcoo Rot' and brought about by lack of fruit and vegetables. The slightest scratch from barbed wire could induce them and lead to serious inflammation and it was noticeable that when morale was low, as in the yeomanry after Katia, their incidence was greatly increased. As much as half the strength of a regiment had on occasions to go on sick parade for dressings, but cases had to be very severe for men to be excused duty, for when too many men were unavailable to look after horses the work thrown on to the others became intolerable. There was one mystery sickness which sometimes struck men in the desert. The Australians called it 'Barcoo Spew'. Men apparently in perfect condition would suddenly be overcome with nausea sometimes followed by actual vomiting for half-an-hour or so and completely

*During the height of the summer of 1916, about 75% of the Australasians developed temporary heart trouble. (Gullett, 107).

recover immediately afterwards. This was probably a form of food poisoning. Anti-cholera and anti-typhoid inoculations were compulsory for all ranks and smallpox vaccination took place whenever a case occurred in the vicinity.[9]

After the fall of Jerusalem Chetwode found it necessary to warn troops that it was

> 'full of the most virulent forms of venereal disease and a considerable number of cases have already occurred through men . . . consorting with women of Jerusalem and the neighbouring villages. In addition to the venereal disease, these women are filthy in the extreme in their persons and there is the greatest danger of men contracting typhus and other diseases from them as well . . . .
>
> 'Every man who wilfully consorts with these women commits a grave crime against his country by rendering himself unfit to fight her battles as a soldier.'

Until towards the end of the campaign, malaria was kept well under control except where marshy areas were taken over from the enemy and before the usual treatment could be carried out, the Turks being apparently virtually immune. The unhygienic conditions of their camps was chronic and cholera was only one of the diseases which were rife. An Australian unit once came across two abandoned sick Turks beside whom was a notice reading in English 'Attention! Cholera, with the compliments of the German Ambulance Corps.' Though the Germans did their best to introduce sanitary discipline to their allies, they found it an uphill task. However, in two field hospitals which the Australians captured in August, 1916, there were found 'all the instruments, fittings and drugs that science could supply'.[10]

The Australians very soon abandoned all formal clothing. Idriess reported that 'few, from the colonel down, possess tunics . . . . Some were dressed in riding-breeches, some in shorts; most wore leggings, others puttees. All were in sleeveless flannels, shirts or singlets. . . . Some of us were shaven, others not. All wore dilapidated hats, most still flaunting a tuft of the once proud emu feather or wallaby fur and often ventilated by a bullet hole.'[11]

* * *

The question of transporting the wounded and sick in desert conditions was a major one, for sand carts and wheeled ambulances (consisting of hooded wire mattresses) borne on broad-tyred wheels were out of the question in heavy sand, while the cacolets, made up of a deck chair contrivance or a man-length box swung either side of a camel proved to be appliances of hideous torture.* The Australians included in their regimental establishments mounted stretcher-bearers who in the early Sinai days were the only efficient means of carrying the wounded. They soon contrived simple sand-sleighs made of galvanized iron sheets turned up at the front and drawn by two horses. More sophisticated sleighs were later introduced by the New Zealanders and adopted by the yeomanry. They solved the problem highly satisfactorily. Early in 1917 the yeomanry regimental saddlers devised a new type of cavalry stretcher. It consisted of two light bamboo poles four feet long joined by a piece of canvas three feet by nineteen inches. It could be carried attached to the sword and proved invaluable when full-sized stretchers were not available. Later on in the mountainous country of Palestine they were carried with ease through the narrow clefts in the rocks.[12]

The New Zealanders were responsible for introducing dental units and before long each mounted brigade was provided with one. Simple cases and plate-work were dealt with in the field, thus avoiding temporary evacuation of large numbers of men. In one brigade nearly 500 cases were treated in one month.[13]

* * *

Inaction in highly uncomfortable conditions bred, of course, intense boredom. 'This cursed desert monotony is nerve-racking,' wrote Trooper Idriess in July, 1916. 'We are the Lost Legion. Some of the light horse have actually managed to smuggle right away to Port Said and there stow away on transports en route to France. Many of us have tried hard to get away legitimately. Curse this inaction.'[14]

*After the battle of Romani two Australians with broken thighs rode their horses for upwards of seven miles so as to avoid the excruciating torment of the camel cacolets. One of them survived. (Gullett, 185)

9

'Good progress was at first made, but late in the day the main attacking force was taken in flank by a British cavalry division and lost so heavily that it was decided to break off and fall back.'

COLONEL KRESS VON KRESSENSTEIN
of the battle of Romani

'Murray's design to fight upon his own chosen ground at Romani was to be fulfilled to a degree rare in warfare.'

Australian Official History

'There can seldom have been a battle the general course of which could be more clearly foreseen than that of Romani.'

WAVELL in *The Palestine Campaigns*

'The enemy who were in 1000s on foot, could see our horses before we could see them in the half light and it was awfully difficult to find cover for [the horses].'

CHAUVEL to his wife, 8 August, 1916

'In the pursuit after Romani, sand made a sound cavalry pace impracticable. If the enemy parties on foot had a start of a few hundred yards, they had very little difficulty in getting away.'

Australian Official History[1]

Chauvel – Cox – Ryrie – Antill – Royston – Chaytor – Romani – the pursuit to Bir el Abd

During this mid-summer period of comparative inactivity in the desert, Murray was constantly pressed to send more and more formations off to France. By the end of June only four out of the twelve infantry divisions which had been in Egypt in January were still there. He even feared for his mounted troops. 'I am assuming,' he told the War Office in May, 'that you are leaving the three Australian light horse brigades and the New Zealand brigade with me. Otherwise I shall be deprived of the only really reliable mounted troops I have.'[2] Fortunately he was not thus deprived and it was upon his Anzac Mounted Division that he was chiefly to rely in the coming operations.

He was exceptionally fortunate in having the fifty-one-year-old Chauvel in command of it. He was a wiry little man, the grandson of a retired Indian army officer who owned a New South Wales cattle station. He had served in the Queensland Mounted Rifles in South Africa, commanding later one of those composite mobile 'circuses' which sprang up towards the end of that war (see Vol. 4, p. 215). In August, 1914, he was on his way to London to act as his country's representative on the Imperial General Staff. He was at once ordered to take command of the 1st Light Horse Brigade which was on the point of landing in Egypt. In Gallipoli he commanded first a brigade and for a short time the 1st Australian Infantry Division. He was perhaps the least dashing of the truly distinguished leaders of mounted soldiers in history. Shy, gentle of speech and quiet of bearing, he was a zealous disciplinarian but a very just one, deeply respected but in no way a soldier's idol. Nevertheless he was particularly solicitous for the lives of his men and horses. General Birdwood (see Vol. 4, pp. 72, 73), who was the General Officer Commanding all the Anzac forces throughout the war, thought him slow and lacking in decision. General Sir Ian Hamilton even went so far as to write that he was not a leader of men.[3] But these views were expressed before the Sinai campaign got under way. After the battle of Romani, a trooper wrote home that Chauvel was 'well liked by both officers and men. A chap feels pretty safe with a leader like him. I saw him riding backwards and forwards under heavy fire . . . and it seemed that he did not know what danger was.' A young subaltern attached in 1917 to Chauvel's headquarters wrote of him: 'He is a wonderful man, as wiry as half a dozen ordinary men. He works from early morning till very late at night [and] is a great favourite amongst all with whom he comes in contact.' Allenby and he got on exceptionally well. The day before Jerusalem fell, he wrote of the Commander-in-Chief:

> 'He is the most energetic commander I have yet come across. I like him immensely and he appeals to my Anzacs tremendously. He is just the kind of man we wanted here. He knows what he wants and sometimes explains it in no measured terms but he generally gets it done. The great thing is he gets about amongst the troops, looks in at hospitals, etc., has a cheery word for the wounded and does not have a fit if he's

> not saluted [showing that he had mellowed considerably!], all of which appeals to the Australians and I think to the other troops also.'

Sometimes, though, he got on Allenby's nerves. His habit of day-dreaming during conferences led to the Bull's clenched knuckles whitening 'in a build-up to an eruption of rage'. In matters directly affecting the Anzacs he usually got his way. The only important matter in which he failed to persuade was the appointment of Howard-Vyse (see p. 277), a non-Australasian, as his chief of staff, an arrangement which in the event worked well – for a time.[4] He was the very antithesis of the British and European cavalry general: always cool, utterly unflamboyant, meticulous in preparation, precise in execution and a dedicated student of military history. No other cavalry commander had such a deep understanding of the pre-eminent purposes of mounted troops. Reconnaissance, scouting, patrolling, probing ahead, protecting flanks and energetic pursuing were for him their chief functions. He was one of the few really outstanding commanders of mounted riflemen, never forgetting that in modern war much more of the actual fighting had to be done on foot than in the saddle and that his light horsemen were in truth infantrymen made mobile by their horses. Many of his officers and men had served in Gallipoli and knew well what it was like to be a foot soldier. Though, as will be shown, there were occasions when full-blooded charges were undertaken by his troops, they were very seldom embarked upon except against semi-demoralized infantry and almost invariably with strong fire support. Both Murray and later on Allenby knew that Chauvel was entirely dependable and that he would never take unjustified risks. They appreciated, too, that though not perhaps a typical Australian, he understood the strengths and the weaknesses of his compatriots to a tee.*

Throughout the campaign Chauvel's General Staff Officer Grade I (GSO 1) was Lieutenant-Colonel (later Brigadier-General) John Gilbert Browne of the 14th Hussars (of the history of which he later became joint author). He was of Australian birth and family and had served in the Boer War, narrowly escaping with his

*The mischievous picture of Chauvel given by Lawrence in *Seven Pillars of Wisdom* is (as is so often the case in that literary masterpiece) entirely discountable.

life at Van Wyksvlei (see Vol. 4, p. 205). He was able and trustworthy. So was the New Zealander, Lieutenant-Colonel Guy Powles, Assistant Adjutant and Quartermaster General (AA and QMG), who also remained with Chauvel throughout the war. He later wrote the history of the New Zealanders in the campaign. Major William Stansfield was the Division's Army Service Corps commander and Lieutenant-Colonel Rupert Major Downes was its Director of Medical Services. Upon both men's exceptional organizational powers Chauvel relied throughout the campaign.

When he became commander of the Anzac Mounted Division Chauvel was succeeded by the fifty-three year old Lieutenant-Colonel Charles Frederick Cox in command of the 1st Brigade. Unlike Chauvel he was not a professional soldier. Known variously as 'Fighting Charlie' and 'Damn-all Cox' he had been a railway clerk in Australia when he enlisted as a trooper in the New South Wales Lancers. As a captain he commanded the handful of lancers who went to England for the Jubilee and for training in 1897 and whom he took to South Africa on the outbreak of the Boer War (see Vol. 4, pp. 82, 137). There as a Major he had won a CB. Well over six feet in height, he was a very typical old-style cavalryman, who left much of the conduct of his brigade in the hands of his staff, but who had the good cavalryman's sure instinct in the thick of battle.

Brigadier-General Ryrie who commanded the 2nd Brigade weighed over sixteen stone but was a superb horseman nevertheless. He was a contemporary of Chauvel, but could hardly have been more different. Chauvel admired his unruffled courage as did the whole of the division, but he found him 'a Member of Parliament first and a soldier afterwards'.[5] He was an exuberant extrovert who made a point of knowing as many of his men by name as possible and who lived as his troopers lived, sharing their discomforts. His competence as a brigade commander was considerable.

Brigadier Antill is a shadowy figure, whose inexplicable decision to retire with his 3rd Brigade when it was most needed at Romani (see p. 64) made it easy for Chauvel to let him go soon afterwards at Birdwood's invitation to command an infantry brigade. He was succeeded by Colonel John Robinson Royston who was in temporary command of the 2nd Brigade at Romani. A South African farmer-soldier, aged fifty-seven, who had fought with the Imperial

Light Horse in the Boer War he later raised his own regiment, Royston's Horse. Known as 'Galloping Jack', he was probably the best loved figure in the whole division in spite of not being an Australasian. He, like Ryrie, was a huge man without it seemed the slightest regard for danger or knowledge of what it was to be tired.

* * *

With what troops the exigencies of the war in France had left him, Murray set about establishing a solid position around the railhead and wells of Romani. His marked superiority in mounted troops was somewhat offset by his inferiority in the air where the newly arrived two-seater Aviatiks of the Germans were faster and more manoeuvrable than the BE2Cs which formed three-quarters of the British force. There were only two Royal Flying Corps squadrons in Egypt at this date, one of them belonging to the Australian Flying Corps and even their best machines, the De Havillands, of which there were but three, were five miles an hour slower than their rivals. One Australian compared the British to the German aeroplanes as a draught horse to a thoroughbred.[6]

* * *

When in July Kress decided that he was in a position to advance again,* it seems that he intended to reach and entrench a position within gun range of the Canal so as to be able to block it and thus interrupt traffic through it. He must have been aware that the EEF had been much reduced in size and he resolved to take advantage of that fact. His total force, including a good Anatolian infantry division, numbered about 15,000 men, of whom perhaps 12,000 were riflemen, supported by thirty guns and thirty-eight machine guns. The heavy guns had been brought forward by running them along planks placed end to end which were repeatedly picked up and carried forward, or by packing small trenches with the prickly bush of the desert. Nearly 5,000 camels and 1,750

*He had been awaiting the arrival of what was known as 'Pasha I'. This formation included a squadron of aeroplanes, two Austrian heavy mountain batteries, four artillery batteries, two trench-mortar companies, five groups of anti-aircraft artillery and five machine-gun companies.

horses accompanied the columns.* It took twenty-five days to reach Katia from Shellal, marching only at night and halting from time to time to build strong entrenchments. It was an astonishing feat at the very height of the summer, which could only have been achieved with air superiority.

* * *

One of the first indications that the enemy was on the move in large numbers came from a brigadier-general's personal reconnaissance in an aeroplane. This must surely have been the first time in history that a senior officer had flown a 'Tac R'. Brigadier-General Edward Walter Clervaux Chaytor was the forty-eight-year-old New Zealander who commanded the New Zealand Mounted Rifles Brigade. Chauvel trusted him above all his subordinate commanders.[7] It was said of him that he was one of those rare soldiers who did everything so surely, thoroughly and yet so quietly and with such apparent ease that it might be said no task set him was big enough to test his full capacity.[8] On 19 July Chaytor reported, by dropping a message from the air, that he had seen large concentrations of Turks at Bir Bayud. Thus alerted, Major-General the Hon. Herbert Lawrence (later to become Haig's Chief of Staff in France), in command under Murray of the northernmost Canal Defence Section in which the Romani position was included, took steps to strengthen it.

The Romani position was a naturally strong one for it was practically impossible for the enemy to approach it from his right flank. He was therefore committed to an attempt to envelop the British right, aiming to cut the railway to the west of Romani. This would entail negotiating the intricate mass of tall dunes to the south and east of the British position where there was no water, where the sand was too soft and heavy for trench making and where even Turkish infantrymen could hardly march at more than

*Much of Kress's equipment had been specially made in Germany for the expedition. The machine-gun and mountain-gun packs were of a design much superior to that of the British who learned much from them. Their camel pack equipment and saddles 'were quite the best thing of their kind' while a manual printed in Arabic on the care of camels 'contained nothing but the soundest advice' which came in very useful for the Camel Transport Corps. (Foster, 142)

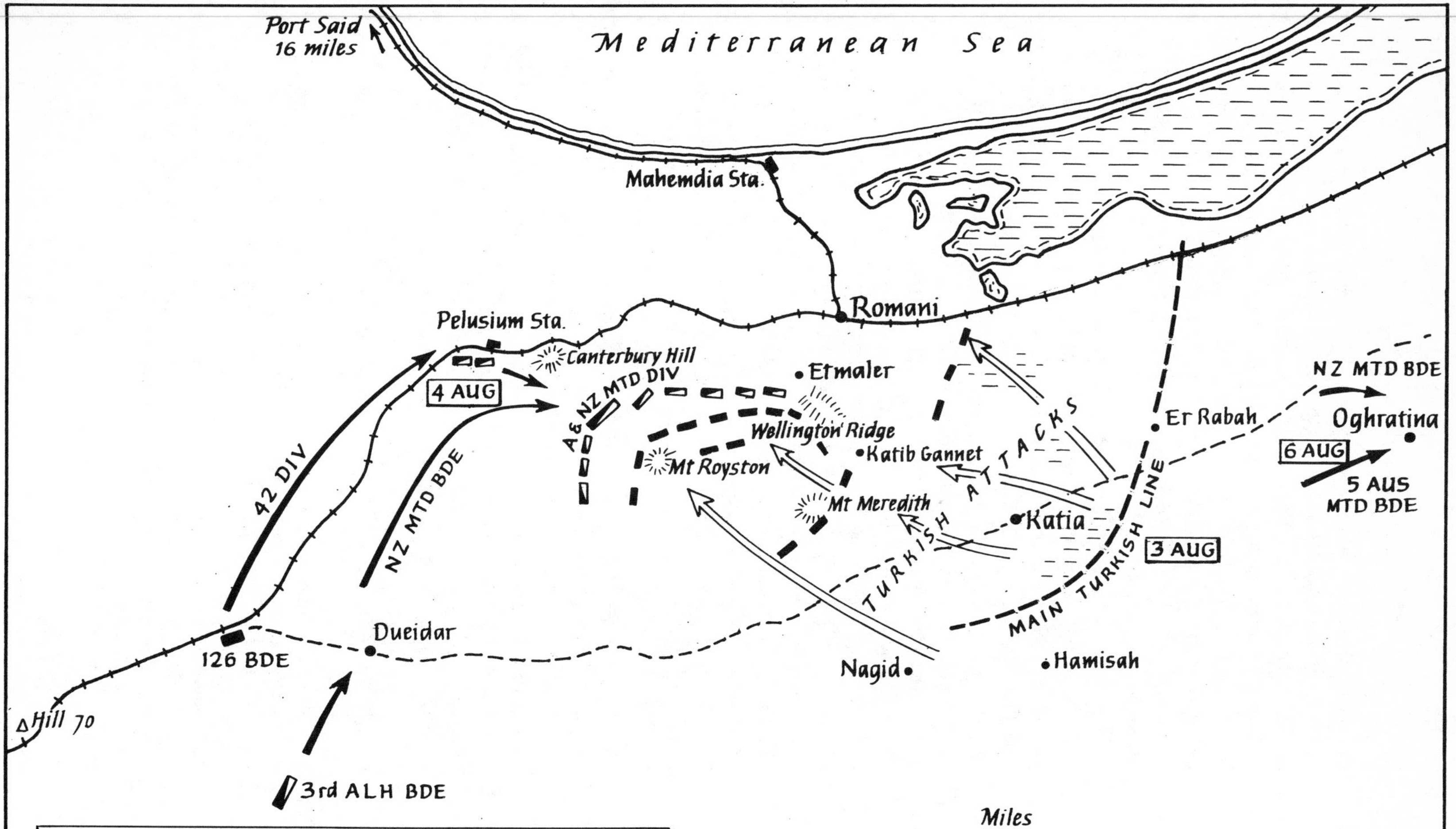

BATTLE OF ROMANI, AUGUST, 1916

one mile an hour. That Murray and Lawrence were hoping that Kress would conform to this pattern is less surprising than that so experienced a commander should actually have done so. The British plan was to allow him to involve his main strength in the dune area and then to counter-attack in the centre with the infantry while sending the mounted troops to get round his left. If the mounted troops could then pursue, Kress's force ought to be cut off and annihilated.

The plan worked well enough except for the last part of it, as will be shown. Part of the reason for this was the unsatisfactory nature of the command structure at this time. Murray sitting in Ismailia maddened Lawrence at Kantara by tending to tell him exactly how to fight the coming battle at Romani; but worst of all the command of the mounted forces was divided. While the slow, methodical and menacing advance of the Turks proceeded, Lawrence formed a new force called Section Mounted Troops by withdrawing Chaytor's New Zealanders (less the Wellington Regiment) from Chauvel's division and adding some yeomanry and two horse artillery batteries all under Chaytor. As late as 26 July Murray began to organize a separate Mobile Column under Brevet Lieutenant-Colonel Clement Leslie Smith, VC, of the Imperial Camel Corps (see p. 18), while Lawrence kept under his direct command Antill's 3rd Brigade at Ballybunion, well to the west of Smith's force.* All this meant that Chauvel had under his command at Romani only the 1,500 or so men of the 1st and 2nd Brigades (the latter of which had the Wellington Regiment attached to it), and that the rest of the mounted troops were, in effect, under three different commanders. Lawrence hoped to coordinate and control all four from Kantara which was twenty-three miles from Romani. He had disregarded Murray's advice that he should move his headquarters nearer to the battlefield.

By 24 July, five days after Kress's line had advanced to the Oghratina oasis area, where he spent ten days digging in and

*The headquarters of Section Mounted Troops under Chaytor were located at Hill 70, six miles from Lawrence's headquarters at Kantara. Beside Chaytor's brigade were the 5th ALH Regiment (replacing the Wellington Regiment), two squadrons each of the Warwicks and Worcesters and one of the Gloucesters as well as the Leicestershire and Somerset batteries.

Smith's force was made up of three Imperial Camel Corps companies, the 11th ALH Regiment (less one squadron) and the City of London Yeomanry (less one squadron).

awaiting his guns, Lawrence had two and a half infantry divisions either manning his entrenchments, along the railway or some way behind in reserve. His redoubts, immediately to the east of Romani, ran south for six miles to the sandhill of Katib Gannet. In and behind them were thirty-six guns, including four sixty-pounders.

During the long wait before Kress came within striking distance of Romani, the 1st and 2nd Brigades, commanded respectively by Royston in Ryrie's absence in England and by Lieutenant-Colonel John Baldwin Meredith in Cox's absence on sick leave,* moved out in turn before dawn each morning, reconnoitred the enemy's successive positions and returned to camp after dark. In the course of these probing patrols numbers of minor engagements took place;† casualties occurred and prisoners were taken. These talked freely, revealing pretty precisely the numbers, identity and quality of the enemy force.

The whole process was very exhausting for both horses and men. Throughout this period Chauvel and his brigade commanders studied the details of the ground over which the enemy was expected to advance. He gave new names to some of its more prominent features such as 'Mount Meredith' (230 feet) and 'Mount Royston' (220 feet). From Dueidar to Katib Gannet, to the south and west of the fortified posts, there was a further line of listening and cossack posts sent out by Chaytor. 'These posts,' wrote Trooper Idriess, 'ride out every night – wee bunches of men creeping close to the Turks. The Listening Posts listen! If they hear the Turks coming they fade back into the night without firing a shot. The Cossack Posts, on sight or sound of Turks, open a furious fire, leap on their horses and gallop back for their lives.'[9]

*As an Australian M.P. Ryrie had obtained leave to attend an Empire Parliamentary Conference in London. Chauvel was not pleased. (*Chauvel*, 76).

Meredith was a New South Walian born in Ireland, aged 52, who had served as a medical officer in the Boer War. He had commanded the 1st Light Horse Regiment from August, 1914 and later commanded the 4th Light Horse Brigade from its formation in February, 1917, until being invalided home in September.

†'The sand muffled all noise made by the Turkish infantry, but the Australians' horses frequently neighed and brought the men under fire at close range.' (Gullett, 128)

On 3 August Kress took advantage of the Australian brigades' ill-advisedly regular routine and followed close on the heels of the 2nd as it withdrew that evening. He hoped to take what was known as 'Wellington Ridge' by surprise overnight. Chauvel, however, suspecting that 4 August, which was the last day of the Mohammedan feast of Bairam, would be the day of the attack, ordered the 1st Brigade to stay out in the outpost line so as to cover, over a distance of about three miles, all the entrances to the sand hill plateau south of the infantry redoubts. Of the existence of the brigade in this area the enemy was quite unaware. Indeed its position was actually on the ground which Kress had chosen as the forming-up place for his assault. When his advance guard discovered it, it halted, waited for further orders and for the arrival of his main force, thus wasting valuable time. Having first opened fire at about 10.30 p.m., it was not until just before 1 o'clock in the morning that the Turkish riflemen attacked, yelling alternatively 'Allah! Allah!' and 'Finish Australia!' The dismounted Australians replied along the whole line. At first there was a low quarter-moon and both sides could see ahead for short distances. It took another hour or so before the Turks got within thirty or forty yards of the light horsemen, who by then were aiming at the flashes of the Turkish rifles, for the moon had just set, complete darkness descending on the field of battle.

Throughout the night it raged, the enemy making slow progress against the greatly outnumbered Australians of the 1st Brigade, who held grimly on behind the smaller dunes and tussocks. Before long they were so hard pressed that they had to fall back on their led horses carrying with them their wounded. They then took up another position from which to carry on the fight.

> 'The bullets,' wrote Lieutenant-Colonel G.H. Bourne, a Brisbane bank manager in command of the 2nd Regiment, 'were making little spurts of flame all round us, owing to the phosphorus in the sand. Here we experienced for the first time the moral effects of turning our backs on the enemy, and the question arose in our minds as we rode, "Can we re-form?" The order "Sections About – Action Front!" was given as we reached the position and was splendidly carried out. This high test of discipline gave us renewed confidence.'

Casualties mounted on both sides as the battle became a troop-leaders' fight, for in the darkness little co-ordination was possible.* At about 3 a.m. the Turks captured Mount Meredith, charging with the bayonet to get there. They lost no time in establishing machine guns on its summit. Not long afterwards the 1st Brigade was forced back over Wellington Ridge, but in spite of repeated efforts throughout that scorching day of 4 August the enemy never managed to establish himself on its crest. The Turks were now within 700 yards of the 2nd Brigade's camp at Etmaler.† Royston, riding along the firing line at this crucial moment, spoke to Lieutenant-Colonel William Meldrum,‡ commanding the Wellingtons (known to the Australians as the 'Well and Trulies').[10] 'You can give them no more ground,' said Royston, 'or we shall lose the camps.' 'If they get through my line here,' replied Meldrum, 'they can have the damned camps.'[11]

It was not until dawn at about 4.45 a.m. that the massed Turkish infantrymen could see how slender was the line of the light horsemen. However, daylight gave the Australians a rare target at close range of which they took fullest advantage. It also made it possible for the artillery of both sides to come into action for the first time. The horse artillery employed to the full from now on the mobility which their pedrails§ conferred on them. Further, Chauvel, Meredith, Royston and the regimental commanders were now able to exercise some degree of control over the battle's

*Far to the south-west, shortly after midnight, the 5th Regiment (attached to Chaytor's New Zealand Brigade) had marched from Dueidar in an attempt to discover the enemy's left flank. At Nagid just about dawn it did so, being fired upon by two Turkish battalions. The regiment then retired to a point three miles on the road back to Dueidar. There it lost contact with Chaytor's headquarters and was therefore of no more use that day. (Gullett, 154)

All the times given are very approximate.

†The 7th Regiment was so close to its base that in the course of the afternoon, the camp cooks, under heavy fire, served the men with tea as they lay in their shallow, scooped-out holes along the firing line. (Gullett, 156)

‡Meldrum was a barrister, solicitor and farmer, who, in 1896, won the New Zealand chess championship. He died in 1964, missing his 100th birthday by five months.

§Invented by B.J. Diplock in 1902, the pedrail was a device intended to replace the wheel of a self-propelled vehicle for use on rough ground. 'The tread consists of a number of rubber-shod feet which are connected by ball-and-socket joints to the ends of sliding spokes. Each spoke has attached to it a small roller which in its turn runs under a short pivoted rail controlled by a powerful set of springs.' (*Webster's New International Dictionary*) progress. Chauvel had wisely held back most of the 2nd Brigade until he could see exactly what was happening to the 1st Brigade. He now gave Royston his head, ordering him to extend the line from the 1st Brigade's right to the west. But it took some time before he could make much impression on the situation. Here was the crisis of the battle. Mount Royston, less than three miles from the railway, was at last taken. At this point, as an officer of the 6th Regiment wrote: 'The ordinary man in the ranks might have been pardoned for imagining Jacko [as the Australasians often called the Turk] had scored a bit of a win . . . . His artillery roared, aeroplanes, unopposed, bombed our camps, led horses and railhead depots to their own sweet will [*sic*]; ineffectively, nevertheless, thanks mainly to the heavy sand.'[12]

The sole reinforcement which Chauvel received at this time was 'D' Squadron of the Gloucesters which was covering the railway at Pelusium Station. Its commander, Major Charles Edward Turner, noticed that infantry were moving about on Mount Royston. Without orders, in the best military tradition, he marched his squadron towards it and quickly became engaged. His prompt action at this critical juncture delayed the enemy's outflanking movement for more than an hour. At about 6 a.m. the Leicester Battery of horse artillery came into action near Etmaler and drove the Turks off Mount Meredith. It then shelled the ground south of it with what was seen afterwards to be great effect.

Throughout this anxious time Chauvel, who had been riding from one brigade to the other under the Turkish shrapnel, was searching the horizon for reinforcements, especially Chaytor's New Zealanders from Hill 70, twelve miles away. In the meantime the Turks were still threatening to get round his open right flank, though their pressure was beginning to slacken through lack of water in the appalling heat.

At this time a battalion commander from the 52nd Division on the 1st Brigade's right, without orders, took over ground so as to enable Chauvel to extend to his right. Later, when Chauvel wanted to relieve his desperately tired light horsemen so as to give them a rest before counter-attacking, he was refused further infantry help on the grounds that the 52nd Division itself, all of whose troops, except for being shelled in their redoubts, had seen no action, was about to counter-attack! A single commander on the spot would have been able to solve all such problems, but Lawrence was twenty-three miles away. Two 18-pounder batteries from the 52nd

Division, as well as the horse artillery, concentrated at this time on the hods and depressions round Mount Royston.

At about 7 a.m. the direct line between Chauvel and Lawrence was cut. They could only communicate slowly and with difficulty by a circuitous route through Port Said and occasionally by wireless. Before that, at 5.35 a.m., Lawrence had ordered the rest of the yeomanry to move from Hill 70 to Mount Royston, but, frightened of throwing in his reserves too early, and still unsure whether the enemy was making for the railway directly to the west, Lawrence took some persuading by Chauvel before, at 7.25 a.m., he ordered Chaytor to move on Canterbury Hill to cover the railway.* However Chaytor, learning by heliograph that Mount Royston had fallen, replied: 'Advancing to attack Mount Royston'. Royston at once galloped up to Chaytor and put him fully into the picture. It was said that Royston used up fourteen horses this day and that in action he looked ten years younger.†

Chaytor's attack began at last between 1 and 2 p.m. in the fierce midday heat‡ and he was soon joined on his right by the yeomanry

*The British plan had been to use the New Zealanders and the yeomanry (5th Mounted Brigade) against the enemy's rear, but the speed and weight of the attack forced both brigades to be diverted to strengthen the line of resistance to the north. (See Foster, 132).

Antill's 3rd Brigade at Ballybunion was now ordered up to Hill 70. Smith's Mobile Column was ordered by the commander of No. 2 Canal Defence Section (south of Lawrence's No. 3 Section), to move on Mageibra, but Lawrence asked for it to be directed on Hamisah. Murray intervened at this point to try to co-ordinate Antill's and Smith's movements, but to such little effect that Smith's men never even began to come into action this day.

†Royston would race up to his men 'with yards of blood-stained bandage from a flesh wound trailing after him. "Keep moving gentlemen, keep moving," was his constant advice to his officers. And to the men, "Keep your heads down, lads. Stick to it, stick to it! You are making history today." To a hard-pressed troop on the naked [right] flank he cried: "We are winning now. They are retreating in hundreds." "And," said one of the light horsemen afterwards, "I poked my head over the top and there were the blighters coming on in thousands." ' (Gullett, 157).

‡When at about 3 p.m. Chauvel telephoned Royston's headquarters for news of Chaytor's attack, he was told: 'Colonel Royston is wounded and has gone for another horse.' Chauvel at once rode off in chase and personally ordered the colonel to a field ambulance to have his wound dressed. (Chauvel to Gullett, c.1921, quoted in *Chauvel*, 79).

from the north-west. For the next four gruelling hours these four brigades, two of them much weakened and thoroughly exhausted, continued the fight as infantry against stubborn opposition along the whole front which had become stationary, the artillery of both sides pounding away.* Soon after 6 o'clock Mount Royston was retaken with some 500 prisoners. The Turks had broken at last. The whole British line now began to sweep forward, gathering a further 1,500 prisoners as it went, and it did not halt till darkness intervened. The New Zealanders and the yeomanry then retired to Pelusium to water. Firing continued until midnight, the Turks from time to time sending up star shells in case there should be a night attack against them.

* * *

In the course of the battle the men of the 1st Brigade had been heavily engaged for most of a twenty-hour period and the 2nd for not much less. Fighting as they were so close to their camps, many of them, as has been shown, were able to enjoy 'their customary hot tea and full rations'.[13] The occasional watering of the led horses had also proved possible. The horses of the 6th Regiment, however, had not been watered on the night of 3 August. By the night of the 5th they had gone waterless for sixty hours. Trooper Idriess was not alone in remarking with what extraordinary fortitude the horses stood the shelling. In Gallipoli he 'used to imagine that horses would go mad when under shrapnel fire, but here they were with it bursting right amongst them, our own guns blazing away under their very noses, and only by the lifting of a few startled heads, just a little uneasiness as shells screamed

*At 5 p.m. the yeomanry and the New Zealanders made a determined if minor counter-attack. Brigadier-General Wiggin's report stated that 'with the NZs on his left [Lieutenant-Colonel Yorke] led two squadrons [of the Gloucesters] at a big ridge, the southern spur of Mount Royston, and took it at the gallop.' Yorke's report gives details of the duel between four mountain guns and the troop of Lieutenant F.A. Mitchell, which ended in the guns' capitulation. (Fox, 103–5).

At this time and at others the yeomanry offered their helmets to the New Zealand machine gunners to be used as 'flame extinguishers' to hide the flashes as darkness came on. The helmets were held over the muzzle and the gun fired through them. The life of a helmet was not a long one, but its work was very effective. (Powles, 36).

exceptionally low overhead, did they betray perturbation at all. Then they would settle quietly down again.'

The Medical Officer of the Worcesters found it sad work bringing back the seriously wounded, 'tied to their horses. . . . At the dressing station cases were dressed and placed under shelters formed out of horse-blankets and swords.' At Pelusium that evening he

> 'found what seemed in the dark to be indescribable confusion – mounted troops coming in to water, infantry detraining and marching out, busy Army Service Corps depots, camel convoys loading up, ammunition columns on the move, wounded arriving in sand-carts and large columns of prisoners being mounted in. After long delays our horses were watered . . . . It was extraordinary how the horses could smell the water long before they reached it; my horse got very excited before I had any idea that we were near water, and then made a rush for it.'[14]

* * *

At 6.30 in the morning of 5 August Lawrence ordered Chauvel, as he surely ought to have done the previous day, to take command of all the mounted troops (except Smith's remote Mobile Column) and to pursue. The brigades were so scattered that it took till 10.30 before the pursuit could begin. By far the freshest if least experienced of the brigade's was Antill's 3rd on Chauvel's right flank. From that brigade the 9th Regiment, soon joined by the 10th and supported by the Inverness battery, made contact at Hamisah, and attacked with great vigour, first galloping mounted and then with the bayonet dismounted. They took 435 prisoners including ten officers, as well as seven machine guns.[15] Their own casualties were trifling. When all seemed to be going particularly well, in spite of continuing heavy artillery fire, Antill took the amazing and never-explained decision to withdraw his brigade to Nagid. Chauvel was thus deprived of the essential co-operation of his only unwearied troops, equal in numbers of riflemen to his 1st and 2nd Brigades combined, at the very time of his advance towards Katia. In this he was supposed to be assisted by the infantry of the 52nd and 42nd Divisions, but their rate of progress was painfully slow* and they failed throughout the pursuit to keep up anywhere near the light horsemen. Without their support Chauvel with the 1st and 2nd and New Zealand Brigades attacked the strongly defended Katia position, charging with bayonets fixed in two lines, squadrons in line of troop columns, and shouting as loudly as their parched throats would allow. It was the first time that the Australasians had attempted a cavalry charge, but unfortunately there was a swamp in front of the Turks' entrenchments and, as soon as it was reached, the leading horses floundered to an ignominious halt, 'bogged to their knees'.[16] There followed an unpleasant moment while the second line of horsemen piled up behind the first, giving the enemy a fine target. Orders to dismount were yelled out, the horses were galloped back to cover and the men proceeded to toil through the quagmire. They got to within 700 yards of the enemy, but could advance no further. On Chauvel's left the Yeomanry Brigade was checked by heavy fire at Er Rabah. On his right the 5th Regiment and the Aucklands were equally held up by artillery fire. Soon after sunset Chauvel was forced, therefore, to order a general withdrawal to Romani. As the Official History puts it: 'Of the expected demoralization in the Turkish ranks there was no sign. Their fire was hot and well directed, while their artillery out-gunned the supporting Ayr and Somerset Batteries.'[17].

Next morning, 6 August, the New Zealanders and the yeomanry found the Katia position abandoned. The Turks had fallen back on another of the fortified positions which Kress's foresight had prepared – at Oghratina. Throughout the day, another excessively hot one, no progress could be made against it.† During the whole of 7 August there was continuous fighting, but the enemy's artillery succeeded in holding back the light horsemen. In the course of the night the enemy evacuated the Oghratina position and by evening

*The infantry's delayed start was largely due to inexperience in loading the camels with their fanatis, while their slow progress, especially that of the 42nd Division, was due to numbers of their men being raw troops unused to desert conditions.

†The infantrymen were so exhausted that hundreds of them became stragglers, many in a delirious state, so much so that on the following day the Bikanir Camel Corps, a detachment of the yeomanry and even aeroplanes had to be employed in searching for them.

4. Starting out on a desert journey

5 .Camels bearing water in the desert (see p.33)

6. Two Cacolet Camel
Sitting-up cases from t
battle of Romani (see p

7. 'Our Water Suppl

of 8 August had retired to the entrenchments at Bir el Abd from near where he had emerged three weeks earlier. By now Chauvel's five brigades, exclusive of horse-holders, numbered less than 3,000 rifles, some regiments putting in the firing line as few as 180. The line in the morning of 9 August extended over some five miles which meant that there was less than one rifle to every three yards. At 5 a.m. the New Zealanders drove in the Turkish outposts, but four hours later the Turks, who had received considerable reinforcements, counter-attacked and by 10.30 all progress came to a halt. Further counter-attacks and massive artillery fire continued until, at 5.30, Chauvel was forced to order a withdrawal to Oghratina. This proved far from easy. For dismounted horsemen to disengage from action against very superior numbers at short range was never a simple operation. Troop alternated with troop and squadron with squadron. Carrying the wounded with them, the irregular line fought its way back to the led horses. Men's elbows became blistered with constant contact with the scalding sand as they held their burning rifles.*

The British losses this day came to over 300 of which eight officers and sixty-five other ranks were killed. No doubt Chauvel took a considerable risk in undertaking this further attack, but if it was a mistake it was about the only one he made in six days of intensive fighting. On 12 August Kress withdrew to El Arish from where he had started what was to be his final attempt to threaten the Canal defences.

* * *

That Romani was a decided victory for British arms cannot be doubted, but equally certain is it that Kress managed to escape the trap into which he had marched. Nevertheless, he lost much matériel and about 9,000 men – a third of his force – of whom 1,250 were buried by the British, and nearly 4,000 made prisoner. But he had saved all his artillery except one mountain battery and all but nine of his machine guns. His leadership, his staff work and the amazing endurance and resilience of the Turkish infantrymen both in the advance and in the retirement were of the highest order.

*In the 8th Regiment an elderly trooper and his three sons made up one section. In the fighting this day he was acting as their horse-holder. An enemy shell killed three of the four horses but he escaped unhurt. (Gullett, 182)

The total British losses were 1,130, of which over 900 came from the Anzacs.*

* * *

The lessons which had to be learned from the battle were numerous. Chief among them was the question of divided command. 'Criticism should be directed,' wrote Chauvel in 1921, 'at Murray rather than at Lawrence. The battle should either have been Murray's or, if it had to be Lawrence's, he should have placed all the troops available for it at Lawrence's disposal.'[18] It has been said that Chauvel's pursuit was too direct, that insufficient efforts were made to work round the flanks.[19] Yet to have placed his much weakened and inferior force across the enemy's rear without close infantry support would surely have been to invite disaster, even if the supply situation had allowed of it. One of the more important lessons learned was that for a long-distance pursuit a far greater camel train than was employed after the enemy began to withdraw from Romani would have to be provided in future. There had not been sufficient camel transport for the mounted division and one infantry division to operate for more than two days' march from railhead. This had meant withdrawal at night for watering and supplies, followed by advance again in the morning, which vastly added to the exhaustion of men and beasts.

On 8 August Murray telegraphed to London: 'Considering that we are operating in the Sinai Desert in the month of August, I think you may feel assured that we are doing all that is physically possible.'[20] This and Chauvel's remark that 'it was the empty Turkish water-bottle that won the battle'[21] are fitting commentaries on the first decisive major battle of the war. Romani marked the end of the Turkish campaign against the Canal. The initiative now passed to the British. As for the Australasian light horsemen, their pluck, dash and endurance gave them well justified self-confidence.[22] At Romani they began to show those qualities which over the next two years were to make them into perhaps the finest

*Of the total, twenty-two officers and 180 other ranks were killed, eighty-one officers and 801 other ranks wounded and one officer and forty-five men taken prisoner or missing.

846 horses died, were killed, destroyed or missing between 3 and 12 August. (Blenkinsop, 164–5)

mounted troops that ever served the British cause. When during the lull after the battle agitation was afoot to get reinforcements for the Division sent to France as infantry, Murray wrote to the War Office: 'I wish to make it clear that I cannot spare a single man. These Anzac troops are the keystone of the defence of Egypt.'[23]

10

'I am losing a division. The policy at home is to take all they can possibly from me, leaving me to hold on for the summer. I think they are right, though it's bad luck on me. At the same time as they withdraw my troops they ask me to be as active as I can so as to draw as many Turks as possible down into the theatre and perhaps give them something to put in the papers during the dull season. It is working on the principle of we give you as much as we can, but should like, and expect, you to guard Egypt and to gain us some success.'

MURRAY to Chetwode, 24 January, 1917

'The famous epigram of Tacitus on the Romans – "they make a desert and call it peace" – might aptly have been inverted for this British advance – "they turn the desert into a workshop and call it war".'

WAVELL in *The Palestine Campaigns*

'Magdhaba was a troop leader's triumph. Once the wadi was captured, the broken nature of the terrain rendered concerted action on any large scale practically impossible.'

LIEUTENANT A.C.N. OLDEN,
10th Australian Light Horse

'Just received your news about Rafah. Splendid. Mounted troops are the thing. *Unofficially* I think you can do a great deal with them yet . . . . I am inclined to think it is a case of *l'audace* for you.'

ROBERTSON (CIGS) to Murray,
10 Jan., 1917[1]

Cabinet's policy changes – Mazar – Maghara – arrival of Chetwode – Magdhaba – Rafah

After Romani, Egypt and the Canal were in effect made safe for the rest of the war. What was to happen next depended chiefly on how the war was going on other fronts. Murray was bombarded with contradictory orders from London. All that was consistent in them was the permission given him to advance to El Arish as a necessary part of the defence of Egypt. By the end of December, 1916, as will be shown, this had been achieved. On the 9th, Lloyd George, two days after he had succeeded Asquith as Prime Minister, had asked Murray to send proposals for action beyond El Arish and to state what additional troops he required to carry them out. 'It was pointed out to me,' wrote Murray in his despatch, 'that the gaining of a military success in this theatre was very desirable.'[2] This change in policy was prompted by the failure of Brusilov's Russian offensive, the costly deadlock on the Somme and the swift overrunning of Britain's newest ally, Romania.* Yet, only six days later, the War Cabinet refused Murray the extra troops he wanted and reinstated the defence of Egypt[3] as his primary mission. Less than a month later, he was told to postpone any idea of an advance into Palestine until the autumn and, on 22 January, 1917, to send one of his infantry divisions to France.[4] Meanwhile Kress was pressing for a complete evacuation of the desert zone, but Djemal, on political grounds, decided to keep garrisons there.[5] With the way in which these were dealt with, this sub-chapter is chiefly concerned.

In October, 1916, Murray had moved his headquarters from Ismailia to Cairo from where he was better able to oversee his numerous other commitments, ranging from Salonika, through the Western Desert, to a revolt in the Sudan, unrest in Egypt and the Arab rising in the Hejaz. He placed all troops east of the Canal under temporary Lieutenant-General Sir Charles Macpherson Dobell, an infantry officer of Canadian origin whose experience of so large a force – larger and more complex than the average corps – was non-existent, but who turned out to be, according to Wavell, 'a sanguine and energetic commander always anxious to engage his enemy'. During September, October, November and December no operations of importance were undertaken. Routine patrolling and much digging for water and trench-making occupied the troops. Training proceeded, too, especially when the summer heat declined. The pipeline crawled forward always some way behind the railway.

*Brusilov's offensive began on 4 June, the Somme offensive on 1 July. Romania entered the war on 27 August, Bucharest fell on 6 December.

> 'It was extraordinary,' wrote an officer of yeomanry in late December, 'the pace that the railhead moved, provided it was not held up by actual hostilities. First came a troop of yeomanry doing advance guard, then the surveyors of the line, often themselves under fire, then some thousands of Egyptian natives belonging to the Labour Corps, who either built up an embankment or else dug a cutting, and finally the construction train itself. The natives performed their work almost entirely by using little spades and fig-baskets, the sand being scooped into the latter and removed as quickly as possible.'[6]

Two raids against Turkish fortified outposts were made by the mounted troops. The first was aimed at the powerfully entrenched garrison of Mazar, forty-four miles east of Romani (see map p. 34). Estimates of the enemy strength varied from 500 to 2,200, but it was never established, for Chauvel with the 2nd and 3rd Australian Mounted Brigades decided to retire as he had been ordered to do should he be unable to overrun the stronghold in the first rush. Numerous things went wrong with this raid. First, the hod in which the brigades encamped on the day before the dawn attack on 17 September provided such feeble concealment that a German aeroplane spotted and machine-gunned them, destroying the element of surprise. Next the Australian companies of the Imperial Camel Corps which were to have been in support were delayed by narrow gullies between dunes which forced single file on them, while the horse artillery, misled by a native guide (a sergeant in the Sinai police), also failed to turn up in time. Worst of all the ambitious scheme for watering the brigades in the desert – which at that time 'in its magnitude was probably without parallel'[7] – was a total shambles. 14,000 gallons had been brought twenty-six miles and deposited at the appointed place by 700 camels each carrying twenty gallons, but organization and discipline broke down when the regiments arrived to drink the water after their retirement. The wild, disorderly scramble at the troughs resulted in the water being exhausted before a large part of the brigades' horses had arrived. The 5th Regiment, for instance, was able to water only one of its squadrons.[8] Many horses went without a drink for thirty hours. Casualties on both sides were not heavy. The lessons learned from Mazar were salutary and soon learned. On 19 September the Turks evacuated the place and the 1st and 2nd Brigades went back to the Canal for a rest.[9]

* * *

The second raid took place between 13 and 17 October. The 11th and 12th Australian Light Horse Regiments and the 1st City of London Yeomanry, supported by 300 men of the Imperial Camel Corps and one section of the Hong Kong and Singapore Battery in support, marched from Bayud (where wells had recently been sunk from which enough water could be drawn for the whole force) to Maghara where a strong enemy post was stationed. The country through which the horsemen marched is the most desolate and difficult of all northern Sinai and, to sustain the 1,100 rifles, it was necessary to provide 3,900 men, including native camel-drivers, with 2,300 horses and some 7,000 camels. On 15 October the 11th Regiment led the way and, in spite of a thick fog, emerged directly in front of the Turkish position. The leading troops immediately charged at the gallop, though in the fog they had only the enemy's rifle fire to guide them. They took a few prisoners. Major-General Alister Grant Dallas, whose career had been chiefly in the 16th Lancers and who was later to command the 53rd Division at Gaza (see p. 96), had orders exactly similar to Chauvel's at Mazar. Because of the persistent fog which concealed the objectives, he decided to withdraw and consider the operation as a reconnaissance only, just as Chauvel had done; but before his verdict could be communicated to the troops, they had dismounted and ascended halfway up the precipitous hill upon which the Turkish fortification stood, covered by the fire from the mountain guns and their machine guns. The enemy, who had been completely surprised, refused a hand-to-hand fight and fled into their main position. It is said that the squadron leaders were confident of carrying it without heavy losses and were disappointed at being recalled. Casualties on both sides were slight. The elaborate administrative arrangements seem to have worked reasonably well.[10]

In November the railway reached Mazar and by the end of the month mounted patrols reached to within three miles of El Arish. From now onwards patrols were more or less constantly in touch with the Turkish line.

* * *

On 7 December Lieutenant-General Sir Philip Walhouse Chetwode 7th Bt, fresh from the Western Front, arrived to take over command of the advanced guard of Eastern Force, known as the Desert Column.* Before his arrival, long-distance patrols boring for water in the Wadi el Arish, the 'River of Egypt', had failed to find any. In consequence it was decided to try to take El Arish, the largest village in the Sinai desert, at dawn on 21 December. The elaborate preparations for this proved unnecessary, for the place was found to have been evacuated and the water supply mercifully intact. It was quite a feat for the division to have marched twenty-three miles over unknown country by night and taken with precision pre-arranged positions encircling the village before it became visible. But what would chiefly be remembered about that night march was the sudden change from deep sand to firm soil.

*It is interesting to note what Meinertzhagen (see p. 134) had to say of Chetwode at the time when Allenby replaced Murray:

> 'Chetwode is an excellent soldier but must be driven. If he acts by himself his every action is bluff and he is a very nervous officer, attributing to the Turk all sorts of wondrous strategy and tactics. Apart from this he is a soldier with sound ideas [see his plan for Third Gaza, p. 125]. What he lacks is the initiative and courage to carry them out as planned. That is just where Allenby will find a very useful and talented servant.' (15 July, 1917, Meinertzhagen, 219-20).

When in April, 1918, Wavell became Chetwode's Brigadier-General, General Staff, he summed him up thus:

> 'He had about the best and quickest military brain I have ever known, an extremely good tactical eye for ground and a great gift for expressing a situation clearly and concisely, either by word or on paper. I think he just lacked as a commander the quality of determination and drive, certainly compared with a man like Allenby. He was aristocratic, rather a snob in some ways and inclined to be satirical at the expense of those not out of the top drawer . . . . He gave quick and sensible decisions and did not worry about detail . . . . It was not safe to bring him work after dinner; he hated it and would be ill-humoured and tiresome . . . . He slept little and woke early but was at his best and brightest in the early morning.' (Connell, 136).

Born in 1869, Chetwode had served in the Boer War (see Vol. 4, 218), having joined the 19th Hussars at the age of twenty. In France he had commanded first the 5th Cavalry Brigade and then the 2nd Cavalry Division. In 1920 he became Deputy CIGS, in 1923 C-i-C, Aldershot Command and in 1930 C-i-C, India, a post he held until 1935. In 1933 he was promoted Field-Marshal and in 1945 he was created Baron Chetwode. He was the father of Lady Betjeman, wife of Sir John. By marriage he was a relation of the present author, who, as a Lance-Corporal in uniform walked with him during World War II in the main street of Chalfont St Giles. He remarked: 'This is probably the first time a Field-Marshal has walked in public with a lance-corporal!' This reflects less his supposed snobbery (see above) than his lack of formality.

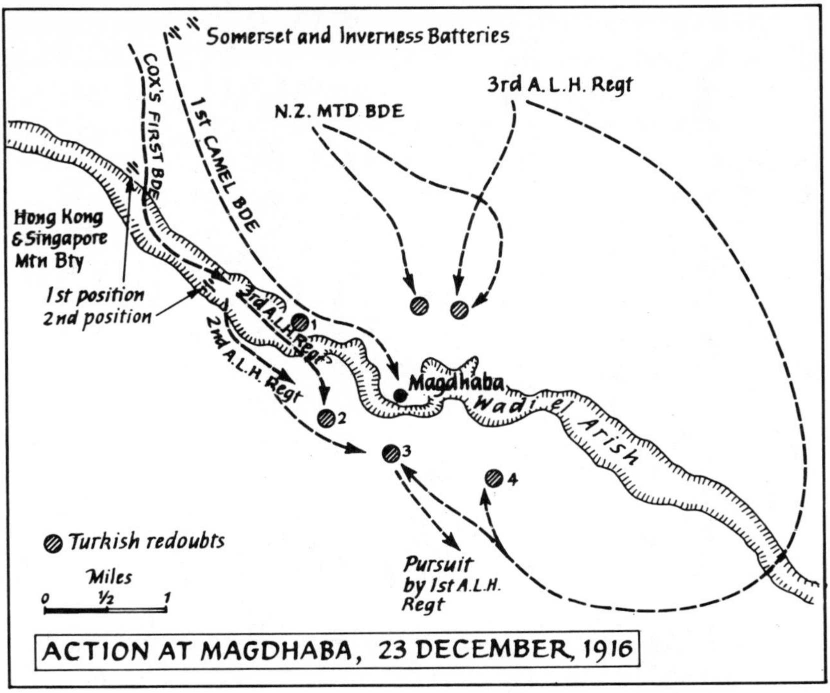

'The hard going for the horses,' wrote Cox, 'seemed almost miraculous after the months of sand; and, as the shoes of the horses struck fire on the stones in the bed of the wady, the men laughed with delight. Sinai was behind them.'[11]

Chetwode now decided upon an immediate assault on Magdhaba, twenty-six miles inland from El Arish, where the Turks were speedily improving a strong defensive position around the village. From this they could present a serious threat to the right flank of any further advance. Having arranged for a supply column to arrive at El Arish by 4.30 p.m., supplemented by the landing of further stores from the sea, Chetwode landed from Port Said on the beach near El Arish on 22 December. Given so little time for preparation, the night march started with a classic traffic jam. When this had been sorted out, the going proved splendid. On firm and level ground for the first time since they had arrived in Egypt, many of the horses 'over-reached and stumbled',[12] but they soon adjusted to the new, delightful conditions. Marching forty minutes in the saddle, leading the animals for ten minutes and resting for ten, as was the normal practice, the 1st Brigade on the right, the Camel Brigade in the centre with the New Zealanders on their left and the 3rd Brigade* far off on the left flank, the division first saw the camp fires of the enemy at about 3.50 a.m. Their existence showed that no attack was expected that day. At dawn the smoke from the fires hid the enemy's position for a time, but ten aeroplanes of the 1st Squadron of the Australian Flying Corps flew low over it at about 6.30 and received a hot reception. In one of them was Lieutenant-Colonel Powles (see p. 58). On his return run he dropped a message almost literally at the feet of Chauvel and his staff. It read: 'The bastards are there alright!'[13]

The enemy's position, forming a rough circle about 3,250 yards in diameter and consisting of five redoubts, numerous rifle pits and four mountain guns was very skilfully concealed. These were made all the more difficult to distinguish because the 1st Brigade as it advanced along the Wadi El Arish put up a dense cloud of fine dust. This, together with the ever-present mirage, made good observation and distance judging impossible, especially for the artillery. Nevertheless, soon after 10.00, as the brigades closed in, the Hong Kong and Singapore Mountain Battery and later those of the Somerset and Inverness Horse Artillery batteries (directly under Chauvel's hand) were soon at work. Progress at first was good, but the Turkish fire was heavy and the difficulty of spotting their positions, even with the help of numerous air sorties, soon slowed the advance down. Some of the regiments got to within 1,000 yards before dismounting. Indeed the advance guard of the Wellington Regiment actually pushed forward to within 400 yards 'where they dismounted in a covered position'.[14]

At about this time one of the airmen who had been dropping situation reports landed his aeroplane next to Chauvel's headquarters with the news that there were signs of a general Turkish retreat. Immediate orders were given to Cox's 1st Brigade to ride directly on Magdhaba along the flat ground between the wadi and to the right of the cameliers. Preceded by ground scouts, the 2nd and 3rd Regiments, led by Cox, trotted for a mile or so before the enemy's mountain guns fired shrapnel at them. He at once ordered them to extend into 'artillery' formation and to increase pace to a gallop. 'For a minute or more the light horsemen enjoyed the

*These two brigades were placed under Chaytor.

The division was complete, except for the 2nd Brigade and the Ayr and Leicester Batteries, which were resting. All the times given are very approximate. The sources vary enormously in this respect.

excitement of a cavalry charge,'[15] but after half a mile they galloped into heavy machine-gun and rifle fire. Cox realized at once that the airman's report was false* and to save his brigade from destruction he swung his regiments, still at the gallop, into a deep, blind tributary of the wadi on his right. There, at 1,900 yards from the redoubt which faced him, he dismounted them. Soon after 2 p.m. the 3rd Regiment at long last took the redoubt, charging with the bayonet. This was the turning point of the battle.†

Between midday and 4 p.m. all the brigades had been hotly engaged, their horses concealed on the right in the wadi – 'an ideal place to leave horses'[16] – and elsewhere behind low mounds and bushes. The experience of the 9th Regiment of the 3rd Brigade fighting dismounted was probably much like that of the other regiments:

> 'When about 1,000 yards from the enemy position,' wrote Lieutenant T.H. Darley, 'snipers and Lewis guns [being employed for the first time in action] were pushed forward to cover the advance, which was made by alternate rushes, troop by troop, each troop supporting the advance by rapid fire. The heavy and accurately placed fire of the enemy began to take effect and a number of casualties occurred, but by 2.30 p.m. the line had been advanced to within 500 yards of the position and drew the attention of the enemy gunners who opened a brisk fire with shrapnel.
>
> 'The line was now straightened up and reserve ammunition brought up for the Lewis guns. At 3.25 the line again advanced by rushes of twenty-five yards, whilst the batteries kept up a brisk fire on the redoubts. On arrival at 150 yards from the redoubts, the line laid low for a spell and at 3.45 bayonets were fixed ready for the final rush. At a given signal

*It was later discovered from a statement by the Historical Section, Turkish General Staff, that numbers of Arab soldiers had decided to retire from their position in a body. (MacMunn, 254).

†Some of the regiments just before setting off had received their 'Christmas billies' from the Australian Comforts Fund. These they attached to their saddles, but as they advanced at the gallop before dismounting for action, the billies worked loose. In consequence the approach to the battlefield was strewn with packets of chocolates, tins of milk and Christmas puddings. (Gullett, 230).

> the whole line leapt to their feet and, rushing forward with wild cheers, carried the outer trenches, many of the enemy being bayoneted before the remainder surrendered.'[17]

It will be seen from the timing given here that the battle had not, after the initial spurt, been going with the speed that had been hoped for. Indeed soon after 1 p.m. Chauvel, having learned that his engineers who had been trying to find water at Lahfan fourteen miles away had failed to do so, proposed to Chetwode by telegraph that the fight be broken off and even before getting his permission sent off to the brigade commanders the following order: 'As enemy is still holding out and horses must be watered [they had not been watered for well over twenty hours], the action will be broken off and the forces withdrawn.' Cox received this message just as the 3rd Regiment was about to take No. 2 Redoubt. 'Take that damned thing away,' said 'Fighting Charlie', 'and let me see it for the first time in half-an-hour.'[18] Chetwode soon afterwards telegraphed Chauvel urging him not to abandon the fight, even at the cost of some of the horses. By the time this message arrived, Chauvel was able to get through on the telephone to Chetwode to assure him that the battle was about to be won.

Well before this, aerial patrols reported that the enemy's retreat had begun. Royston, commanding 3rd Brigade, at once ordered the 10th Regiment with two sections of the Machine Gun Squadron to make a wide detour so as to cut off the Turkish withdrawal to the south and south-east. Led by the twenty-two-year-old Major Horace Clement Hugh Robertson, the regiment galloped across the open plain in extended order under machine-gun and rifle fire, 'holding their bayonet-tipped rifles like lances and cheering lustily'. At the edge of the wadi the retreating Turks surrendered in terror, yielding up ninety prisoners. Leaving a party to collect these, the three squadrons dropped into the wadi and, as Lieutenant Olden of 'C' Squadron reported,

> 'rapidly extended along its bed until "C" Squadron on the left had swung right round to the south-eastern side of the [No.4] redoubt. Flinging themselves from their horses, "A" and "B", with two troops of "C" Squadron, emerged from the west side of the wadi and pushed on dismounted towards the redoubt proper, whilst [the 2nd Regiment] of the 1st Brigade was seen attacking it from almost the opposite

> direction. [When the 2nd Regiment reached the redoubt they galloped into it, shooting from the saddle.][19] Two troops [about seventy strong] had in the meantime ridden further round and had almost reached the Camel Corps, when suddenly they observed an opening leading straight to the redoubt [occupied by at least 300 Turks]. [They] charged right through the enemy's outer line and reached the very heart of the redoubt. The balance of the regiment rushed in support towards the objective, but the Turks, finding themselves attacked on all sides, and having suffered heavy losses, surrendered at a quarter-past four by hoisting numerous white flags.'[20]

Throughout these spirited attacks, the squadrons went forward in a succession of mounted rushes, galloping from cover to cover, dismounting, engaging for a time in rapid fire and then riding forward again. The machine gunners rode with them and gave very effective covering fire. Royston, who kept up with the regiment, at one point galloped up to a Turkish trench and was instantly faced by five enemy rifles. He raised his cane and shouted at the riflemen in Zulu (since he knew, of course, no Turkish), 'whereupon the Turks, impressed with the demonstration, dropped their rifles and held up their hands.'[21]

In the last charges made by all three brigades their units had become mingled and, indeed, on at least one occasion regiment had fired on regiment. After the last isolated party of the enemy had surrendered, it took some time to water the intensely thirsty horses at the crowded but intact wells, to collect the 1,282 prisoners* and to re-assemble the units. The Anzacs' and Camel Corps casualties were amazingly small. Five officers were killed and seven wounded, while seventeen other ranks were killed and 117 wounded.† Chauvel thought that the Turks' poor shooting was

*Ninety-seven of the enemy were buried next day. The wounded numbered about 50. (Murray, 104; Gullett, 227, though, gives the figure of 300). The four mountain guns, 1,250 rifles, 100,000 rounds of ammunition and ninety-one horses and camels were taken.

†A high proportion of these was incurred by the two Camel Corps companies which had to charge the redoubt north of the wadi, in conjunction with some of the light horsemen, over very flat country devoid of cover.

Twenty-seven horses were killed and thirty-two wounded. (Blenkinsop, 166).

one reason. Increasing skill in minor tactics exhibited by all the troops was certainly another, especially the adroitness with which the Lewis gunners kept up with the forward troops, thus keeping down the enemy's fire. Magdhaba was the first action in which each man carried two bandoliers, which added much to the fire strength of the regiments. This innovation became standard practice for the rest of the war.

After the battle Chetwode signalled to Dobell: 'Conduct of action was left entirely to General Chauvel. I consider that he conducted it most admirably under most difficult circumstances and if you concur hope you will bring him to favourable notice of the C-in-C.'[22] Indeed Magdhaba was a considerable victory. Scarcely any of the enemy escaped. The Turks had a great advantage in that much of the ground around their position was devoid of cover, while small bushes and excrescences in the ground hid their exceptionally well positioned trenches and pits. On the other hand Chauvel had twelve guns to the enemy's four. When later Chetwode addressed the New Zealand Brigade he said 'that the mounted men at Magdhaba had done what he had never known cavalry in the history of war to have done before. They had not only located and surrounded the enemy's position but they had got down to it as infantry and had carried fortified positions at the point of the bayonet.'[23] This suggests that Chetwode had not grasped that the Anzac Mounted was not a cavalry division and that light horse and mounted rifle training and organization were specifically designed for just such fighting. After all, they were armed with bayonets not swords. Magdhaba supplies a classic example of the correct use of mounted riflemen.

As the devoted stretcher bearers collected the wounded (whose locations were marked by small fires), and brought the most serious cases into the well appointed Turkish hospital, placing the others in the cacolets for their excruciatingly painful journey back to El Arish, the regiments started on their return march.* Very few men had had any sleep for the last three nights. In spite of the piercing cold and the stifling dust clouds, which almost blinded the horses causing them to collide with each other, most of the men

*A convoy of 400 camels met them at 4.30 a.m. with urgently needed food and drink. At daybreak next day, Christmas Eve, each prisoner received one tin of bully beef, 1 lb. of biscuits and one quart of water. 'It is doubtful,' remarked Lieutenant Darley, 'whether they had ever received such a generous ration during the whole of their desert campaign.' (Darley, 63).

slept in the saddle. Many experienced singular hallucinations: strange animals and streets of well-lit houses were amongst those reported. Chauvel, who had slept during only one of the last four nights and had ridden more than sixty-three miles in thirty-six hours, suddenly galloped off to a flank with another officer, returning quietly to slip back into their places in the column a mile or two later. Both said that they had seen a fox and thought that they were hunting![24]

* * *

After Magdhaba there was left in Egypt only one other formed body of the enemy. At El Magrunein just south of Rafah, and twenty-nine miles from El Arish, 2,000 Turks occupied a strong entrenched position next to the Egyptian-Palestinian frontier. Kress wanted them withdrawn, but Djemal insisted that they stay. Murray determined as soon as the railway reached El Arish and the extensive work on the station and siding had been completed that this 'small detached garrison within striking distance of my mounted troops'[25] should be surrounded and captured. Chetwode resolved to lead the raid himself with a force consisting of the Anzac Mounted Division, less one brigade, the 5th (Yeomanry) Mounted Brigade with 'B' Battery of the Honorable Artillery Company in support, the Imperial Camel Corps Brigade and the No. 7 Light Car Patrol (six machine-gun-carrying Fords) which did valuable work during the action. The approach march was admirably executed* and soon after dawn on 9 January the position was in the process of being almost completely surrounded, the mobility of the troops being exploited to the full.† It proved to

*On the left the Gloucesters 'marched in column of squadrons, line of troop column' with 'B' Battery marching parallel to the regiment between it and the sea. (McGrigor, 334).

†'The silence of the dawn was broken by the crowing of a single "rooster", which was replied to by a spontaneous burst of cheering and laughter from the whole [3rd] Brigade.' (Olden, 107).

In error, the orders for the 5th Brigade, which provided the advance guard, instructed the bearer sub-division of its Field Ambulance Column, with its sand-carts and sledges, to *precede* the leading squadron of the Worcesters. In consequence, soon after dawn, the enemy's first shots were fired at this non-combatant unit as it proceeded solemnly in the van of the whole Desert

(continued over)

be considerably more difficult to get at than had been the Magdhaba entrenchments. 'When daylight broke,' wrote Chetwode in his despatch, 'the ground was seen to be almost entirely open and devoid of cover, while the immediate neighbourhood of the works was almost a glacis. I confess I thought the task was almost beyond the capacity of dismounted cavalry to carry through.'[26] And so it proved.

Except for the capture of the Police Post at Rafah which was galloped by the Canterbury Regiment, yielding six German and two Turkish officers and sixteen other ranks, and a second gallop by the Auckland Regiment over some out-lying trenches, yielding two more German officers and twenty Turks, there was no mounted action. The horses and camels had to be left in most cases well over 1,000 yards from the skilfully constructed redoubts of the enemy position atop a 200-foot hill. From 10 a.m. onwards a day-long dismounted action followed. Progress across the open turf land was painfully slow and by 2 p.m. had become almost non-existent. Thus it remained for nearly two hours, the machine guns and the artillery keeping down as much as possible the Turkish fire while the thin encircling line of rifles, only on average one to about every four yards,[27] lay as close to the ground as possible.*

At the beginning of the action a detachment of the 8th Regiment and parties of the Wellingtons had been sent to watch for possible enemy reinforcements known to be at Khan Yunis and Shellal. At about 3.45 some two and a half Turkish battalions were reported to be advancing on the Anzacs' rear. Air reconnaissance confirmed this but put the number at 2,500 men. Chetwode discussed the situation with Chauvel on the telephone and decided at 4.25 to

---

Column! It was soon rescued by the squadron galloping up in front of it. (Teichman, 100; Godrich, 55).

It is interesting to note that officers were ordered to leave behind everything that was not essential, 'as horses,' wrote Lieutenant McGrigor of the Gloucesters, 'must be relieved as much as possible'. He even left his 'shaving tackle' behind. 'One carries one's emergency rations [a tin of bully beef and some biscuits],' he added, 'on one's person.' (McGrigor, 333).

*To try to prevent the guns firing on their own troops, as had happened at Magdhaba and was difficult to avoid when they were formed in a circle, black and white discs were issued to all regiments. These were waved to show the artillery their positions. (Darley, 67).

break off the fight and withdraw. He at once pulled out the 5th Brigade, and himself began to ride back to El Arish. This decision was reinforced by a mistake made in the planning of the battle. He had decreed, against his brigadiers' wishes, that all wheeled transport which was not absolutely necessary should be left behind at the assembly point of Sheikh Zowaiid. This meant that the reserve ammunition was some ten miles in the rear. Of course he had not expected that the action would be so protracted and therefore so prodigal of ammunition, but in the event not only were four of the New Zealanders' machine guns put out of action for want of their reserve ammunition, but the Inverness Battery also ran out of rounds and had to be withdrawn as early as 3.30.* About 410,000 rounds were expended by the machine guns and rifles and 1,637 by the batteries.[28] For only some 225 enemy killed and under 200 wounded, and with not more than 2,700 riflemen and 114 Lewis and other machine guns employed, this was an unusually large expenditure, but it is explained by the unusual conditions of the action.

As at Magdhaba, it was the determination of the officers and men that snatched victory from the jaws of defeat at this critical moment. The brigades had only just received the retirement orders when the New Zealanders and the 3rd Brigade, led by their brigadiers, Royston yelling out, 'Come on lads! We've got them!'[29], went in with the bayonet, running uphill without a vestige of cover. Some of the New Zealanders were seen to be firing their rifles as they charged.[30] The concentrated machine guns gave effective covering fire and just before dusk vital parts of the enemy's defences fell. Again as at Magdhaba the rest of the Turks soon surrendered. The engagement was over. Virtually the whole of the garrison was made prisoner.† Among those captured was the Oxford-educated German major who was in command of the

*The intention was that the reserve ammunition should be sent on after daybreak, but it mostly failed to arrive. The quartermaster of the Wellingtons was so concerned at the small supply of rounds for his regiment that during the early hours of the battle he went forward to discover that it was desperately calling for ammunition. He seized a cable wagon, emptied its contents, filled it with boxes of rounds and galloped with it to the New Zealanders just in time for their final assault. (Powles, 77–8).

†Thirty-five officers, including the Turkish commander and one German, and 1,600 other ranks, including ten Germans. Four mountain guns, four machine guns and 137 animals were also captured. (MacMunn, 270).

machine guns. He was 'a magnificent, well-dressed figure'. A young yeomanry lieutenant accosted him with a 'good morning', to which he replied in perfect English: 'Good morning to you – wasn't that a damned good scrap?'[31] The enemy's reinforcements, fewer in number than had been feared, seeing the display of their comrades' white flags, were easily persuaded to return whence they had come. The Desert Column's losses were 487, including seventy-one killed of whom three were officers: thrice the casualties at Magdhaba. Most of the wounded were hit by rifle fire, the shooting of the Turkish gunnery being very erratic. Leaving detachments to collect the wounded and bury the dead, Chetwode's force once again faced the long march back to El Arish. 'We had,' wrote Trooper Godrich of the Worcesters in his diary, 'twenty-five miles to march home. We watered the horses at Sheikh Zowaiid, the first drink that they had had for thirty-six hours. I shall never forget that march: after a gruelling day, preceded by a long march and a terribly cold outpost, we had all reached the limit of human endurance; practically everybody slept in the saddle and left it to the horse to get him home.' Most of the troops did not arrive at base before 6 a.m. on 10 January. There they were welcomed by some Scottish infantry who turned out of their tents to pump water for the horses.[32]

The success of the raid was especially remarkable when it is considered that a mounted force shrinks to hardly more than half its actual strength when fighting dismounted and that its lack of depth and of fire power makes it far from being the ideal means, unlike infantry proper, of effecting a protracted advance against strong works and against time. Both at Magdhaba and at Rafah the Camel Corps Brigade proved its worth for, unlike the light horsemen, it was able to deploy its men in greater depth and in greater numbers. 'Our aeroplanes,' wrote Captain O. Teichman, Medical Officer of the Worcesters, 'were wonderful during the whole day, continually spotting for the batteries by dropping smoke-balls over the enemy and bringing in fresh news of the Turkish dispositions and reinforcements.'[33] For the first time in the campaign wireless was used from aircraft to direct the artillery.[34] The German aeroplanes were also active throughout the day 'dropping light bombs on every available target'.[35]*

*Chetwode's chief staff officer was Colonel Arthur Lynden Lynden-Bell, known everywhere as 'Belinda'. In mid-February, 1917, he wrote to Chetwode:

(*Continued overleaf*)

> 'You may have trouble shortly with more bombing of the line in consequence of our going for the enemy's aerodromes, but the chief [Murray] did not consider it right that hostile aircraft should be going over Mahemdia and other places in our lines without anything being done in retaliation. As you know, the arrangement was that if the enemy did not drop bombs we would not drop bombs on them, and that if he came over without bombs he was to be met and fought in the air. This business, however, of fighting in the air is extraordinarily difficult for our aircraft, as apparently the enemy's planes pop in suddenly from the sea. The Chief has therefore decided to continue going for the enemy's aerodromes at Bir Saba [Beersheba] and Ramleh till he has knocked them out, after which the old policy will be resumed of not bombing the enemy unless he bombs us.' (18 Feb., 1917, Chetwode).

Sergeant Bill Walker of the Worcesters was among those decorated after Rafah. He was awarded the DCM, according to a member of his troop, 'for carrying on with the Troop after our officer had turned and fled: the only example of cowardice our regiment experienced through the war.' (Godwich, 57).

11

'The intention was to repeat the manoeuvre of the Magdhaba and Rafa actions on a larger scale.'

WAVELL in *The Campaigns in Palestine*

'At 9 a.m. on 26th March, an aviator reported to Army Headquarters that the British were advancing on Gaza with two infantry divisions, and with three cavalry divisions.'

KRESS VON KRESSENSTEIN in *Between Caucasus and Sinai*

'At 3 p.m. the order "Right Wheel Gallop" was given and we steered dead straight for our objective . . . The charge was in every way a splendid achievement, but owing to the failure of the infantry we were forced to retire at dusk.'

CORPORAL A.W. FLETCHER, Lincolnshire Yeomanry

'A captured Syrian doctor who was in Gaza on March 26th asked Ryrie: "Why did you pull out from Gaza on the first attack?" Ryrie, with characteristic bluntness, replied: "You can damn well search me!" The Syrian added that when Tala Bey (the garrison commander) discovered in the morning that the British had withdrawn, he "laughed for a long time".'

The Australian Official Historian

'Places over which the cavalry had careered freely during the first battle became a network of trenches and a tangle of barbed wire before the second.'

LORD COBHAM, Worcestershire Yeomanry[1]

The first and second battles of Gaza

On 11 January, 1917, Murray was given the final go-ahead for an autumn campaign in Palestine.[2] He realized that to make this effective and speedy he ought to try to gain the Gaza-Beersheba line, the twenty-five-mile-wide gateway to Palestine, as soon as possible. The advantage of possessing the springs in the Wadi Ghazze, five miles south of Gaza, was obvious. If Gaza itself, a town with a population of 40,000, could also be taken, a partially metalled road system, fit for all types of transport, would immediately become available from railhead, while the enemy would undoubtedly be compelled to abandon Beersheba. By 6 March the enemy had evacuated all the country south of the Gaza–Beersheba line, but an attack upon Gaza had to be delayed till railway and pipeline had reached Khan Yunis where there was an almost unlimited water supply.*

In the meantime, a major reorganization of Dobell's Eastern Force was undertaken. The infantry now consisted of only three divisions instead of the five which Murray had demanded. Not all of these were of the highest quality, nor were they up to full establishment. Another, the 74th (Yeomanry) Division, was in the process of being formed from the dismounted yeomanry brigades in Egypt. His mounted force, on the other hand, was augmented. The partial liquidation of the Western Desert campaign had released a number of regiments while the War Office had agreed to the reconstitution of the 4th Australian Light Horse Brigade. This was made up of the 4th, 11th and 12th Regiments which had been unbrigaded since returning from Gallipoli. It was commanded by Meredith, who had done so well at Romani (see p. 64). A second mounted division named the Imperial Mounted Division was now formed. A British regular cavalryman, Major-General Henry West Hodgson, who, aged twenty-one, had entered the 15th Hussars in 1889, was given its command. Sound but uninspired, with an engaging personality, he had been largely instrumental in bringing the operations against the Senussites to a successful conclusion, but his appointment ruffled Australian feathers no end.[3]

The brigades within the new division and Chauvel's Anzac Mounted Division were re-arranged with Chauvel's consent, though some of his officers objected both to the re-grouping and to

*Lance-Corporal Godrich of the Worcester Yeomanry was 'trekking quietly along' on a reconnaissance towards Gaza on 23 March when he came to the top of a ridge to behold a view that 'nearly took our breath away'. It was the army's first 'peep of the "Promised Land" '. The frontier between Palestine and Egypt had been reached and the desert left behind. 'Down in the valley,' wrote Godrich in his diary, 'lay a large village of white houses surrounded by thousands of trees in bloom and beyond that miles of young barley.' The village was Khan Yunis. 'The horses enjoyed it, for they were eating grass and green barley every moment they could, their first feed of green stuff for very many months.' (Godrich, 60).

the new division's name. Murray's object was to mix the Australasians with the yeomanry. This was a recognition that the days were past when, as Murray had written to Robertson four months previously, the yeomanry 'always got lost and the Anzac cavalry had to find them and bring them home. Now,' he had continued, 'they are learning to find their own way about.'[4]* They were, in fact, becoming desertworthy and dependable to a degree which would have astonished an observer at the time of Katia. Murray thought, nevertheless, that the yeomen would benefit and learn from working and fighting beside the Anzacs, which was really a considerable compliment to the latter.

Chauvel's division now comprised the 1st and 2nd Australian Light Horse, the New Zealand Mounted Rifles and the 22nd (Yeomanry) Mounted Brigades, while Hodgson's division consisted of the 3rd and 4th Australian Light Horse and the 5th and 6th (Yeomanry) Brigades, all forming the Desert Column under Chetwode. In fact, both divisions by the time of the attack on Gaza were one brigade short. The 1st was resting in the Canal area and the 4th was not yet ready for operations. The Gloucesters of the 5th Brigade were attached to Money's Detachment, a small specialized body of mixed arms. The Imperial Camel Corps Brigade was attached to the Desert Column for the operation.† By

*As late as mid-August, 1916, a lieutenant of the Gloucesters wrote: 'It does make me angry the way those in authority out here think the yeomanry more or less hopeless. Some remarks I have heard since I have been at GHQ have fairly made my blood boil. They never get credit for anything.' (McGrigor, 222)

†*Anzac Mtd Div.:* 1 *ALH Bde* (Cox): 1, 2, 3 ALH Regts; 2 *ALH Bde* (Ryrie): 5, 6, 7 ALH Regts; *NZ Mtd Rifles Bde* (Chaytor): Auckland, Canterbury & Wellington Mtd Rifles; 22 *Mtd Bde* (temp. Brig.-Gen. F.A.B. Fryer): Lincs, Staffs & E. Riding Yeo. Regts.

Imp. Mtd Div.: 3 *ALH Bde* (Royston): 8, 9, 10 ALH Regts; 4 *ALH Bde* (Meredith): 4, 11, 12 ALH Regts; *5 Mtd Bde* (Wiggin): Warwick, Glos. & Worcs Yeo. Regts; *6 Mtd Bde* (temp. Brig.-Gen. T.M.S. Pitt): Bucks, Berks & Dorset Yeo. Regts. Each division was served by four artillery batteries: *Anzac Mtd Div.:* Leics, Somerset, Inverness & Ayr Btys, RHA; *Imp. Mtd Div.:* Notts & Berks Btys, RHA, and 'A' and 'B' Btys, HAC.

Imperial Service Cavalry Bde (temp. Brig.-Gen. M.H. Henderson) remained under Dobell. It consisted of the Mysore and 1st Hyderabad Lancers.

Also under the GOC were five anti-aircraft sections, seven tanks, not employed on this occasion, two batteries of light armoured cars and one motor MG bty, as well as Engineers and Signals (including a pigeon section).

25 March, when preparations for the *coup de main* against Gaza were complete, the mounted troops, including the Camel Corps Brigade, numbered some 11,000. Of these about 6,000 rifles were available in the firing line.*

Dobell's intention was that Chetwode with one infantry division and part of another† should assault the town from the south and south-east. At the same time Chauvel was to prevent the enemy's retreat from it and he, Hodgson and the Camel Corps Brigade, extended over an arc of some fifteen miles, were to reconnoitre to the north, east and south. They were to hold off the 15,000 reinforcements and forty-five guns which the Turks were believed to be capable of bringing from between ten and seventeen miles away. Dobell's order added that Chetwode was then 'to attack the enemy's force occupying Gaza'.[5] In Gaza itself the garrison was believed to number about 3,500 rifles and twenty guns.‡ The Desert Column's artillery totalled ninety-two pieces of which thirty-two were the 18-pounders of the Royal Horse Artillery.§

As at Magdhaba and Rafah, the capture of Gaza had to be

*For the coming battle five days' rations were provided. For this purpose twenty-six saddle horses were trained to carry improvised packs made of canvas in the form of saddle bags, each side being large enough to take two twenty-five-pound tins of biscuits and twenty-four tins of bully beef. Each regiment now had forty-nine pack horses. The improved packs caused much trouble, so much so that an extra man had to be left with every four horses. This entailed a considerable loss to the firing line. This was already in some regiments below strength. A squadron of the 9th Regiment, for instance, when dismounted could only put some sixty rifles in the line. (Darley, 76).

The number of squadrons in the yeomanry regiment was actually reduced by one 'as we have so many packs!' (McGrigor, 421).

†53rd (Welsh) Division and one brigade, 54th (East Anglian) Division. Dobell kept the rest of 54th Division under his own hand as support to the mounted troops. 52nd (Lowland) Division formed the general reserve, as also did one brigade of the uncompleted 74th (Yeomanry) Division.

52nd Division's Divisional Troops included headquarters and a squadron, R. Glascow Yeomanry; those of 53rd Division, a squadron, Herts Yeomanry; those of 54th Division, headquarters and a squadron, Herts Yeomanry.

‡Chetwode's estimate was only 2,000. Kress had very recently reinforced the garrison. (See MacMunn, 415, Kress, 506 and below).

§According to Wavell a total of only forty-two were employed in the main assault. (Wavell: *Pal.*, 82).

accomplished in a day. Water, food and ammunition could only be carried for a twenty-four hour battle. It was exceptionally unfortunate, therefore, that in the early morning of 26 March the infantry were held up by thick fog, very rare at that time of year. The approach march, which included the crossing of the wide Wadi Ghazze with its forty feet deep banks, went well, but the fog descended at about 4 a.m. and was at its thickest an hour later, just before dawn. It began to lift from the higher ground soon after 7.15 a.m. and had virtually dispersed by 9 a.m. but not until 11 a.m. from some depressions.

> 'It was on the movements of the 53rd Division,' wrote Wavell in *The Campaigns in Palestine*, 'that the fog imposed really serious delay; two priceless hours [MacMunn puts this at not more than an hour] were lost in waiting for it to clear. Even so, the brigades had reached their first assembly points by 8.30 a.m.* and were then within 5,000 to 6,000 yards of their goal, the Ali Muntar position [a prominent hill which was the key to Gaza]. Their attack was not, however, launched till near midday [some four and a half hours behind schedule]. The rapid staging of an infantry attack over unknown ground is a high test of the training of staff and troops. To the onlooker, it often seems incredibly slow. So many processes have to be carried out, the reconnaissance of commanders, the elaboration of the plan and issue of detailed orders, the ranging of artillery, the disposition of the means of inter-communication, of medical aid and so forth; and these processes have to be repeated in each successive lower formation. A chief of even the most fiery and energetic temperament cannot ensure rapidity of execution unless his own staff and those of the units under him are fully practised. The 53rd Division had not been in action since Gallipoli, and was, maybe, "short of gallop".'[6]

*Browne, a very reliable source, says that the Anzacs did not reach Beit Durdis until 09.30, forty minutes late, due to the fog. The pace was affected by the need for the flanking detachments to keep close in to the column and by the movement of successive positions being consequently restricted. (Browne, 231).

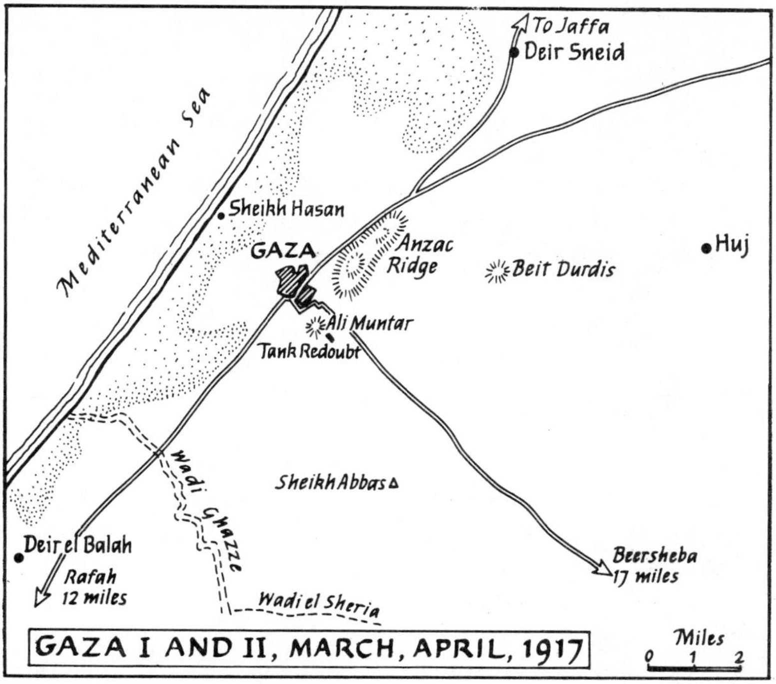

GAZA I AND II, MARCH, APRIL, 1917

The mounted troops on the other hand were certainly not short of a gallop. With visibility down to less than twenty yards, the guides of the vanguard provided by the 7th Australian Light Horse Regiment of the 2nd Brigade arrived at their crossing place on the Wadi Ghazze very nearly on time. Their compass work was excellent and just before 8 a.m. they were at Sheikh Abbas. As soon as the fog cleared, German aircraft became active both in reporting positions and in machine-gunning. No great damage was done. The ground in front of the rapidly advancing horsemen was gently rolling plain, much of it planted with barley and 'actually intersected by English-looking brooks fringed with kingcups'.[7] For these troops, long experienced in night marching, the fog served to screen their penetration into the enemy's country. While it lasted the Turkish outposts were denied a sight of them as they rode on, until the 7th was fired on by an enemy patrol. This was at once galloped down. A few hundred yards further on the crew of two German aeroplanes were awakened by the firing. The men of the 7th galloped up, but just failed to stop the planes taking off to report the light horsemen's presence.

Further north, men of the 2nd Brigade came upon and captured what Trooper Idriess called

> 'a queer little crowd . . . in two funny little coaches that might have come out of Queen Anne's reign. There was a body guard of mostly Arab cavalry all done up like wedding cakes. But the occupants of the carriages were smart-looking men in natty uniforms. They were the Turkish divisional commander and his staff, moving in to Gaza to take up his command . . . . I shall never forget the disgusted general. His *à la mode* Kaiser moustache was continually a-twitch as vainly he turned his back on the unkempt troopers who threatened him at every angle with all breed of cameras. He struck at some with his cane; at which one sunburned villain remonstrated: "Aw, be a sport, general!" When he was taken to the Old Brig. [Ryrie], he complained bitterly of our disgusting coarseness. He demanded some of us at least should be shot. The Old Brig. looked as if he was going to explode, but saved himself and roared instead: "Well, it *was* damned funny, wasn't it?!" '[8]

After dispersing numbers of small Turkish patrols, the mounted troops had closed the cordon around Gaza by about 11.00 a.m. For an hour and a half they waited, wondering what had happened to the infantry attack. When at last it went in, they could witness with complete clarity the hail of enemy shell, machine-gun and rifle fire ripping into the Welsh foot soldiers as they advanced by rushes without the slightest cover until faced by the impenetrable cactus hedges which surrounded the town. Progress was pitifully slow and casualties mounted alarmingly.

Dobell and Chetwode, whose headquarters were side by side on the bank of Wadi Ghazze, suffered agonies of frustration at the lengthy delay. To make it worse not only Major-General Dallas, commanding 53rd Division, but apparently all his staff officers as well, had gone forward, leaving no responsible officer at his headquarters for some two hours.[9] Four increasingly peremptory orders to push the attack vigorously fell on inattentive ears.* Just after the infantry attack started Chetwode signalled to Chauvel and Hodgson: 'Both cavalry divisions will reconnoitre immediately

*Part of the delay was caused by slowness in laying cable to the artillery from divisional headquarters.

with view to closing in on enemy at Gaza to assist infantry if ordered. One brigade only from each division will be employed, leaving two brigades to continue observation. No news by aeroplane of any enemy movement from any direction.'[10] While this order was being transmitted, a captured Turkish sergeant stated that the Gaza garrison, instead of numbering two battalions only as supposed and as an earlier prisoner had reported, in fact consisted of six. Chetwode, with Dobell's consent, therefore sent a further order at 1 p.m.: 'Most desirable that mounted troops should press in on Gaza to assistance of infantry .... General Chauvel is placed in command of both mounted divisions.'[11] For some unexplained reason this order, which instructed him to employ the whole of Anzac Mounted Division in the attack, though sent by wire, took an hour to reach Chauvel. To enable him to carry it out much reorganization had first to be undertaken. His division's mounted patrols had to be released for their new task by the Imperial Mounted Division moving northwards, while the latter's place was taken by the Camel Brigade. Chauvel had also to move his headquarters so that he could oversee the operation. It was 3.15 p.m. before he issued his orders to his brigadiers. The 2nd Brigade was to attack from the coast to the main road, the New Zealand Brigade from the road to the ridge north-east of Gaza (known later as Anzac Ridge) and the 22nd Brigade from due east of the town. At 4 p.m., before all his patrols had been relieved, Chauvel decided to launch the assault. Just before that moment Chetwode suggested that Hodgson should add one of his brigades to assist Chauvel's division. Royston's 3rd Brigade was detailed for the purpose.

The Turks, concentrating on opposing the infantry's attack from the south, were taken by surprise. They hardly expected to be attacked from the north and east so late in the day. Consequently, though the attacking line was necessarily very thin, the advance was at first remarkably swift. For instance the 5th Regiment, which had been ordered to cooperate with the New Zealanders, rode forward with bayonets fixed at the gallop.

> 'Major [William] Chatham with "C" Squadron', wrote another officer of the regiment, 'galloped down the Jaffa–Gaza Road, which runs through an olive grove, closely followed by "B" Squadron under Major [Frank Graham] Newton and the remainder of the regiment. As soon as we

appeared in the open on the road, the regiment was heavily shelled . . . . Fire was opened from the cactus hedges and native villages to the right by remnants of the Turkish battalions that had been dispersed earlier in the day. Lieutenant [Frederick Mitchell] Waite, who was on the right of the advance, at once wheeled his troop to the right, jumped the hedges still mounted, and got amongst the Turks. [He] used his revolver until shot through the body in five places, his men firing their rifles from their horses.'[12]*

Trooper Idriess describes what happened next as the regiment reached

'those giant walls of prickly pear. The colonel threw up his hand – we reined up our horses with their noses rearing from the pear – we jumped off – all along the hedge from tiny holes were squirting rifle-puffs, in other places the pear was spitting at us as the Turks standing behind simply fired through the juicy leaves. The horse-holders grabbed the horses while each man slashed with his bayonet to cut a hole through those cactus walls. The colonel was firing with his revolver at the juice spots bursting through the leaves – the New Zealanders had galloped by to the left of us, the 7th Light Horse were fighting on our right. Then came the fiercest individual excitement – man after man tore through the cactus to be met by the bayonets of the Turks. . . . It was just berserk slaughter . . . . The Turkish battalion simply melted away; it was all over in minutes.'[13]

There was much snap shooting in the maze of cactus. An officer of the 5th said afterwards: 'The Turks ran in and out like rabbits, and we shot them as they ran.' One farrier-sergeant was seen sniping from the saddle while his mount nibbled at the grass. An officer shouted at him 'Why not get off?' 'I can see them better,' he replied, 'from here' and went on shooting.[14]

As dusk approached, the urgency of taking the town was obvious to all. Even the men with the led horses forced themselves

*Waite was awarded the Military Cross and Shoeing-Smith Thomas Jones and Trooper Septimus Laws Gofton were awarded the Military Medal for their exploits on this occasion. (Wilson, 97)

through the single-file passages cut through the hedges immediately on the heels of the riflemen. One party of them took twenty prisoners and shot others. Lieutenant G. Snow of the 7th Regiment, a Victorian engine driver in peacetime, at one point shot down numbers of Turks by running into the open with a Hotchkiss which he placed on the shoulders of one of his troopers.[15]

The New Zealanders (or 'En Zeds' as the Australians called them) meanwhile pressed on equally fast.* By 5 p.m. the Canterburys had worked along the ridge and actually attacked the Ali Muntar defences from the east. Having joined up with the infantry, they finally gained the position at about 6.40 p.m.,† at which time the Turks had at last withdrawn from it. The Wellingtons, when they had cut through the hedges, eventually captured two 77 mm. Krupp guns with their ammunition limbers. These had been put out of action by the Somerset Battery. Held up by enemy fire from several houses, the regiment immediately formed an extempore gun crew. One of the guns was pointed at the most troublesome house less than seventy-five yards away 'and Corporal Rouse who was "O.C. Gun Detachment" looked through the barrel until the target was well in view, inserted a shell, closed the breach and fired the gun. Result, large hole in the house and twenty terrified Turks covered with debris ran out and surrendered.' Another shot from the gun caused such destruction that Corporal Rouse was heard to remark that 'the New Zealanders had made, at any rate, a new street in Gaza'.[16] At the same time the yeomanry regiments of 22nd Brigade which had advanced at a gallop along the track from Beit Durdis also made contact with the men of the 53rd Division in the streets of Gaza.

As the moon rose after a scorchingly hot day, during much of which a fierce khamsin had blown, the Turks had everywhere fallen back into the centre of the town. As Wavell has put it: 'The victory so hardly struggled for was won.' To this he added: 'but an hour too late'.[17] At 6.10 p.m. Chetwode with Dobell's approval issued a telegraphic order to Chauvel. He was to break off action at

*To each of the three regiments were attached four machine guns, the other four being held in reserve.

The Aucklands had only three troops up, the rest not having yet come in from their observation duties well away to the north and east.

†Sunset at this date is almost exactly at 6 p.m. (MacMunn, 307).

once and retire across the Wadi Ghazze.* To understand why he thought such an apparently unwarranted order necessary the situation from the enemy's point of view must be examined. By 16 March Kress knew from air reports that Murray intended to advance on Gaza. He immediately strengthened the garrison to seven battalions supported by twelve howitzers, two 10-cm 'long' guns and two field batteries. At the same time he ordered a division from Jaffa to march towards Gaza. One regiment was expected to arrive at the latest by first light on 27 March. When early on the 26th Major Tiller, the garrison commandant, first reported the British attack, Kress told him to hold out to the last man and ordered a division which was near Jemmame to advance towards Ali Muntar and another, equally near to Gaza, to threaten the British rear. Further troops were to march at once from Beersheba. Most of these, Kress hoped, could be in action close to Gaza before darkness fell on the day of the battle. Due, however, to what Liman von Sanders called 'typically Turkish' delays, they had covered scarcely half the distance to Gaza by nightfall.[18]†

Dobell, Chetwode and Murray, who had set up his forward headquarters on his train at El Arish, were unaware of most of this. What they did know is difficult to gather.‡ The first sighting of enemy reinforcements came at 4 p.m. when Chauvel's headquarters reported: 'Three columns are moving towards us from Deir Sneid.' Fifty minutes later Hodgson's headquarters reported '3,000 infantry, 2 squadrons cavalry, advancing from Huj in south-westerly direction'.[19] These seem to have been the only written messages received, but Chauvel is known to have spoken from time

*Powles purports to give the exact wording of the order: 'Owing to the lateness of the hour and the strength and position of the enemy forces pressing in from the north and east and the difficulty of continuing the attack in the dark in the town of Gaza, the GOC Desert Column has decided to withdraw the Mounted Troops.' (Powles, 93. See also, Browne, 230).

†The reinforcements might have started earlier had the fog not delayed the British infantry's attack. (Browne, 232).

‡Just as the withdrawal order was being despatched, Dallas reported that a redoubt north-east of Ali Muntar had fallen, but this did not seem to Chetwode to be sufficient reason to alter his order. It was not until a little later that he learned of the capture of the whole ridge. (MacMunn, 307).

to time to Chetwode on the telephone. There was, however, other vitally important information which ought to have been available to all three commanders. The Intelligence Department in Cairo, which had the key to the Turkish cipher, picked up all the messages between Kress and Tiller and telephoned them to Eastern Force exchange at Rafah within a quarter of an hour of receipt. At least four of these* are known to have reached Rafah before 6.30 p.m., but the first which trickled down to Dobell did not arrive till an hour before midnight! 'The obvious explanation' is given by the Official Historian. Dobell's headquarters were far from the exchange, 'there was heavy pressure on the line forward and proper discretion as to priority of messages was not exercised.'[20] The diary of a major in the Cairo Intelligence Department gives some idea of what those messages consisted:

> 'Things began to hum about midday . . . . Tiller anxiously reports no sign . . . of any reinforcements. Von Kress assures him Divs. 3 and 16 coming to his assistance. By evening Tiller reports British [the mounted troops] have entered town by N. and E. and situation very bad. Von Kress asks him concerning morale of troop commanders and troops. Tiller replies that former refuse to face combat at dawn . . . . At midnight messages from Gaza pessimistic . . . . Troops cannot face any further artillery fire. Kress [says that] he will attack from Jemmame and Sheria at "first morning grey". Tiller says English in the town and unless reinforcements sent there is very little hope.'

By 5.40 a.m. on the 27th all papers were burnt and the wireless station was blown up.[21]†

Chauvel protested most strongly at the withdrawal order, but Chetwode stressed the menace of the Turkish reinforcements' advance against Hodgson, pointing out that, as the 3rd, 5th and 6th Brigades were already engaged with their vanguards, the risk was too great. Chauvel eventually gave in, saying: 'Well, if the 5th Mounted Brigade are heavily engaged too, we had better

*The originals are not to be found (MacMunn, 310. See also Gullett, 289).

†Gullett states that a message was sent from Gaza to Kress saying 'position lost, 7.45.' (Gullett, 289).

withdraw.'[22]* Chaytor when he received the order is said to have insisted on having it in writing before he would obey it. When the 2nd Brigade received it Ryrie told his staff not to obey until every one of his widely scattered men had been collected. This he was unable to do till 'nearly midnight and during that time there were no signs of the enemy'. In fact the Turkish reinforcements with Kress's acquiescence[23] halted everywhere the moment night fell. The 9th Australian Light Horse Regiment, for instance, acting as rearguard to the 3rd Brigade, stayed out some miles east of Gaza till 3 a.m. and were not troubled in the slightest by the enemy until they began to retire, and then only cautiously.[24]

Not only was the enemy unwilling to fight at night, but he remained ignorant of the withdrawal until dawn. This was fortunate, for to effect a retirement in such a situation was a severe trial of skill and discipline. Some of the regiments were far from their horses and it took time to gather them together. The New Zealanders uncharacteristically lost their way, but recovered themselves when they bumped into Hodgson's headquarters. His brigades screened Chauvel's withdrawal and they did not begin to concentrate their regiments until about 2 a.m. All the dead and wounded were safely brought out and most of the prisoners too, except the badly wounded ones.[25]† The 2nd Brigade brought up the rear and remained near enough to succour the infantry, some of which throughout the 27th remained in action on the open ridges south of Gaza. These were not withdrawn in total till the early hours of 28 March.

*He realized that the success of this brigade which was blocking the enemy thrust up the Beersheba–Gaza road would be vital to the extrication of his division. (Chauvel, 105).

Gullett says that when Dobell gave the actual withdrawal order Chauvel exclaimed: 'But we have Gaza!' to which Dobell replied: 'Yes, but the Turkish reinforcements are all over you.' This exchange, though likely to be authentic, is confirmed by no other source. (Gullett, 294).

†These were left in a prominent position, each being given a full water bottle by Chauvel's order. 'Some of the poor devils made an awful fuss at being left . . . and their shrieks followed us as we rode off.' (Anon. *Experiences of a Regimental Medical Officer*, 15, AWM, quoted in *Chauvel*, 105).

8. Mysore and Bengal lancers with the Bikanir Camel Corps in the Sinai Desert, 1915

9. Men of the 10th Australian Light Horse leaving the Suez Canal defensive line on 10 June, 1916, to destroy the supplies of water in Wadi um Muksheib

10. 'We capture and question a lone prisoner as to his intentions of why he is alone and get some useful information; needless to say, he is a deserter'

11. When in action, cover for horses had to be found. The usual practice was for one man to look after four horses

The British lost 523 killed and 2,932 wounded, mostly lightly.* The Turks lost 301 killed, 1,085 wounded and 1,061 missing. Of this total of 2,447, fifty-seven were Germans or Austrians.[26]

A considerable number of post mortems on the battle have been penned. Here it may suffice to give Chauvel's considered opinion. He wrote to Birdwood on 25 July: 'The infantry treated the fog as an obstacle and waited until it had cleared instead of looking upon it as a god-send and making every use of it as we did.' This is perhaps not quite fair, since the infantry had to advance against an enemy force in position, while the mounted troops were faced with nothing more than open country. Chauvel also blamed Dobell for not employing 'the whole available force', meaning more of the infantry and artillery. Yet it seems certain that supply arrangements for such an addition were difficult if not quite impossible.

> 'I think,' continued Chauvel, 'the withdrawal was wise as many thousands of reinforcements were moving on us from three different directions, some less than six miles away, and the town itself was full of Turks, so that we mounted troops . . . would have had to face round at dawn† to meet these fresh troops with an unbeaten though no doubt demoralised enemy immediately behind us, unless . . . we had been able to clean up the town in the night which was unlikely.'[27]

What Chauvel does not mention are some other major reasons why the battle did not go the way that Magdhaba and Rafah had. Neither Dobell nor Chetwode, with perhaps less excuse, at any time left their headquarters to communicate in person with Dallas or Chauvel. There was no shortage of horses or motor cars and the going was excellent, nor were the various headquarters any great

*512 were put down as missing, mostly prisoners and nearly all from the infantry. (MacMunn, 315). The mounted troops' casualties were comparatively slight, the Anzac Division, for instance, had only six men killed, forty-six wounded and two missing. Twenty-three horses were killed and thirty wounded. Some 150,000 rounds of small arms ammunition were expended. (Browne, 231).

†Browne, 233, calculates that the Imperial Mounted Division and the Imperial Camel Corps Brigade together could not have mustered more than 5,600 rifles to fight off superior numbers attacking from four widely separated directions. This assumes that the whole of the Anzac Division would have been engaged in occupying Gaza, which was a probability.

distances apart. Instead, in the words of Chauvel's biographer, 'both elected to conduct the battle with telephone and signal pad, like a training exercise for junior staff officers.'[28] Surely Chetwode should in person have chivvied Dallas and have learned the up-to-date situation from Chauvel. For instance, he was excessively worried about the horses not being able to water and gave this as a weighty reason for the withdrawal order. Had he conferred with the mounted divisions' commanders he would have learned that nearly all the regiments had managed to obtain water in the course of the day.* There is much evidence that the system of communications throughout the battle was for the most part inefficient.† Further, the staffs were very inadequate both in quality and numbers. Chetwode and Dobell shared what there were, the former's being enough for a division only, not a corps, and the latter's being improvised from one of the Canal Defence Sections. As Allenby was later to prove (see p. 121), there were plenty of trained staff officers sitting in Cairo offices.

Murray's despatch gave a ridiculously inaccurate and astonishingly complacent account of the battle, representing it as a victory in which the Turks only just avoided annihilation. His popularity with all ranks fell to new depths. Djemal is said to have sent a message to Murray saying, 'I have defeated you in the field, but you have beaten me in the communiqué.'[29]

* * *

*There is ample evidence of this: MacMunn, 300, 305; Browne, 227, 230, 232; Teichman, 120; Gullett, 294; Adderley, 97, *et al.*

†The historian of the Royal Corps of Signals stresses the paramount importance, where headquarters are split into more than one echelon, of messages being addressed to the appropriate echelon, or alternatively that they

> 'do not get held up at the wrong echelon, as was the case with the intercepts of the Tiller wireless messages [see p. 101 above]. It is also essential that the moves of headquarters are carefully timed and coordinated so that they are not inaccessible at a critical phase of operations . . . .
>
> 'There were the misunderstandings on the telephone . . . . In 1917 field telephones were not very efficient instruments and were awkward to handle. It is doubtful whether either of the two generals had had much telephone experience or fully realized the pitfalls, and it is not surprising therefore that in the stress of battle they should have fallen into error.' (Nalder, 181, 182).

An important reason why First Gaza, as the battle came to be known, failed was the paucity of artillery support. For Second Gaza, which took place three weeks later, between 17 and 20 April, the number and weight of guns were much increased. Gas shells were used for the first time in the theatre, and fire from warships was also employed. So, too, were eight recently arrived tanks, all of which were blasted out of action.* This accretion of strength became possible because the railway had been advanced to within seven and a half miles of Gaza. Equally, virtually the whole of Dobell's force could now be deployed: three infantry divisions with their full complement of artillery and another in reserve as well as all the brigades of the two mounted divisions.

If First Gaza was a costly near victory, Second Gaza was an even more costly total defeat. It was intended by Murray and Dobell, both excessively, almost insanely optimistic, to be a Western Front type battle in the grand style. It was fought against a greatly increased garrison in extremely well orientated entrenchments and bastions.† These extended intermittently for some eleven miles to the east, about half way to Beersheba, thereby preventing the possibility of a repetition of the encirclement of Gaza by the mounted troops as at the earlier battle. There never was the slightest hope of any element of surprise and with one single exception every attempt by the infantry, the Camel Corps Brigade and the largely dismounted mounted troops to reach the enemy's trenches was shattered and frustrated. The assault frontage of about 15,000 yards, in spite of the augmentation of the artillery, was supported

*An officer of yeomanry describes how the tanks 'arrived in the outpost line, where dead silence and no lights were the order, making enough noise to wake the dead, exhausts red hot, and preceded by a man on foot who yelled directions to the driver . . . . Their presence must have been obvious to the Turks. They were to undertake a task very different from anything yet asked of tanks, for the distance they had to go before they could close with any enemy, the heat and the lack of decent maps made the proposition one of a nature quite foreign to anything yet encountered by tanks on the Western Front.' (Powell-Edwards, Lt-Col H.I. *The Sussex Yeomanry and 16th (Sussex Yeomanry) Battalion, Royal Sussex Regiment, 1914–1919*, 2nd edn, 1921, 77).

†'Turkish reinforcements,' wrote Trooper Idriess of the 5th Regiment, 'are pouring into Gaza day and night, and to think that one little horse brigade and one little NZ brigade galloped right into the back doors of the town and drove the Turks to their very mosque!' (Idriess, 261).

by no more than 150 guns. Further, the supply of gas shells and the number of tanks were quite inadequate.

The Anzac Division on the right flank was little engaged, while the brigades of the Imperial Mounted Division were used almost entirely as infantry. They incurred 547 casualties while the Anzac's numbered only at most 187.* The Camels' casualties numbered 345, while the infantry lost 5,328. The Turks' losses seem to have amounted to a little over 2,000. They took 272 prisoners while the Turkish prisoners numbered about seventy fewer. The British loss in horses and camels numbered 2,129.[30]† The official historian sums up this expensive but otherwise uninteresting battle thus:

> 'A dogged advance against imperfectly located entrenchments and in face of fire from hidden artillery, without adequate support from that arm on the side of the attackers . . . . There were wide gaps in the Turkish position, but the redoubts were well sited for mutual support and permitted the retention of reserves for counter-attack outside the danger zone. Since the Turkish infantry did not flinch from counter-attack, the result was never in doubt.'[31]‡

During the night of 19 April and the following day the infantry and many of the mounted troops, exhausted though they were, dug sturdily in expectation of a counter-attack. This Kress arranged to launch along the whole front, but Djemal once again refused his assent.[32]

* * *

*This figure is given by Browne, 239. Gullett, 334, quoting Dobell's report to Murray, puts it at only 105.

†It is interesting to note that 335 camels died of heat apoplexy between 24 and 26 April, showing that the khamsin, which started during the battle, blew for at least eight days of blistering heat. None of the horses or mules died from this cause. (Blenkinsop, 186).

‡The yeomanry at Second Gaza truly came into their own. From now onwards the Australasian jeering ceased. Trooper Idriess says that the yeomanry 'fought in a way that has made the crowd accept them as brothers'. (Idriess, 277).

'We were given a treat,' wrote Lance-Corporal Godwich of the Worcestershire Yeomanry on 3 May.

> 'We were taken down to railhead and thoroughly disinfected or "de-loused" as the wags called it. The process was as follows: two railway wagons and an engine were the baths. We undressed by the wagons, tied our clothes in a bundle and the bundles were then packed into the wagon nearest the engine, the doors closed, then steam was forced through everything at high pressure. While this was going on, as many as possible got into the other wagon and had a good bath. It was an extremely funny sight to see dozens of men in their birthday suits, except that we all kept our big sun-helmets on. After half an hour's steaming the kits were thrown out of the steamer. We opened them to the sun for two or three minutes, dressed, saddled up and rode back to camp feeling very much cleaner and comfortable. This disinfecting was a jolly good idea. We all welcomed it, but it was not done frequently enough. I think I went through it four times in three and a half years in the desert. Probably the infantry had it more often, because we mounted troops were not often near the railway.'[33]

12

'The two Gaza battles were as stimulating to the Turks as they were mortifying to the British .... Each army held good defensive ground. Each was confident of its capacity to fling back an assault. By the end of April the campaign had reached a stalemate.'

The Australian Official History

'21.5.17 Cavalry warfare is about over I think .... They can't say we haven't done our share – we have taken every inch of ground this side of Kantara ... and I should think I have ridden on an average the whole distance at least three times – the infantry have simply followed us up.'

LIEUTENANT R.H. WILSON,
Royal Gloucestershire Hussars Yeomanry

'Helped by the Russian inertia in the CAUCASUS and the summer check to active operations in MESOPOTAMIA, the Turk appears to have decided to cry "check" to us definitely on his present line. I do not think he had first intended to do so, but emboldened by our failure to annihilate his advanced force at GAZA by a coup de main and by his success in staving off further attacks and influenced no doubt by the moral and sentimental advantages of holding JERUSALEM, he has succeeded in overcoming the objections of CONSTANTINOPLE and is making every preparation to make our further advance as difficult as possible.'

CHETWODE in 'Notes on the Palestine Operations', 21 June 1917[1]

Six months' stalemate starts – Chetwode succeeds Dobell – Chauvel succeeds Chetwode – Chaytor succeeds Chauvel – re-organization of mounted divisions – arrival of Barrow – expedition to destroy Asluj-Auja railway

After Second Gaza both sides set to in earnest to fortify and entrench their lines. 'There comes nothing out of Eastern Force now but demands for trench mortars and handbooks on trench warfare,' wrote a Staff officer in Cairo.[2] A glance at the map will

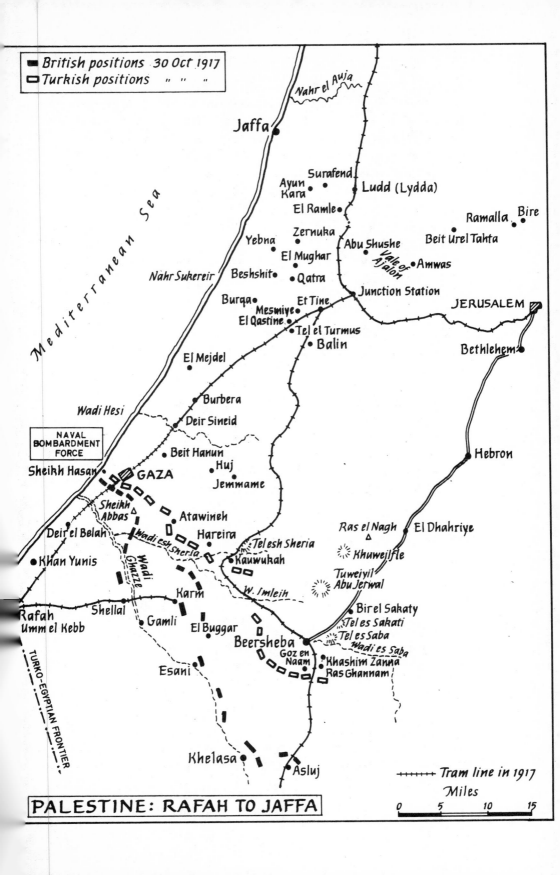

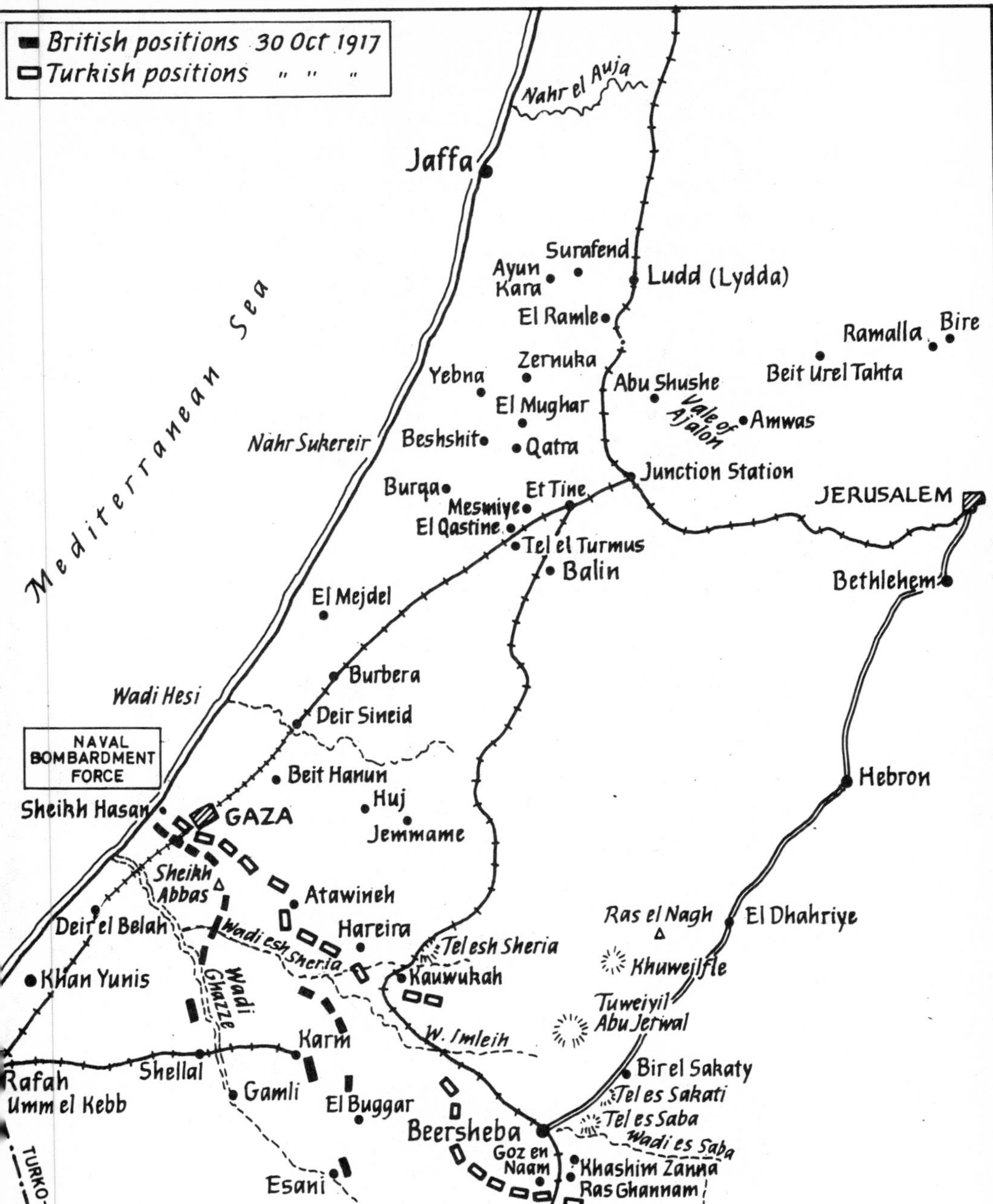
British positions 30 Oct 1917
Turkish positions " " "
Nahr el Auja
Jaffa
Mediterranean Sea
Surafend
Ayun Kara
Ludd (Lydda)
El Ramle
Ramalla
Bire
Beit Urel Tahta
Zernuka
Yebna
Abu Shushe
Vale of Ajalon
El Mughar
Amwas
Nahr Sukereir
Beshshit
Qatra
Junction Station
Burqa
Et Tine
JERUSALEM
Mesmiye
El Qastine
Tel el Turmus
Balin
Bethlehem
El Mejdel
Burbera
Wadi Hesi
Deir Sineid
NAVAL BOMBARDMENT FORCE
Beit Hanun
Hebron
Sheikh Hasan
GAZA
Huj
Jemmame
Sheikh Abbas
Atawineh
Deir el Belah
Hareira
Ras el Nagh
El Dhahriye
Wadi esh Sheria
Tel esh Sheria
Khuweilfle
Khan Yunis
Kauwukah
Wadi Ghazze
Tuweiyil Abu Jerwal
Karm
W. Imleih
Rafah
Shellal
Bir el Sakaty
Umm el Kebb
Gamli
Tel es Sakati
El Buggar
Tel es Saba
Beersheba
Wadi es Saba
Goz en Naam
Khashim Zanna
Ras Ghannam
Esani
TURKO-EGYPTIAN FRONTIER
Khelasa
Asluj
Tram line in 1917
Miles
0
5
10
15
PALESTINE: RAFAH TO JAFFA

show that the strongly entrenched positions taken up by both sides, unlike those on the Western Front, were only for a part of their length close to each other.* Those nearest to the town were some 400 yards apart and those on the extreme British right were divided by as much as nine miles. It will also be seen that both flanks to the east were open. For the British this meant that only lack of water constricted the manoeuvring power of their overwhelmingly superior mounted forces.†

The state of standstill was to last for the next six months. It was only relieved by mostly ineffectual bombing raids and by constant patrolling. A corporal of one of the 3rd Brigade's Scout Troops, known by their rivals in other brigades as 'Royston's pet boys', describes how these patrols were conducted:

> 'We would line a ridge, leaving our horses below us in the care of the horse holder for the day, and keep watch. Then a few shots would be fired. The Turks first, as much as to say "you have come far enough". After a while we seemed to come to an understanding that if you don't shoot we won't either. However, they were waiting for our retirement. As soon as we began to withdraw, they would gallop smartly forward and open a brisk fire, following from ridge to ridge, and repeating the performance until we regained our lines. We would do the same thing. We called it good fun and often played tricks on them, pretending to withdraw and then open fire. In this way we would keep them guessing. We were only a troop of thirty in strength and they could have been about the same.'

*The Turks had made Gaza into a strong modern fortress. Between Gaza and Beersheba their line consisted of five strong localities, the furthest east being around Beersheba which was about four and a half miles from the Hareira group. The rest of the localities were separated from each other by gaps of 1,500 and 2,000 yards. 'They are not,' wrote Chetwode in his 'Notes on the Palestine Operations' (see p. 126), 'in all cases mutually supporting as regards rifle and machine-gun fire, but they are mutually supporting as regards the artillery fire from neighbouring groups of works.' (Chetwode, 2–3).

†This open right flank had one important disadvantage: it made it easy for enemy spies to reach the rear of the British lines. Most of them were natives but at least one daring German penetrated to the camps in the disguise of a British or Australian officer.

The Australians on some of these minor 'stunts' covered their bayonets with cloth so as to prevent them glistening.[3]

On one occasion a squadron of the 7th Regiment captured two enemy troops by means of a clever ambush. As the Turkish cavalrymen fled they were overtaken and, apparently grabbed 'by the collar at the gallop' and then dragged 'each man in turn from the saddle' and all made prisoner. Young Lieutenant Bob Wilson of the Gloucesters won a Military Cross at Gamli on 15 October when 'in charge', as he put it, 'of a little ambush' where his troop 'played Merry Andrew with a few of their cavalry and pinched some lances and brass stirrups – irons and bits, besides a lot of other stuff.' Both his sergeant and his corporal received Military Medals for this action. Two days earlier a patrol consisting of Corporal Morgan and three troopers of the Gloucesters had been lured into a trap by a Bedouin and then confronted by two troops of Turkish cavalry. All except one trooper were captured. One of the two prisoners taken by Wilson proved to be one of the Turks who had captured Corporal Morgan's patrol. The regiment had got its revenge![4] On 26 June a corporal of the Worcesters 'had the satisfaction of pushing his sword through the back of a flying Turk at full gallop'. Though the regiment had been armed with the sword for 123 years this was the first known occasion of its employment in battle by a member of it.[5] Every night the regiments in the line had to provide listening posts. These went out well in advance of the ordinary outposts. 'One of us,' wrote Lance-Corporal Bostock of the 10th Regiment's Scout Troop, 'would hold the two horses while the other moved about fifty yards further on and lay with ear to the ground. If we were too close to the horses their noise would interfere with our hearing. We took it in turn to listen.'[6]

Both sides, as the months wore on, received reinforcements. By the beginning of June the Turks numbered about 33,000 rifles, 2,200 sabres, 130 machine guns and 120 pieces of artillery.[7] Murray's recurring demands for more infantry could not be fully met,* but under pressure from Lloyd George, the Cabinet was persuaded not to be content with the security now obtained for Egypt and the Canal, but to increase, so far as resources allowed,

*A new, weak division was formed on 21 June 'from units sent from abroad or hitherto in Egypt'. Another division started arriving from Salonika on 14 June. (MacMunn, 358)

their commitments and efforts on the Palestine front. The failure of the French offensive in Champagne strengthened the Prime Minister's resolve to try to find a way round the expensive deadlock in France. The Turkish concentration for the recapture of Baghdad also spurred them on to take pressure off the Mesopotamian front.* In June and July the 7th and 8th Mounted (Yeomanry) Brigades were therefore withdrawn from the Salonika front and added to Eastern Force, together with a considerable accretion of artillery.

Dobell, who it seems had suffered mild sunstroke, was relieved by Murray soon after Second Gaza and Chetwode replaced him at Eastern Force headquarters, while Chauvel took over command of all the mounted troops formed into the Desert Mounted Corps (Descorps). He was the first officer of a Dominion force to command a corps. He in turn was replaced in command of the Anzac Mounted Division by the New Zealander, Chaytor. Lieutenant-Colonel Meldrum relinquished his command of the Wellingtons to take over the New Zealand Mounted Rifles Brigade. The value of the mounted troops was now fully recognized, for by late June Chauvel had ten brigades to dispose of. These were reorganized into three divisions (each comprising three brigades instead of four as previously):† Anzac Mounted (Anzacdiv.),

*The chief events elsewhere at this time were the start of intensive submarine warfare on 1 February, Maude's capture of Kut on 24 February, the beginning of the Russian Revolution on 9 March, the capture of Baghdad on 11 March, the entry of the United States into the war on 6 April, the start of Nivelle's abortive offensive in France on 16 April, Lawrence's capture of Akaba on 6 July and the Italian defeat at Caporetto on 24 October.

†Five days after Second Gaza, Chetwode wrote to Lynden-Bell:

> 'I have been thinking a good deal about the question of the Desert Column cavalry. There are now two heavy divisions of four brigades each. All cavalry experience tells us, and the other day confirmed it, that four brigades are more than one man can handle in an open fight on a broad front. The French have two, the English have three. Cavalry work naturally in threes and the fact of having two four-brigade divisions makes it very difficult for the Corps Commander to keep a reserve in his own hand as he has to withdraw individual brigades who have to be commanded by the senior brigadier with no staff to run it. I know that the question of Staffs would be very difficult, but I feel that it is worth thinking about and that two three-brigade divisions and one two-brigade, the latter possibly strengthened by any odd cavalry such as the Westminster Dragoons and Indians, would result in a better working instrument.' (25 Apr., 1917, Chetwode)

Australian Mounted (Ausdiv.), Imperial Mounted (renamed at the Australians' request and still commanded by Hodgson), and Yeomanry Mounted (Yeodiv.) which consisted of nothing but yeomanry regiments.* To command this Major-General Barrow (see Vol. 4, 455) came out from France in June.†

The unpleasing prospect of a whole summer to be spent on the Palestine frontier enduring the inescapable heat, the dust, the flies, the lice, the recurrent khamsins, the unvarying diet and above all the boredom, was not made more acceptable by the low morale which inevitably followed two lost battles. Fortunately both Chetwode and Chauvel directed their considerable energies with marked success to making as bearable as possible the harsh, protracted defensive period ahead. Chetwode determined to make his defence as aggressive as possible, employing at any one time the minimum of men in the trenches. He eschewed a linear front line such as prevailed in France and substituted a sequence of separate, mutually self-supporting localities requiring comparatively few men to man them. Vigorous training‡ and regular rest periods could thus be provided behind the line. When the mounted divisions were in it they lived in bivouacs made of light wooden hurdles covered with grass mats erected over rectangular pits. These provided some protection from the sun and from enemy bombs. When they took their turn on the seashore they were accommodated in tents.

* * *

**Anzacdiv.*: 1 and 2 ALH Bdes, NZMR Bde (XVIII Bde RHA (12 18-pdrs)); *Ausdiv.*: 3 & 4 ALH Bdes, 5 Mtd Bde (XIX Bde RHA (12 18-pdrs)); *Yeodiv.*: 6, 8 and 22 Mtd Bdes (XX Bde RHA (12 13-pdrs)). Two regiments of 7 (Yeomanry) Mtd Bde and the Essex Bty, RHA, became Corps Troops. The infantry divisions now had greatly augmented artillery support (MacMunn, 357–8). Three fully equipped French battalions and 500 Italians joined in the course of June.

†Barrow's GSO 1 was Lieutenant-Colonel William James Foster, an Australian from Melbourne and an officer of the Australian Permanent Force. After the war he did a course at the Staff College and in 1927 commanded the 2nd Cavalry Brigade at Tidworth, the first Dominion officer to command a regular army formation in Britain.

‡In late May the Mark III rifles were handed in and exchanged for the new Mark IV. At the same time the Mark VI small arms ammunition was withdrawn and the new Mark VII, which had a considerably higher velocity, replaced it.

Leave to Cairo, 'a city', according to Corporal Fletcher of the Lincolnshire Yeomanry, 'blessed with grandeur unequalled in the world yet packed with all the lust and vice conceivable', was perforce very limited: one officer and ten other ranks every week per regiment. Frequent concerts were given in the Anzac Hostel which had been set up in 1914 and taken over by the YMCA in 1916. It could sleep 550 men.*[8] Home leave was, of course, absolutely out of the question. Many men had not seen home for well over two years. The next best thing after leave to Cairo was the rest camp set up at Khan Yunis.[9] There the band of the New Zealand Mounted Rifles was much in demand, as also was the 74th Division's 'Palestine Pops' concert party.[10] Inter-brigade boxing matches were frequent and 'B' Battery of the Honorable Artillery Company 'gave some excellent Pierrot performances'.[11] Race meetings and gymkhanas which had been started before First Gaza now became a prominent feature of military life. Enclosed paddocks, totalisator facilities, jumps and marked courses were provided and large sums were made for charities.[12] Regimental sports were organized on a grand scale. In the Gloucesters, for instance, wrestling on horseback, 'a harem race' (whatever that may have been), a mounted tug-of-war and a mule race for officers took place in late July.[13] Trooper Idriess describes another form of sport:

> 'Scorpions are maniacal fighters. When two are tumbled into a tin they pull one another to pieces. Some of the boys are quite fond of their pets and carry their prize-fighters with them in a tobacco tin. They bet on the results of fights. The stakes are mostly wood to boil a quart pot, tobacco, chocolate (if any parcels have arrived), and money in the case of the affluent. A number of scorpions are "squadron leaders", others even have regimental reputations should they have survived numerous fights .... The proud owners are gorging them up with fresh meat and blood; we don't know where they got either from, but there's a rumour that one man has been draining Charlie Cox [the brigade commander]'s fowls. "Fighting Charlie" would shoot him. Before the fight, each

*The YMCA opened 120 centres between Kantara and Damascus over three years. It had an excellent reputation for producing canteens as close to the front line and as early as possible. (Hogue, 206) "trainer" tickles his poisonous pet with a straw. The little beasts clash their pincers in maniacal fury . . . . Should both be torn to pieces, the fight is a draw.'[14]

More civilized recreation was provided by the 5th's padre, Chaplain Maitland-Woods, who, says Idriess, 'has got the boys archaeologist mad. In their precious spare time they are digging all along the wadi [Ghazze] and finding queer old stone houses and buried tombs and things so musty with centuries that even the padre does not know what they are. The boys disdain anything Roman for that is too modern.'[15] At Shellal under a fourth century mosaic floor bones thought to be those of St George were uncovered. The padre sent a memorandum to headquarters reporting his discovery. 'Can find no record of this man,' came back the reply. 'Send full name, number, regiment and identity disc of Trooper St George [or, in another version, Saint]'! The Australians were enthusiastic card players. Lieutenant McGrigor of the Gloucesters wrote in October that he had an Australian post on his left 'within a quarter of a mile; they sent over a message to my N.C.O. asking if any of the men would care to join in a game of Nap!!!* What an unique crowd these fellows are really!'[16]

On the coast where the divisions took turns when out of the line there was much bathing† and in some places fishing. In one regiment fishing lines were bought from the regimental funds, 'but it was found that Mills grenades were much better'. One yeomanry officer bought out of his own pocket 100 pairs of sandshoes for his men. 'They are a great comfort in camp as one gets so tired of boots.'[17] Sent up the line from Cairo was a special delousing railway truck into which clothing was put and 'steamed for an hour'.[18] (See p. 107) Extra food and small luxuries arrived from time to time, especially for the Australasians. In June the 1st

*A card game named after Napoleon I, dating from 1879, 'in which each player receives five cards and calls the number of tricks he expects to win; one who calls five is said to "go Nap" and to "make his Nap" if he wins them all.' (*Shorter O.E.D.*)

†The regimental orders of the Gloucesters included one day the following: 'False teeth, loss of, while bathing. The attention of all concerned is drawn to the danger of losing false teeth while bathing unless they are removed before entering the water.' (McGrigor, 486).

Regiment received 'fifty-one cases of comforts' and a special issue of beer (which had to be paid for), some of which came from Japan.[19] Later on, the meat issue became frozen rabbit from Australia 'which was very good'.[20] A particularly pleasing addition to the troops' comfort was the earlier capture from the Turks of a large quantity of waterproof bivouac sheets. These were made of duck (a cotton fabric, lighter and finer than canvas). They were five feet square and had buttons and button holes along each edge so that they could be joined together to form shelters from the sun. With them came pointed poles in sections with which they could be erected. They were not of much use in wet weather as their ends were open. Later on similar sheets were issued by the Ordnance department.[21] On the coast there was a plentiful supply of dates and figs as well as of quails which the Arabs caught in nets.[22]

As always life was less dreadful for the mounted men than for the infantry. But for both, the sameness of the food and the lack of fresh vegetables severely affected their health. Septic sores which often developed into serious inflammation and caused agonies for men sitting on their saddles were an evil which hardly an officer or man escaped. Scratches from barbed wire were a frequent cause. Later on in the campaign a superior German-made ointment was captured. 'We tried it on our septic sores,' wrote Trooper Idriess. 'The men crowding around that wagon reminded me of the flies crowding our sores .... The ointment is grand. Within twelve hours most of the small sores grow scabs and soon after stop from running. The ointment burns like blazes.'[23] The sores epidemic became so bad that 'at times the majority of the men in a regiment would be swathed in bandages'.[24] Further, the heat and the strain meant that as many as a third of the men in one unit suffered from dilation of the heart, and sunstroke was common. So were the 'sand-fly fever', which left those attacked limp and exhausted for days on end, and 'Nile boils', referred to as 'the Order of the Nile'.[25] The health of the British, though, was vastly superior to that of the Turks. Kress states that 'at this period the army was losing in sick and wounded 3,000 to 4,000 men per month, and about 25 per cent of its strength – 10,000 men – were always in hospital.'[26] This manpower drain, though possibly exaggerated, was partially caused and much exacerbated by the inadequacy and inefficiency of the railway system which resulted in a constant shortage of supplies. 'We know,' wrote Chetwode, 'he is short of rolling stock and has practically no coal.'[27] There was also a high rate of desertion, particularly during the lengthy journeys which brought in reinforcements from the north.

At the time of the capture of Magdhaba in December, 1916, the Turks had extended their Ramle to Beersheba railway line almost to the Wadi el Arish. Now that the British were established before Gaza this was said to pose a threat to the lines of communication. In fact, it had not been used since January. The enemy was believed to be picking up the rails for use on the Et Tine to Gaza branch line which he was constructing. It was vital therefore that he should be prevented from doing so. A few days of intensive training in railway demolitions by men of Anzac Division and the Camel Corps, under their engineering squadrons, resulted in over thirteen miles of the line between Asluj and Auja being demolished at the rate of a mile in every twenty minutes.[28]* The expedition, which was a total success, was undertaken in a *khamsin* between 15.00 on 22 May and 22.30 on 23 May. At its head was the divisional commander, Chaytor – on a bed placed in a sand cart! He had been ill for days past, but was determined to lead his men.[29] The demolition and covering parties travelled fifty-eight and sixty-four miles respectively, losing not a man and encountering only a few hostile Arabs. It was a remarkable example of the use of the mounted forces.

Its most important result, though, had nothing whatsoever to do with the railway demolition. In the everlasting search for water, Chauvel's Chief Engineer, Temporary Brigadier-General Reginald Edmund Maghlin Russell, had been for some time researching in the records of the Palestine Exploration Fund. He discovered that Asluj and Khelasa had been populous centres in ancient times. Local Arabs confirmed that water existed in both places. He calculated that by a fortnight's work this could be developed sufficiently to support two divisions in an 'attack on Beersheba from the east without moving across the enemy's front.'[30] The significance of this was to be vital in five months' time (see p. 128). The enemy at some time did much to destroy the Asluj water supply. The pumping plant had been removed and explosives used to blow in the sides of the forty-foot-deep wells. For five days, just before Allenby's great advance commenced, the 2nd Australian

*Several tons of explosive were packed in kerosene tins, while the clips for attaching the charges to the rails were made from the steel bands in which hay for the horses was brought up from Kantara. (Gullett, 352)

Brigade was sent to repair and develop this vital supply, installing oil engine-driven pumps and long rows of canvas horse troughs. While the men were at this job, Allenby himself paid them a visit and 'complimented them on their energy'.[31] 'We thought quite a lot of him,' wrote Trooper Idriess, 'coming out all this distance and seeing with his own eyes what is being done.'[32]

13

'A new G.O.C. in C. out here .... Everyone is delighted. Few had a good word for our last one.'

LIEUTENANT SIR RANDOLF BAKER Bt,
Dorset Yeomanry

'[Allenby's] face is strong and almost boyish. His manner is brusque almost to rudeness, but I prefer it to the oil and butter of the society soldier. Allenby breathes success and the greatest pessimist cannot fail to have confidence in him .... He looks the sort of man whose hopes rapidly crystalise into a determination which is bound to carry all before it.'

CAPTAIN RICHARD MEINERTZHAGEN,
Cairo, 15 July, 1917

'Sinai had been cleared of the enemy and all danger to the Suez Canal removed. The physical difficulties of the road to Palestine had been subjugated. No enemy was behind; Egypt herself, if there were sullenness among a section of her people, was quiet and prosperous. The Arab Revolt had been fostered and supported until it had become an important factor in the war. All was prepared for Britain, if she so desired, to put forth in this theatre efforts far greater than before and to concentrate against that Ally of the Central Powers who here stood in arms against her, strength enough for decisive victory to be ensured.'

The Official Historian[1]

Murray replaced by Allenby – effect of Allenby's arrival – Allenby's first review

On 11 June Murray received a cable from London informing him that he was to be replaced by fifty-six-year-old Lieutenant-General Sir Edmund Allenby who was on the point of sailing for Egypt fresh from his command of Third Army and the battle of Arras. He arrived in Cairo on 28 June.* In his final despatch he wrote of his

*Lloyd George first offered Smuts the command, but he demanded an adequate force to land from the sea to cut the Turkish communications. This

(continued over)

predecessor: 'I reaped the fruits of his foresight and strategical imagination, which brought the waters of the Nile to the borders of Palestine .... The organization he created ... stood all tests and formed the cornerstone of my successes.'[2] All this was true and had Murray not allowed his multifarious responsibilities other than those connected with Eastern Force to prevent him from directing the Sinai campaign in person, he might just conceivably have become the famous commander which his successor became. However, his failure to stand up to French when he was his Chief-of-Staff in France in 1914 and his tendency in that post to faint at moments of crisis make this unlikely. The parallel between his case and that of Wavell in the next war is obvious. He was not of Wavell's high calibre, nor for that matter of Allenby's, but his country and especially the War Cabinet had reason to be grateful to him, especially for the unselfish way in which he had reinforced the Western Front with division after division. By contrast his firmness in insisting upon the retention of his mounted troops was entirely admirable. The Official Historian sums up the position correctly: 'Like the commanders of many other British "advanced guards" sent to open a campaign with insufficient resources, he was superseded because he had failed to achieve the success expected.'[3]

The new Commander-in-Chief's advent boosted morale immediately. One of Murray's failings was his remoteness. He was virtually never seen by the front line troops. 'Between the Canal and Gaza,' complained an Australian light horse brigade-major, 'I never set eyes on Murray.'[4] Corporal Fletcher of the Lincolnshire Yeomanry expressed the other ranks' view: 'Never was a change so popular amongst soldiers. Each individual cried aloud his pleasure when the announcement was made.'[5] The Medical Officer of the Worcesters heard that 'the "man-power" general had arrived and that after combing out the "indispensables" in Cairo and Alexandria, he was about to do the same at the base at Kantara; it is

---

was refused and he therefore declined the offer. (Churchill, *'16–'18*, 335).

Allenby brought with him Major-General Sir John Stuart Mackenzie Shea (see p. 175) who had commanded one of his divisions in 3rd Army. Allenby had removed him from his command for criticizing the orders he had received, but when Robertson at the suggestion of Lord Derby, Secretary of State for War, invited Allenby to employ him in Palestine, there was ready agreement, showing how little Allenby bore grudges. Shea became what was probably the best of all the infantry divisional commanders of the campaign. (Wavell, 186).

considered by some that he should be able to collect at least a division from that place!' Numbers of 'elderly and obese' cavalry officers were sacked and replaced by younger, leaner men.[6] Allenby's ruthlessness and speed in this respect contrasted vividly with Murray's comparative complacency. Trooper Idriess found among his fellow Australians 'a wonderful spirit of optimism over the arrival of the new general . . . . He has come right out to the line. We hear that he is a cavalry man too. New 'planes, more men, more guns are coming with him.'[7] Indeed they were, for he had demanded not only the 5,000 men required to bring his existing force up to strength, but also two more infantry divisions with a full complement of artillery, three more squadrons of aircraft (consisting chiefly of the new Bristol Fighters, superior in speed and manoeuvrability, especially climbing power, to any of the German planes),* twenty-six anti-aircraft guns and much else. Generally speaking he got what he wanted, but many of these reinforcements could not arrive in time for operations before October.

Hardly a week had passed after his arrival in Cairo before he paid a five-day visit to the front and arranged to remove his headquarters from Cairo to Umm el Kebb near Rafah. 'You know,' he is supposed to have said, 'General Headquarters' roots in Cairo and Ismailia are like alfa grass. They are getting deep into the ground and want pulling up. Moreover, Staff officers are like partridges: they are better for being shot over.' His first tours were made 'in a particularly disreputable Ford truck. He sat,' says Wavell, 'perched up on the front seat alongside the driver, an Australian, who was clad only in a sleeveless vest and very attenuated shorts.† The picture of these two, with one of the personal staff bumping painfully in the body of the truck behind,

*Even now the home authorities were apt to be shortsighted in the distribution of new aeroplanes. 'When a new type was produced they usually insisted on equipping all the squadrons on the Western Front before any of the new machines were sent East. It took many squadrons to produce much effect in France, while a single squadron could change the whole balance of air-power on the Palestine front in a few days.' (Wavell, 210)

†Though he thought shorts 'indecent and abominable', his chief reason for banning them for mounted troops was because there had been cases of blood-poisoning caused by legs rubbing against the heated horses. He insisted at first on riding-breeches and leggings, but later allowed men to ride in long 'slacks' of khaki drill. (Allenby, I/14-2, 32, LHC, quoted in James, 138; Gullett, 646)

remained long in the memory of those who witnessed it.' When asked why Liman von Sanders (see p. 219) lost battles, Allenby replied, 'He sat too much in his office: did not go about his front to see things for himself.' Lieutenant Bob Wilson of the Gloucesters tells of Allenby's first formal review. 'It meant a lot of "spit and polish" . . . . We sat on our horses for three hours without moving an eyelash with drawn swords which ultimately weighed five tons – whilst he rode round. This was after three hours of forming up and getting into shape – a battle is a picnic compared to a show of this sort.'[8]

An officer of the 9th Regiment found that he was 'constantly seen at all points of the front line, and if not actually in sight, his presence seemed to be felt.'[9] So frequent were his lightning visits that a certain lower commander arranged for a GHQ signals officer to send him whenever a descent was imminent a warning message: 'BL' ('Bull Loose')! That morale had been restored by late August is attested to by Bob Wilson who wrote home that 'the transport, supply, medical and sanitary arrangements are absolutely perfect; everybody is merry and bright.' As for the Turks, they were soon to be mightily impressed by 'Allah Bey'.[10]

Of the senior British generals who were cavalrymen, Allenby is in some ways, if not the most attractive, the most interesting. That he was very different from French and Haig in character and outlook is abundantly clear.* Whether he was a better, rather than only a more fortunate commander will never cease to be a matter for debate. That, unlike them, he left behind only mostly uninformative letters and no diaries or memoirs adds to the difficulty of arriving at a secure judgement. Like Montgomery a quarter of a century later, he was fortuitously placed in an exceptionally propitious situation, though at the time nearly everyone thought that he had been shunted into a backwater. That he made an outstanding success of the task set him is not in doubt. He has been lucky, as well, in being the subject of two rather too hagiographical books by Wavell.†

*For instance he did not share Haig's abhorrence of Lloyd George. After the war 'when the former Prime Minister had offended the army by his criticism of Haig, Allenby was asked to add his voice to the protest. "Attack Lloyd George?" he answered. "But I like the little man. He won the war, though for heaven's sake don't tell him so." ' (LHC, Allenby Papers, 6/IX.15)
†Wavell and Wavell: *Pal.*

It is hard to say that he performed especially well either as commander of the Cavalry Division or, later, of the Third Army on the Western Front.* In neither post did he reveal anything approaching greatness. It seems very unlikely that had he remained in France he would not now shine out in history with more effulgence than do, for instance, Rawlinson, Byng and Plumer. Had any of them, or even Gough and Haig's other Army commanders, been handed a large independent command with an open flank and adequate forces, there is reason to suppose that they, too, would have grasped the opportunities bestowed on them with something not far below equal ability.

Nevertheless, the overwhelming show of self-confidence, the daring and the dexterity of deception and manoeuvre which can be displayed in the Palestine command were of the very highest order. He was probably the best of the bunch for discharging the unique mandate with which he was entrusted. Though he was never 'popular' in the sense that Montgomery was, his judgement when it really mattered was good; his single-minded, if often ruthless, dedication earned him the almost complete confidence of his staff, his inferior commanders and, above all, of his men, and the *brusquerie toute anglaise* of which Picot, the French Consul-General,[11] complained, did his cause no real harm, either during or after the war. His famous *brusquerie*, due probably to a basic shyness not incompatible with his self-confidence, certainly unnerved many of those on its receiving end. As Lawrence James, his latest biographer puts it, he never saw 'anything incongruous about playing the censorious, fulminating commanding officer one moment and the friendly general, anxious to get to know his men next'. Lawrence James then gives an example: 'Once, after spotting an unconcealed ammunition dump, Allenby descended from his car and fell angrily upon the officer responsible, whom he discovered in his tent, shaving. After an apoplectic dressing-down, Allenby paused, extended his hand to his "semi-paralysed" victim and quietly said, "Well, good day. I am very glad to have made your acquaintance." '[12]

*There is evidence that, after the battle of Arras, Haig was sufficiently unhappy with Allenby's conduct of it to ask that he should be replaced. (Edmonds' unpublished memoirs, Edmonds Papers, III/2/15, quoted in Newell, 365).

Maurice Hankey, no mean judge of men, after seeing Allenby in September, 1915, came to the snap opinion that he was a stupid man. (Roskill, S., *Hankey, Man of Secrets*, I, 1969, 297)

Allenby's post-war career was distinguished during his seven years as virtual ruler of a much disturbed Egypt by his firmness in dealing with the manifold and intricate situations with which he was faced. It does not surprise one to learn that his political masters in London did not always find him malleable.

14

'Allenby's hawks are circling as if they own the sky.'

TROOPER IDRIESS, 16 July, 1917

'I had decided to strike the main blow against the left flank of the Turkish position, Hareira and Sharia. The capture of Beersheba was a necessary preliminary to this operation, in order to secure the water supplies at that place and to give room for the deployment of the attacking force on the high ground to the north and north-west of Beersheba, from which direction I intended to attack the Hareira-Sharia line.'

ALLENBY'S despatch, 16 December, 1917

'It was an unorthodox battle, for the two wings of Allenby's army which struck these alternating blows were fifteen to twenty miles apart, linked only by a screen of one mounted division and open to a counterstroke in the centre. But Allenby had rightly judged that the nature of the ground and of the enemy made the danger of such a counterstroke an acceptable one.'

WAVELL in *Allenby: A Study in Greatness*

'Large dumps sprang up everywhere; here were piles of forage, there stacks of rails for rushing forward. Little engines fussed about drawing truckloads of shells.'

An officer of yeomanry in mid-September, 1917[1]

Chetwode's plan for Third Gaza accepted by Allenby – Eastern Force abolished – Chauvel succeeds Chetwode – Chetwode commands XX Corps – Bulfin commands XXI Corps – reconnaissances, transport, water and deception preparations for Third Gaza

Even before Murray's departure, Chetwode, ably assisted by his chief staff officer, Major-General Guy Payan Dawnay, had set about preparing an appreciation and a plan for future operations.* This he presented to Allenby in June. At the foot of his file copy he wrote: 'The above plan was adopted by Gen. Sir E. Allenby on his arrival in Egypt and was carried out – commencing with the capture of Bir Saba [Beersheba] on Oct. 31st & ending with the fall of Jerusalem on Dec. 9th 1917.'[2] Allenby's debt to Chetwode was fully acknowledged. 'My plan,' he wrote in his despatch of 16 December, 'was based on the scheme he put forward .... To his strategical foresight and tactical skill the success of the campaign is largely due.'[3] Allenby accepted Chetwode's estimate of the troops required: seven infantry and three mounted divisions. On 12 August he abolished Eastern Force and substituted three corps (the whole still known as the Egyptian Expeditionary Force): the Desert Mounted Corps under Chauvel, promoted lieutenant-general, the first Australian to attain that rank;† the XX Army Corps under Chetwode‡ consisting of four infantry

*Early in the second week of April Dawnay had put forward an alternative to the frontal attack which became *Gaza II* on the 19th (Dawnay to Wavell, 29 Dec., 1938, Allenby 6/VIII/35, quoted in Newell, 370).

Chetwode insisted to Allenby that Dawnay's contribution should be fully acknowledged. Allenby in reply wrote to Chetwode: 'I know what a valuable man Dawnay is, and, without belittling your work, I fully recognize the part he has taken in preparing the appreciation and plans.' (13 Aug., 1917, Chetwode).

†The fighting strength of the corps, including its three divisions, the 7th Mounted Brigade and the Imperial Camel Corps, was at this time about 18,700. (Osborne: I).

On 31 August Brigadier-General Richard Granville Hylton Howard-Vyse, known long before he served with Australians as 'the Wombat', reported to Chauvel as his Chief Staff Officer. Allenby, describing him as the 'best staff officer in the British Army' (Chauvel, 123), brought him from France where he had been GSO, 5 Cavalry Division. He had joined the Royal Horse Guards in 1902 and very unusually for an officer in that regiment had passed the Staff College. In 1917 he was aged thirty-four and before long he and Chauvel became close friends, the latter appreciating his great ability as a staff officer. (His career ended unhappily in 1918).

‡Chetwode, of course, lost his superior command. As a cavalryman he might have resented being given command of an infantry division. Certainly some senior British officers did. (See *Chauvel*, 118 and 241). Chetwode, however, wrote to Chauvel: 'You and I have worked together in the greatest harmony. We have together helped to write a small page of history.... I shall always be very proud of having had such a fine body of men under my command as your Anzac mounted troops.... I cannot say how much I envy you the command of the largest body of mounted men ever under one

(continued over)

divisions and the XXI Army Corps under Lieutenant-General Edward Stanislaus Bulfin consisting of three infantry divisions.*

Meredith, invalided home to Australia, was succeeded in command of 4 Australian Light Horse Brigade by Brigadier-General William Grant of the 11th Regiment, a surveyor and pastoralist from Queensland. 'He had learned on the wide plains that bushcraft which made him famous in Sinai as a guide on night marches over the maze of sand-dunes.' He was more excitable and impulsive than his fellow brigadier-generals. Royston was granted leave to his home in South Africa on urgent personal business.† He was succeeded in command of the 3rd Brigade by Brigadier-General Lachlan Chisholm Wilson who had been the highly successful commander of the 5th Regiment, but he took over only three days before the assault on Beersheba. He was a solicitor who had learned his soldiering as a trooper in the South African War. Shy, short and silent, 'his quiet figure' according to the official Australian historian, 'concealed the spirit of a great master of horse'.[4]

6 (Yeomanry) Brigade was now commanded by Brigadier-General Charles Alexander Campbell Godwin of the Indian Cavalry, who had served with Barrow in France. 8 Brigade came under the command of Brigadier-General Claude Stuart Rome of the 11th Hussars. Rome, beside being a champion boxer, had captained the Harrow XI against Eton at Lords. Brigadier-General Fryer, who commanded 22 Brigade, had served in the Inniskilling Dragoons.[5]

---

hand§ – it is my own trade – but Fate has willed it otherwise.' (12 Aug., 1917, quoted in *Chauvel*, 118).

§So far as the present author can make out this is not strictly accurate. Marshall Bessières commanded a cavalry corps in 1809 which probably numbered about 9,000 more than Chauvel's 20,000 in 1917. However, when in 1918 a fourth division was added to the DMC Chauvel probably did command 'the largest body of mounted men ever under one command', though in battle the actual number seems not to have exceeded a maximum of 23,500.

**XX Corps:* 10, 53, 60 and 74 Divisions; *XXI Corps:* 52, 54 and 75 Divisions. *GHQ troops:* Imperial Camel Corps Brigade, 7 Mounted Brigade, Imperial Service Cavalry Brigade and 20 Indian Infantry Brigade.

†'I have never seen such a downcast body of men as the regiment who farewelled the colonel by the wells of Asluj,' wrote Trooper Idriess. 'Major [Donald Charles] Cameron,' he continued, 'is colonel now; all hands like him. We are lucky in our officers.' (Idriess, 317).

*   *   *

Chetwode's and Dawnay's plan, in outline, was to feint at Gaza, then to employ most of his mounted troops in a vigorous thrust designed to capture Beersheba 'of the Seven Wells', a large, squalid Arab village, the desert gateway to Palestine, which had since 1914 been the chief Turkish base, and to assault the enemy's left, some ten miles north-west of that place. Thereafter he would be in a position to roll up the enemy line towards Gaza. When the Turks were thus forced to abandon that town, Chauvel would be well placed to intercept or harass their retreat over the great plain of Philistia, a strip of rolling downland fifteen to twenty miles wide, wonderfully well suited to mounted troops. While these operations were proceeding, the enemy's attention was to be fixed on Gaza 'by every available means, including a heavy bombardment and a determined holding attack.'[6] The success of the plan depended upon deception, surprise and swiftness of execution. The chief difficulties were transport and water. Between the Wadi Ghazze and Beersheba there were fifteen miles of barren plains and dry wadis. The daily requirements of the right flank striking force alone were calculated at 400,000 gallons of water of which a quarter, weighing about 500 tons, would have to be carried over a virtually roadless wilderness, while the horses could not be watered at all until Beersheba had been captured. All the transport available, including that of two of the three infantry divisions opposite Gaza (which was withdrawn from them for the purposes of the operation, including over 30,000 pack camels), was estimated to be only just sufficient to supply Chauvel's troops with ammunition and food up to one march beyond Beersheba. For water they must depend upon the rapid capture intact of at least some of the seventeen wells thought to be in the town.* This had to be achieved in a single

*In October, 1916, a Jewish spy had revealed the location of fresh water wells in the region of Beersheba. Chetwode communicated this information to Allenby when first discussing the plan with him. The spy was Aaron Aaronsohn, who had reached London from the Jewish settlements in Turkish territory. (Chetwode to Edmonds, 26 Sep., 1924, Edmonds Papers, II/1/32; Gilbert, Martin *Exile and Return: the Emergence of Jewish Statehood*, 1978, 129-30). For a brilliant semi-fictional account of Aaronsohn's and his wife's activities, see Smith, Colin, *The Last Crusade*, 1991.

It was known that for the post-Beersheba phase of the operations hardly any water existed until Sheria and Hareira were taken.

day or else the whole operation would fail. Some of the similarities between Chetwode's plan and Roberts's for the relief of Kimberley seventeen years before (see Vol. 4, pp 124-8) are striking. The difficulties were almost exactly the same: transport, water and secrecy.

Allenby's original intention was to make the attack in September, but it soon became clear that the earliest moment when the essential preparations could be ready was late October. The dangers inherent in leaving it so late were great. The enemy might well launch an attack before that time and indeed he was known to be preparing one. The greater risk, perhaps, was that the army would be caught by the heavy autumn rains which would bog down both the transport and above all the mounted troops in the sticky mud of the maritime plain over which they were to operate after Beersheba had been taken. Further, the subsequent attack on Jerusalem would probably have to be carried out in the extreme wet and cold of the mountainous country surrounding that city. Nevertheless Allenby considered that the risk of launching the attack before his arrangements were more or less complete was the greatest of all.

There was a vast amount to be done. The base at Kantara had to be transformed into a great inland port capable of berthing several ocean-going liners simultaneously. Thirty miles of metalled road were laid there. The doubling of the railway to Bir el Mazar was pushed ahead at the rate, during September and October, of a mile a day and the branch to Shellal threw out a twenty-eight kilometre offshoot to Gamli. A nineteen kilometre light railway was laid from Deir el Belah along the southern side of the Wadi Ghazze. Special construction gangs consisting of volunteers from the troops were trained so that the railways could be rapidly expanded in the wake of the advance. At Shellal a natural rock basin, improved by a masonry dam, provided a water reserve of 500,000 gallons and a pipeline was carried forward from there for some distance. At Shellal, too, arrangements were made to fill 2,000 camel fanatis with 25,000 gallons an hour.*

*Springs yielding some 14,000 gallons an hour of slightly saline water were also developed at Shellal, and the pipeline from Kantara was 'connected up with Shellal, and the wells at Dier el Belah were connected up with the trenches south of Gaza .... At intervals along the Wadi Ghazze a total of over 3,000 running feet of masonry and wood troughs were provided for watering horses and camels.' (Blenkinsop, 201)

Many of the wells in this area used for watering animals were infested with leeches. So as to prevent these getting into the troughs close-mesh wire gauze was fitted on to the intake pipe. Alternatively water was strained through fine cloth. (Blenkinsop, 194)

Phase I: Deployment (*October* 24–30)

The Twentieth Corps moved east towards Beersheba, the Twenty-first Corps remaining opposite Gaza. The Twentieth Corps had practically the whole of the transport of the army, the Twenty-first Corps being left immobile. One mounted division covered the gap between the two corps. The remainder of the Desert Mounted Corps moved south to Khelasa and Asluj. From October 27 the Twenty-first Corps, assisted by warships, carried out a heavy bombardment of Gaza.

Phase II: Capture of Beersheba (*October* 31)

The Twentieth Corps captured the main defences of Beersheba while the mounted troops, after a night march of thirty miles, attacked the town from the north-east. Colonel Newcombe's detachment placed itself astride the Hebron-Beersheba road. The Twenty-first Corps continued the bombardment of Gaza.

Phase III: Attack on Gaza (*night of November* 1–2)

While the Twentieth Corps was preparing to attack the left of the Turkish main line the Twenty-first Corps assaulted a portion of the Gaza defences in order to attract the enemy reserves. Meanwhile the flank-guard of the Twentieth Corps became heavily engaged in the hills north of Beersheba, at Khuweilfe.

Phase IV: Exploitation as intended by G.H.Q.

While the Twentieth Corps broke the Turkish left, the Desert Mounted Corps was to pass round this flank and intercept the retreat of the whole Turkish army.

Phase IV: Exploitation as it actually occurred (*November* 6)

Owing to the fighting at Khuweilfe and the water difficulties, the mounted troops were scattered and tired, instead of collected and fresh, when the moment came. As the Turks still held out at Khuweilfe the mounted troops had to pass through a comparatively narrow gap, instead of round a flank. Only four brigades out of ten were immediately available.

Phase V: Pursuit (*November* 7 *onward*)

Owing to the supply and water problem, the Twentieth Corps had to halt after November 6 and transfer all its transport to the Twenty-first Corps, who took up the pursuit along the Plain of Philistia.

THIRD BATTLE OF GAZA

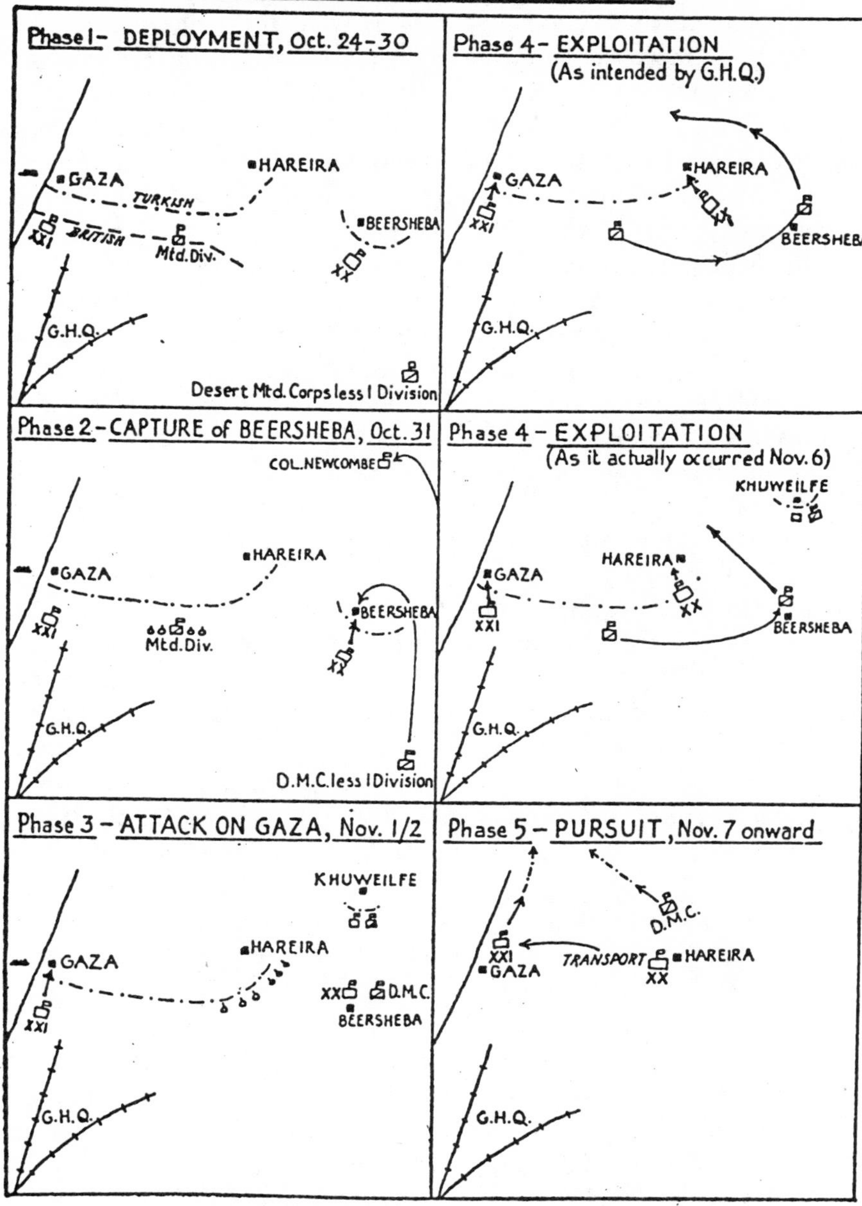

Deep bores were sunk wherever water was divined. Three wells near Khan Yunis were developed to yield 130,000 gallons daily.*

While all these basic preparations were going ahead intensive training proceeded with equal energy. For the mounted troops ever increasing speed in turning out in full marching order at a moment's notice was much practised. One of the horse artillery batteries achieved the record, taking only eleven minutes to turn out fully accoutred and with all its ammunition, rations and stores.[7] A vital aspect of the rehearsals for what was to come was the galloping of trenches and of course there was intensive musketry training. New ground-to-air contact methods were devised and practised for keeping in touch during battle. Each troop was provided with ground flares which were lit to show aeroplanes flying a pre-arranged identification signal where the front line was. Each regimental headquarters was furnished with groundsheets by which they signalled to the pilots who replied by klaxon horn using the morse code.[8] At night red flares for communication with aircraft were used for the first time.[9]

Soon after Allenby's arrival, regular fortnightly reconnaissances towards Beersheba were undertaken by one of the two divisions in the line. They had a double purpose, the chief of which was to familiarize the commanders with the ground over which the attack on the town was to take place. The secondary purpose was to make the Turks believe that these reconnaissances were a routine matter, so that when the actual advance was made they would not think them anything exceptional. The historian of the Staffordshire Yeomanry gives a good account of what these 'stunts' were like:

> 'The route to Beersheba was a desert of blistering rocks, crossed by a succession of shallow wadis which were nothing more than boulder-strewn clefts in the limestone hills. The crossings were narrow and exposed and many of them were commanded by Turkish batteries in the hills, so that the reconnaissance would be carried out in the face of a galling fire, as often as not supplemented by bombs from hostile aircraft. The march would start in the afternoon and continue through the night, and the following day would be spent in

*At this time on the extreme right of the British position, the front was so undefined that British and Turkish patrols frequently watered their horses in the Wadis 'when the other was not about!' (*Twentieth MG Sqn*, 11)

> holding a line of outposts in front of Beersheba. While the cavalry were out they acted as cover for parties of [XX Corps] infantry commanders and Royal Engineers making a close study of the ground over which they were to advance as soon as the main attack was mounted . . . . It was hard and galling work. They entailed a thirty-six-hour march all told and caused much suffering to the horses, since for them there was no water till they returned. The men started off with one full water bottle and one spare was brought up during the day in water carts.'[10]

It was said, according to an officer of the Gloucesters, that these reconnaissances were organized 'so that staff officers could exercise their horses and test their motor-cars in safety'![11] Chetwode and his staff used to be referred to irreverently as 'the Royal Party'.[12]

Allenby's plans to deceive, mystify and surprise were immensely elaborate. During early October sham preparations for a landing on the coast at the mouth of the Wadi Hesi were made. Naval vessels came in close to the shore and took soundings. A fleet of small craft appeared near Gaza. A large-scale camp was laid out in Cyprus and preliminary enquiries were made of Cypriot contractors for supplies for a large force and buoys were laid down to direct transports. Numerous bogus messages were sent by wireless and docks and wharves were designated for embarkation of troops and stores.* Between 24 September and 31 October wireless messages were sent daily from Allenby's headquarters, hinting at the attack on Beersheba being merely a more comprehensive reconnaissance, and at Allenby being at Suez from 29 October to 4 November.†[13]

*This scheme had been thought up by Murray before Romani. It was now revived on a greater scale. Whether it succeeded in its object is not clear. A German aeroplane flying over Cyprus on 17 October reported nothing out of the ordinary, but by then much of the plan may not have been implemented. (Falls, 30).

†Captain Meinertzhagen, disregarding Allenby's veto, 'started a regular plane service in August over the Turkish lines at sunset, dropping small packets of cigarettes with a little bit of harmless propaganda in each packet; after a few days the Turks became accustomed to their cigarettes and would run out of their trenches when they saw the shower of little packets of cigarettes descending . . . . Thousands of packets were dropped, each cigarette heavily doped with opium.' (Meinertzhagen, 223-4)

The most successful act of deception was carried out by Captain Meinertzhagen, in charge of Advanced Intelligence. He compiled a dummy Staff Officer's notebook 'containing all sorts of nonsense about our plans and difficulties'. On 10 October he rode out towards Sheria, was chased and fired at by a Turkish patrol. He dismounted, had a shot at his pursuers and in remounting let go his haversack, field-glasses, water bottle and rifle, 'previously stained with some fresh blood from my horse'. He then made off, dropping the vital notebook, various maps and his lunch. He saw one of the Turks pick these up, 'so now went like the wind for home and gave them the slip'. On reporting the loss to headquarters of the Desert Mounted Corps, Meinertzhagen was cursed for his carelessness and reported to Allenby for his negligence and stupidity. Captured documents later revealed that the enemy had been completely taken in and that they had altered their plans accordingly.[14]*

Though, as the Official Historian has put it, this ruse 'was to have an extraordinary effect hardly to be matched in the annals of modern war',[15] neither it nor any of the other deception operations would have been successful in concealing the concentration of troops and supplies which were essential for the attack on Beersheba without almost complete air superiority. This had been achieved a good month before, the German airmen having been almost driven out of the skies by then. The patrols of 40 Wing, Royal Flying Corps, were so effective in keeping the enemy planes away or so high as to be unable to see much, that during the ten days during which the great flanking movement was underway only a single German aircraft managed to take photographs of the

*A similar stratagem was carried out by a light horseman in which another haversack was dropped during a reconnaissance. It contained 'a half-finished letter to his girl' saying that the mounted troops were 'having a rough time constantly reconnoitring towards Beersheba to hoodwink the Turk who was really going to be attacked at Gaza'. The details of this exploit have never been revealed. (*Chauvel*, 121).

In an official Turkish report, Colonel Hussein Husni, criticizing Kress's action in the battle of Beersheba, wrote: 'He relied on —'s pocket book.' (Falls, 62).

Meinertzhagen was responsible for virtually stopping the infiltration into the British lines of Arab and disguised Turkish spies. He also employed some thirty Jews, Arabs and Egyptians as agents, not a single agent having been employed up to his arrival. By mid-August he was receiving much information about the enemy's strength, dispositions and intentions. By late October he had produced very accurate maps showing these. (Meinertzhagen, 216).

12. This purports to be a photograph of the actual charge at Beersheba, 31 October, 1917 (see Appendix II, p.349)

13. The charge at El Mughar, 13 November, 1917 (see p. 191)

14. This artist's impression, after the capture of the Turkish guns at Huj, illustrates the tasks confronting the victors after an engagement (see p. 173)

area. It was brought down near Khelasa, its occupants being captured.[16]

The enormous task of building up supplies for the flank attack thus went ahead without the enemy being aware of it. At least 20,000 camels as well as thousands of horses and mules, with 50,000 Egyptian labourers, were engaged in the preparations. By day the minimum of work took place, but as soon as darkness fell a veritable 'buzzing hive of surely-directed industry' came into being 'as train followed train and convoys rolled eastwards in the choking clouds of dust'.[17]*

The plan in its final form placed Bulfin's XXI Corps opposite Gaza; next, to its right, came Barrow's Yeomanry Division covering the gap between XXI Corps and Chetwode's XX Corps and acting as a general reserve. To it was attached the Camel Corps Brigade. This left Chauvel (on Chetwode's right) with only seven mounted brigades (each capable of putting only about 800 dismounted men into the firing line) and twenty-eight horse-artillery guns, Barrow coming again under his command only after Beersheba had fallen.

*'From Shellal the labourers hastened along the railway east of the Ghazze, stripped off the brown camouflage which screened their uncompleted work, set about plate-laying and the screwing up of the pipeline.' (Falls, 40)

15

'The tasks of the Desert Mounted Corps are to
(a) Attack Beersheba from the east so as to envelop the enemy's left rear; and
(b) Seize as much water supply as possible in order to form a base for future operations northwards.'

Desert Mounted Corps Operation Order No. 2, 26 October, 1917

'The quintessence of the Beersheba operation lay in a speedy decision, and the whole plan was devised with the object of a quick and crushing success, so that the Turk might have no time to destroy the wells, the chief prize that Beersheba had to offer its captor.'

WAVELL in *The Campaigns in Palestine*[1]

The enemy situation before Third Gaza – preparations and preliminaries: Third Gaza

While the British were perfecting their plan, the enemy was dithering and divided. In May General Eric von Falkenhayn, one of Germany's most outstanding commanders, had been sent out to try to coordinate and enliven the Turks.* In June he reported that the retaking of Baghdad was feasible provided a sufficient force was carefully prepared and provided the Palestine front was secure. Endless arguments between the Germans and the rival Turkish leaders wore on throughout the summer and early autumn. Von Falkenhayn soon saw that Turkish sluggishness and what he learned of the British build-up in Palestine made the attack on Baghdad impracticable. In early September he visited the Gaza front and proposed that the whole of the VII Army which was slowly assembling for the Baghdad venture should be transferred to Palestine with a view to an offensive designed to forestall Allenby's coming onslaught. The plan, which was eventually agreed upon, was to concentrate the VII Army for a blow from the Beersheba area aimed at Allenby's right flank, in conjunction with

*Between 1914 and 1916 he had been Chief of the General Staff. He had been removed after the failure of the attack on Verdun.

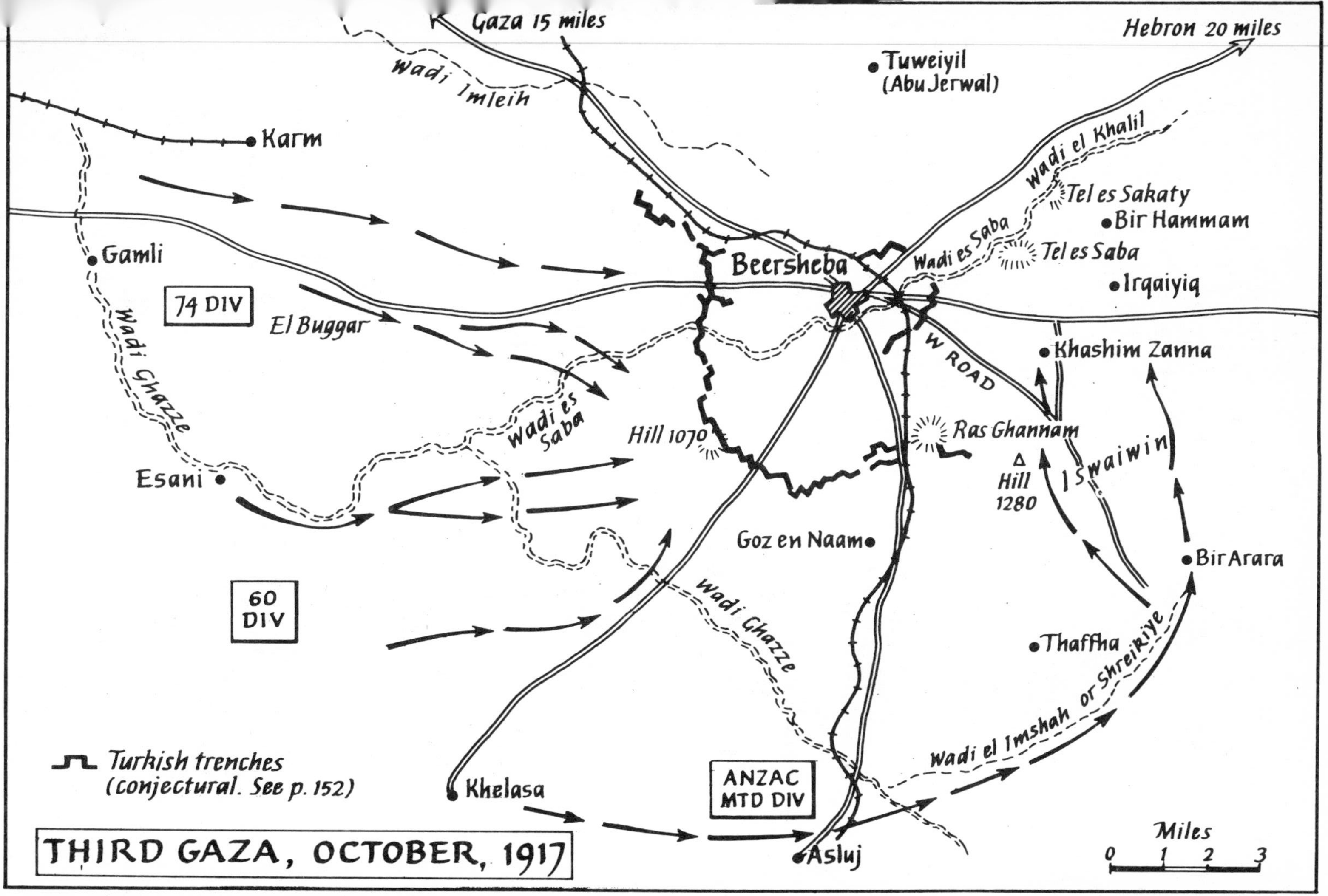

THIRD GAZA, OCTOBER, 1917

the 8th Army already in the line. But, as Wavell has put it, 'before the army could even reach the front . . . the opening of the British offensive caught the enemy commander-in-chief with his plans incomplete, his organization in process of change and himself on the railway between Aleppo and Jerusalem.'[2] It is difficult to be sure about the enemy's numbers at the end of October,* but it is probable that the British had a preponderance of about two to one in infantry, three to two in guns and eight to one in mounted troops.[3]

Z (for Zero) Day (or D Day, as it would have been called in the next world war), was fixed for 31 October, largely because the previous nights would have been blessed by moonlight. Operations began, nevertheless, a week earlier with a methodical, Western Front type bombardment of Gaza. This was progressively augmented in violence and supplemented by naval gunfire as the week wore on. It was the heaviest carried out throughout the war outside the Western Front. While the Turks' attention was thus concentrated on their right flank, Chetwode and Chauvel marched their striking force over to the other flank. The transfer of XX Corps was effected by a night-by-night side-stepping to the south-east. It was during this process, on 27 October, that some yeomanry regiments of 8 Mounted Brigade (temporarily under command of 53 Division) were overwhelmed by superior Turkish cavalry at El Buggar (see p. 24). This was the only time during the whole campaign that the enemy's cavalry executed a proper charge.

Chauvel's seven mounted brigades effected their deployment to the various water-points from which the attack on Beersheba was to be launched over the two nights 27 and 28 October, having begun their concentration on the 21st. Anzac Mounted Division moved to Asluj while Australian Mounted Division moved to Khelasa.† This movement was undoubtedly, as the Official Historian

*On 30 October: *7 Army* (Fevzi Pasha): one infantry division at Kauwukah; one cavalry division and one infantry division based on Beersheba, about 11,000 effectives. Kress says that on 27 October 'six battalions, four batteries and three squadrons' were sent 'to reinforce the garrison of Beersheba'. (Kress, 510) *8 Army* (Kress): five infantry divisions at Gaza and Sheria, with two in reserve, about 33,500 effectives. The total of artillery guns was about 300, and of machine guns about 250.

†It was during this move that, according to Trooper Idriess, 'a 'plane was to co-operate with us. She located us, as we thought. The whole regiment in

(Continued over)

has put it, 'the most anxious of all, because [it] supplied the clue, if the Turks could read it aright, to the whole programme.'[4] The transport column alone was six miles long. It consisted of more than 300 four-wheeled vehicles, some of them lorries, and twenty-four Holt tractors, beside the 7,000 'heavy burden' camels of the water convoy, the 'light-burden' camels carrying three days' rations and numerous horses, mules and donkeys. The evacuated camps near the coast were left standing and lit up at night. By day the troops were hidden so far as possible in the numerous deep wadis. When as sometimes happened enemy aircraft were sighted, a warning whistle was sounded, upon which the troops, though already concealed in the gullies, would throw themselves down on their faces and remain motionless till the danger was past.[5] A subsequently captured document shows that on 29 October the enemy believed that only one infantry and one mounted division were being moved towards Beersheba,* a tribute chiefly to the British superiority in the air.[6]†

On 26 October Hodgson issued to the Australian Mounted his Preliminary Instruction No. 1:

> 'It is to be noticed that the country is built for mounted action, whereas any dismounted attack is handicapped for want of cover. The Divisional Commander hopes that all brigades will endeavour to profit by their knowledge of these facts.
>
> 'To manoeuvre an attack mounted, an *arme-blanche* weapon is necessary. The Divisional Commander suggests that the bayonet is equally as good as the sword, if used as a sword for pointing only; it has the same moral effect as a

---

mass just stood upon a bare, rocky hill gazing up at her circling low. She did not see us at all. Flew back to Asluj and reported that they could find no trace of the regiment. So yet again an aeroplane has proved a dud scout.' (Idriess, 317)

*Cyril Falls in the Official History states that the enemy was 'fairly accurately informed of the British dispositions', that the evacuated camps were discovered to be empty on 28 October and that next day the Turks 'had marked down . . . three infantry divisions east of the Ghazze, the 10th Division approaching the wadi and actually more cavalry at Asluj and Khelasa than were there'. He does not give his sources and Wavell seems to intimate that Falls's statements are based on unreliable Turkish claims. (Falls, 40-1; Wavell, 112)

†But see p. 150 for the bombing and machine-gunning successes of the enemy aeroplanes on 31 October.

> sword, as it glitters in the sun and the difference could not be detected by the enemy.
>
> 'If used in this manner, the point only should be sharpened, to ensure that the men point instead of striking.
>
> 'The Divisional Commander suggests that the bayonet, used thus, will be more effective as an *arme-blanche* weapon than the rifle with bayonet fixed, as he fears that the latter method would render the control of the horse difficult in manoeuvre, and would leave the right arm too tired to give the final thrust.
>
> 'The G.O.C. directs that steps be taken at once to have the points of all bayonets sharpened by the armourers.'[7]*

This order bore abundant fruit five days later (see p. 156).

When the Anzacs arrived at Asluj they found there a small party of Hejaz Arabs with a stiffening of British machine gunners, all mounted on camels, under Lieutenant-Colonel Stewart Francis Newcombe of the Royal Engineers. This was an independent raiding party which was to make a wide detour east of the mounted troops with a view to raising the Bedouin against the Turks, and to harassing any reinforcements which might march along the road from Hebron and any of the Beersheba garrison which might escape along that road in the opposite direction.†

The Anzacs whilst at Asluj were filled with joy when on 25 October a thunder-storm of some violence broke over them and, more important, over Beersheba. The resulting pools of water were to prove copious enough to be extremely valuable on the 31st (see p. 161).

* * *

*Hodgson had, in fact, applied to be allowed to arm his two Australian brigades with the sword a short time previously. Since the Australians had no experience whatsoever of the use of that weapon, sanction, not surprisingly, was not given. (Osborne: I, 350-1).

†Newcombe was the first head of the British Mission to the King of the Hejaz. To Allenby he suggested this side-show enterprise taking his party through the desert country to the east of Beersheba which he had surveyed and the inhabitants of which he knew personally. On 31 October he cut the telegraph to Jerusalem. For his further activities see p. 164.

Though Falls in the Official History hardly mentions it, there was a 'Gaza School' which believed that a third direct attack on Gaza, now that the force available was so much larger than for Gaza II, would have succeeded, thus leading to an earlier, more decisive victory. It was argued by some senior officers that a gap made in the enemy line between the town and the sea would have made it easier for the mounted troops to capture the entire garrison comparatively speedily. How seriously Allenby considered this option is not known. What seems likely is that there was a general repugnance in many minds for another assault after the two earlier repulses.[8] Though in print Liddell Hart supports the Beersheba plan, he wrote to Lloyd George in 1934 that had another frontal attack on Gaza been made instead of it, 'the Turkish army could have been overthrown at this time almost as certainly as it was a year later.'[9] This is purely conjectural and is typical, perhaps, of Liddell Hart's unhappy, unscholarly habit of asserting what can never be proved. Nevertheless, as will be shown, Chetwode's plan depended upon good luck respecting the water supply situation, and, further, in the event it took both time and considerable casualties to effect the fall of Gaza which did not take place until 7 November.[10]

16

'A force quite disproportionate to the garrison [of Beersheba] was to be deployed against it. . . . It was like taking a county cricket eleven to play a village team; but the pitch was a difficult one and there was much at stake.'

WAVELL in *Allenby: A Study in Greatness*

'If there was one lesson more than another I had learned at Magdhaba and Rafa it was patience, and not to expect things to happen too quickly. At Beersheba, although progress was slow, there was never that deadly pause which is so disconcerting to a commander.'

CHAUVEL after the capture of Beersheba

'Till now the Turks had looked upon the British mounted troops as chiefly valuable for reconnaissance or raids in superior numbers on isolated posts, as at Rafah, and had considered that the Turkish infantry had little to fear from them. . . . The commanders of the British mounted troops first learnt from this incident their own strength and that in this theatre there was likely to be brilliant opportunity for the cavalry arm even in frontal attack.'

CYRIL FALLS in the *Official History*

'This charge was a gallant affair and deserves all the praise which has been bestowed on it. It showed yet again how great is the protection from fire effect given by speed of movement.'

WAVELL in *The Palestine Campaigns*[1]

Third Gaza: XX Corps attack – actions of Descorps – charge of 4th Australian Light Horse Brigade – capture of Beersheba

By 30 October, the eve of Halloween, some 12,000 mounted men and as many horses were assembled at Asluj and Khelasa. Great attention was paid to what might be allowed to be carried on the horses for the coming assault and the days of fighting which were to follow it. To each saddle were attached two nosebags holding two days' forage consisting of nineteen pounds of grain. A third day's forage was transported in limbered General Service wagons, three to each regiment. A fourth wagon carried cooking utensils and technical stores, while entrenching tools were carried on pack animals. Neither greatcoats nor blankets were to be taken and each man was issued with a pair of officer's pattern saddle-wallets in which three days' rations, including the iron ration, were carried. These were made up of bully beef, biscuits and groceries. Included, too, were the few articles of clothing allowed.[2]*

Thus equipped, the 7th Australian Light Horse in the afternoon of 30 October moved out as advanced guard. By 8 p.m. Chaytor's Anzac Division had cleared its dismantled bivouacs at Asluj. It had ahead of it a night march of some twenty-four miles. It had not long left Asluj before it was followed by Hodgson's Australian Division which had already covered the ten miles from Khelasa and was to follow in Chaytor's tracks. This meant that Hodgson's men had to march a total of about thirty-three miles to Chaytor's twenty-four. Since their objective was well to the left of Chaytor's, (see below), a more direct westerly route would have been even shorter than theirs. (See map, p. 137) However, it seems that there was 'only one track available along the Wadi el Shreikiye'.[3] The column was some twenty miles in length and the hills on each flank were so rough that it was impossible to employ a moving flank guard. The 7th Mounted Brigade at Esani had only some seventeen miles to cover before taking up position near Goz en Naam. From there soon after 5 a.m. it sent out patrols (mostly dismounted) towards Ras Ghannam where it found the Turkish trenches strongly held. By 9 a.m. it got in touch with XX Corps on its left and 4 Brigade on its right, as its Commander, Brigadier John Tyson Wigan, had been ordered to do.

The infantry of Chetwode's XX Corps, about 45,500 strong, well to the left of the mounted troops, were at much the same time marching over the eight miles which would bring them to within some 2,000 yards of the enemy works. By 3 a.m. on 31 October they were 'marshalled for battle without a hitch of any kind'.[4] By 8.30 a prominent, strongly defended outwork on a commanding height, known as Hill 1070, had been captured. Later on, after an hour and a half's fierce artillery bombardment, the whole position

*The horses were hard and fit, inured to long marches and practised in going for long periods without water. Each carried not less than twenty stone. Many units spent four hours compulsorily resting in their 'bivvys' during 30 October. (*Twentieth MG Sqn*, 18)

between the track to Khelasa and the Wadi Saba had been taken and an outpost line was soon established about 2,000 yards further east. By about 7 p.m. XX Corps had completed its task, its casualties numbering 136 killed and 1,010 wounded. Its captures totalled 419 Turks, six field guns and numerous machine guns.*

Meanwhile one of the more amazing night marching feats of the campaign was under way. The two mounted divisions were negotiating, virtually without a single check and exactly according to the timetable, some of the wildest, most featureless country of southern Palestine. Much of it was unreconnoitred. For security reasons no Bedouin guides were employed and the new maps, though on the whole excellent, lacked details. The weather was airless and sultry and a thick pall of dust lay upon the marching horsemen, reducing the moon to a dull red sphere. Near Thaffha, more than half way to their jumping off positions, the Anzacs' leading brigade, the 2nd, which was to form the right of the attack on Beersheba, advanced northwards to Bir Arara. From there they were to proceed eventually to Tel es Sakaty on the road to Hebron. The rest of the Anzacs and the Australian Division at the same time turned left towards the Iswaiwin area some six or seven miles east of Beersheba. From here at dawn the town in its shallow depression came into view, the hilly country abruptly giving way to the open plain in which runs the broad, shallow Wadi Saba. Thus came to an entirely satisfactory end what was possibly the biggest night march which has ever taken place in war, made entirely across country.

*Falls states that since the XX Corps had now attained all its objectives, it could 'without doubt have captured Beersheba itself before the mounted troops. The objective had, however, been allotted to the Desert Mounted Corps, and Sir Edmund Allenby, who had come up to General Chetwode's headquarters at El Buggar, ordered General Chauvel to capture it before nightfall.' (Falls, 57). Yet such an advance by XX Corps would certainly have disorganized, in Gullett's words, 'the next stage in the operations' by taking away Chetwode's men from their next objective which was the trenches of Hareira to the north-west. Gullett also says that Chetwode's infantry divisions 'could not be moved on the town because of the absence of water supplies.' (Gullett, 392) Wavell says that 'it was desired to reserve all the water in Beersheba for the mounted troops.' (Wavell, 211). This seems to refute Falls's view conclusively. However, according to A.J. Hill, Chauvel's biographer, Chauvel later told Gullett that just before giving his orders for the 4th Brigade charge, he had 'the nagging thought that Chetwode . . . might rush some of his own troops into the town'. (*Chauvel*, 127, 240)

Chauvel established his headquarters on a conspicuous hill near Khasim Zanna, giving him as he said 'a dress circle view of the whole show'.* Such a position for a Corps Commander was unique in World War I and reminds one of nothing more than Raglan's position at the battle of Balaklava.

> 'So prominent a feature,' writes Chauvel's biographer, 'occupied as it was by a headquarters with its inevitable movement of officers and despatch riders and the flash of maps in the sun, could not fail to attract the attention of the enemy, but it was not until Chauvel was eating a hasty lunch with his staff that their peace was shattered by a salvo of shells. Luckily there were no casualties; they were shelled again and towards sunset they were bombed and machine-gunned by enemy aircraft, but without loss. The scene on Khasim Zanna was in marked contrast to contemporary corps H.Q. in France, where commanders and staffs in their châteaux became used to the rattle of the windowpanes during distant bombardments but only in exceptional circumstances saw some of their troops in action. A small touch of colour at Descorps – the code name for H.Q. Desert Mounted Corps – was the red and white corps flag attached to a lance which was borne by a mounted orderly when the commander was in the saddle and was set up in the ground wherever he fixed his H.Q. in the field.'[5]

All was now set for the second stage of the Desert Mounted Corps's programme. At 9.10 a.m. Meldrum's New Zealand Brigade began its advance against a flat-topped tel (or mound) called Tel es Saba, three miles due east of Beersheba. This, which is about 400 by 200 yards in extent and rises steeply to a height of a few hundred feet, was held by some 350 men with one machine-gun company. These were skilfully arranged and 'had two tiers of fire which swept the wadi [es Saba] bed and the bare plain'.[6]

About a mile north of Tel es Saba was another mound, Tel es Sakaty. Ryrie's 2nd Brigade on Meldrum's right was ordered at 9.30 to advance with all speed on this mound so as to be in a position to ward off any counter-attack from the Hebron motor road. With the 7th Regiment in the lead the brigade at first

*Chaytor's headquarters were established nearby.

cantered and then galloped in artillery formation (i.e. widely spread) across open ground under first shrapnel and then machine-gun and rifle fire. Trooper Idriess says that the men smoked as they

> 'broke into a thundering canter holding back in the saddles to prevent the horses breaking into a mad gallop. I think,' he added, 'all men get scared at times like these; but there comes a sort of laughing courage from deep within the heart of each, or from a source he never knew existed; and when he feels like that he will gallop into the most blinding death with an utterly unexplainable, don't care, shrieking laugh upon his lips. . . . Some of our led horses became so madly excited that improperly strapped-on packs rolled under their bellies, but the regiment galloped on and I did not envy the unfortunate horse-leaders who had to dismount with their frantic horses and resaddle those packs all alone on that shell-thrashed plain.'[7]

The enemy's fire proved so intense that, although the brigade had suffered not a single casualty whilst galloping to a point just short of the road, when they reached it the enemy guns, finding the range, now fired point-blank into the 7th's squadrons, causing a number of casualties. Forced to dismount and place their horses under cover, after a stiff fight which lasted until 1 o'clock, the tel was at last made good and so were the wells near it and the line of the Hebron road. For the rest of the day it stayed there protecting the Corps's right flank.

In the meantime Tel es Saba was proving a very hard nut to crack. The surrounding open ground put a near approach out of the question, but it was much broken up by steep-sided wadi beds. These proved invaluable since the Aucklands on the left and the Canterburys on the right were able to advance mounted to within about 800 yards of the tel along the well concealed approaches which they provided. They were covered by the Somerset Battery which opened up at 3,000 yards. But for the next half mile or so, as the enemy machine-gun fire from skilfully concealed positions became concentrated, the two regiments had to struggle on on foot yard by yard, covering fire being supplied by the machine guns. The wadi beds furnished excellent cover for the horses.

Since the enemy machine guns could not be located – it was well into the afternoon before they were* – and since the mounted batteries with their light 13-pounder guns could do little to keep down their fire, Chaytor at about 11 a.m. put in the 3rd Regiment from Cox's 1st Brigade, which was in reserve, to attack the hill from the south-east, covered by the Inverness Battery. However, this regiment's advance up the Wadi Saba proved impossible, so its commanding officer decided to advance along open ground to the wadi's south, with the Aucklands conforming on the north. The 3rd, breaking into a fast gallop under fire, got to within 1,500 yards before having to dismount. The Somerset Battery at this time managed to range on the tel at 1,300 yards. A section from the battery was now moved to the east of the tel to try to deal with machine guns on high ground to the north which were holding up the Canterburys. This regiment soon crossed the Wadi el Khalil and began to threaten the tel from the north. While all this was going on, German aircraft were dropping bombs which caused a number of horse casualties.

Since the assault was taking so much longer than had been hoped, Chaytor felt impelled at about 1 p.m. to commit the 2nd Regiment. It came up on the 3rd's left to stiffen the attack.† The men

> 'advanced at the gallop until they reached a zone of heavy fire, when their dismounting afforded a pretty example of light horse work at its best. So rapidly,' writes the Australian Official Historian, 'were the galloping horses checked, cleared and rushed back by the horseholders and so quickly did the dismounted men resume their advance on foot, that the Turks, under the impression that the regiment had retired on the horses, shelled the galloping animals, while for a time the riflemen were not fired upon. The Australians speedily cleared a group of huts in which riflemen had been concealed and then pushed on towards the wadi. The two advanced

*Lieutenant Hatrick, the Signalling Officer of the Aucklands, was responsible for their eventual location. 'From a concealed position in the front line [he] observed for the Somerset Battery and directed the fire by flag signal.' (Powles, 137).

†At about 1.30 p.m. Chauvel ordered Ausdiv to place 3rd Brigade at Chaytor's disposal. In the event it was not required. (Falls, 57). Its appearance mounted may, though, have contributed to the collapse of the tel's garrison. (Gullett, 392).

> squadrons of the 3rd had by then gained the bank of the wide wadi immediately opposite the Turkish position on the tel; the Wellington Regiment had been thrown in on the right of the Aucklands and the enemy was now under intense converging fire. Every Turk who showed was an easy target and the position of the garrison had become precarious.'[8]

Under cover of all available guns and machine guns, the Aucklands at 2.10 p.m. started their final assault, crawling forward by a series of short rushes from cover to cover. Within half an hour they had reached the eastern flank trenches, captured sixty prisoners and three machine guns. Two of these they turned upon their late owners, thus substantially increasing the covering fire for the decisive dash with the bayonet which they made twenty minutes later. At last the tel was taken, together with two more machine guns and another seventy prisoners. The remaining Turkish troops fled towards Beersheba and north-westwards. 1st and 3rd Brigades chased them and made for objectives north of the town so as to isolate it and complete its capture. But the ground ahead was rough and difficult and time was pressing. The sun would soon set and it would be dark soon after 5 p.m.*

Though Chauvel, with reluctance, had felt compelled, in view of the slowness of the attack on Tel es Saba, to send Wilson's 3rd Brigade to Chaytor's assistance, he had resisted the temptation to hurl more of Hodgson's Ausdiv into what was only the preliminary phase of the struggle. Consequently he had Grant's 4th Brigade and Fitzgerald's 5th Mounted Brigade if not at hand at least as a substantial reserve. As time was so urgently pressing he realized that methodical progress must now give way to a swift mounted charge in an attempt to enter the town from the east. Hodgson, Grant and Fitzgerald were with him at his headquarters. It is said that Fitzgerald pleaded to be allowed to employ his sword-armed

*There is sufficient evidence to be able to assert with confidence that the sun was above or only slightly below the horizon throughout the subsequent charge. Since sunset was at 4.52 (Dr R. Hirst, Curator of Astronomy at Melbourne Science Museum, Jones, 37) the charge must have started at or very soon after 4.30. Accounts, including those of eye-witnesses, vary, but both Falls and Gullett agree on 4.30.

yeomanry regiments, but Chauvel quickly decided that the 4th Brigade, which was appreciably closer – about six miles due east of Beersheba – should undertake the attack. To Hodgson he said: 'Put Grant straight at it'.* At the same time he arranged with Hodgson for Fitzgerald's yeomen to follow Grant's Australians and for the independent 7th Mounted Brigade to come up on Grant's left.†

There now followed a most agonizing period‡ before Grant's regiments could be collected together. One of them, the 11th, was spread over a line of outposts extending towards the 7th Mounted Brigade, some two miles to the south-west, while the 4th and 12th, east of Iswaiwin, were scattered over a wide area in single troops so as to avoid the frequent enemy aerial bombing of both cavalry and

*Grant told Falls that Hodgson 'took him to the corps commander who directed him to "take the town before dark", without giving him instructions as to how the attack was to be carried out and that he himself was therefore solely responsible for the mounted charge.' Chauvel's recollection was that he had given no orders personally to Grant. (Falls, 58). Grant's official report states that 'he received direct instructions' from Chauvel. (Osborne: I, 352).

Chauvel was usually scrupulous in not preferring his Australians to the yeomanry. 'If I did ever favour the light horse,' he said afterwards, 'it was at Beersheba, when, in giving the lead to Grant, I was perhaps influenced by a desire to give a chance to the 4th and 12th Regiments, which up to then had seen very little serious fighting.' (Gullett, 393).

†Whilst these orders were being implemented, Allenby had been in communication with Chauvel by telegram. Earlier in the afternoon General Headquarters had prudently enquired of Descorps whether, should the wells of Beersheba not be available, watering could take place in the Wadi Malah. Chauvel had replied: 'Water situation in Wadi Malah [not identifed: possibly Wadi es Saba] is not hopeful and if Commander-in-Chief approves, it is proposed to send back all troops which have not watered to Bir Arara and Wadi el Imshash, if Beersheba is not in our possession by nightfall.' This reply was at once relayed to Allenby who was away from headquarters at the time and who, *having no knowledge of the original enquiry*, interpreted Chauvel's reply as contemplating withdrawal. 'The Chief orders you to capture Beersheba today, in order to secure water and prisoners.' Thus Allenby reacted in a typically peremptory manner. His order soon became known to the whole army. This created the legend that he was directly responsible for the 4th Brigade's charge. (Gullett, 393-4).

‡There is much uncertainty as to when exactly this marshalling process began. It was a complex operation and can hardly have been completed in the half hour posited by Jones, 30. The 4th Regiment, however, might well have been ready within that time span and the 12th not long afterwards.

infantry.* These two regiments had been quietly resting, off-saddled, since early morning. Though not watered, the horses had been fed and so had the men.† While they were being hastily collected by their two seconds-in-command, Grant with his brigade major, and Lieutenant-Colonel the Hon Murray William Jones Bourchier, commanding the Victorians of the 4th Regiment, and Lieutenant-Colonel D. Cameron, commanding the New South Walian 12th Regiment, both of them graziers by profession, 'galloped forward', in Grant's words, 'to reconnoitre a covered way of approach for the brigade to the point of deployment, and for the direction of the attack. This was necessary as the 3rd Brigade had just previously been heavily shelled in attempting to cross exposed ground.'[9]

The two regiments were eventually assembled behind a ridge about four miles from Beersheba, which was now in full view, and about one and a half miles from the forward trenches. In front of them lay a long but slight slope broken occasionally by tracks cut by the rains and totally devoid of cover – in some respects good but in others dangerous country over which to gallop.‡ The point of deployment was just north of hill 1280, the regiments forming up on either side of the Iswaiwin road (road W on the maps on p. 152).

*'C' Squadron of the 9th Regiment, 3rd Brigade, and a section of the 3rd Machine Gun Squadron were bombed late in the day. Thirteen men were killed and Lieutenant-Colonel Leslie Cecil Maygar, VC died from wounds. Twenty others were wounded and fifty-eight horses became casualties. (Darley, 100). Though this was the worst example of the havoc caused by enemy bombers, other units lost numbers of men and especially of horses. These, particularly when their riders were fighting dismounted, were very vulnerable, packed as they were into every available depression in the ground.

The Palestine Brigade of the Royal Flying Corps comprised at this time: 14 Squadron, allotted to XXI Corps, and 113 Squadron, two flights of which were allotted to XX Corps and one to Descorps. Also employed were 11 Squadron and the Australian Flying Corps' 67 Squadron. Bombing raids behind the enemy lines were undertaken chiefly by a special squadron formed from 20 Training Wing at Aboukir.

†During the advance a number of wells had been exhausted. 'For example, the 4th Machine-Gun Squadron had waited a whole day at Asluj while engineers tried to pump more water. Eventually there was enough to fill the men's water bottles, but none for the horses.' (Jones, 37). Most of the horses had watered between twenty-six and thirty hours before. A few had gone forty-eight hours without a drink. (Jones, 27).

‡Trooper S. Bolton of the 4th remembered that the ground was 'full of little wadis and washaways. The worst one we came to was about thirty feet wide and from six to eight feet deep.' (Jones, fn 31, 37).

Recent aerial photographs showed that a system of cleverly masked trenches lay somewhere outside the town's perimeter.* It was clear that these trenches were protected neither by barbed wire nor by *gardes de loup* – pits dug as cavalry traps.†

The two regiments, carrying their bayonets in their hands,

*There is difficulty in establishing the strength, extent and exact siting of the trench system with which the light horsemen were confronted. That some of it was dug only in the days immediately preceding the British attack is certain. (See Husni Amir Bey, *Yilderim*, trans. Capt. G.O. de R. Channer, n.d. pt 3, AWM, quoted in Jones, 28, 37). Gullett, 397, says that the first trench was 'a shallow unfinished one held by only a few riflemen'. He adds that 'close behind them was the main line, a trench in places ten feet deep and four feet wide thickly lined with Turks.' Chauvel himself (quoted in Elliott, E.G. 'The 4th L.H. Brigade Cavalry Charge at Beersheba', unpublished paper in possession of Jones, and quoted in Jones, 28, 37) speaks of 'two lines of trenches eight feet deep and four feet wide'. Allenby's despatch refers to 'two deep trenches'. (Pirie-Gordon, 2). Trooper Eldridge who rode in the charge wrote in his diary: 'There were trenches everywhere,' while an ambulance man who helped to get the wounded into Beersheba hospital remembered the 'redoubt very strong' and wondered 'how we ever took it'. (Diary of Tpr C. Eldridge, in possession of Jones and quoted in Jones, 28, 37; see also 'Scotty's Brother' *The Desert Trail*, 1919, 117).

In 1923 the United States Cavalry School

> 'published a detailed map of the charge showing complex and unusual systems of Turkish trenches across its path. Its accuracy is confirmed by a scale map of the Beersheba defences, drawn by an officer of the 4th Brigade, which shows a 2,000-metre crescent of earthworks between the Wadi Abu Sha'ai and the Wadi es Saba, dug across the slopes of a low hill called El Shaibat. It shows the formidable 550-metre redoubt facing the 12th Regiment and the succession of trenches and dugouts facing the 4th. The fortifications stretch back about 500 metres, the forward trenches being three kilometres from the edge of the town. . . . There were crucial gaps. But they offered a considerable barrier to a mounted attack and excellent cover for defending infantry.'
> (U.S. Cavalry School *History of the Palestine Campaign*, Kansas, 1923, 91; War Diary, 10/4, 4 L.H. Bde, 31 Oct., 1917, AWM, both quoted in Jones, 28, 37).

The four maps on p. 152 illustrate how hard it is to estimate the scale and the positioning of the trenches.

†These were often constructed by the Turks. They were usually in three closely adjoining rows of circular holes, shaped like inverted cones, about six feet in diameter and four feet deep. At the bottom of each was a spike. (Tallents, 106). They were 'made in wet weather when the clay was soft with some special instrument, as they were all perfectly symmetrical.' (Cobham, 103).

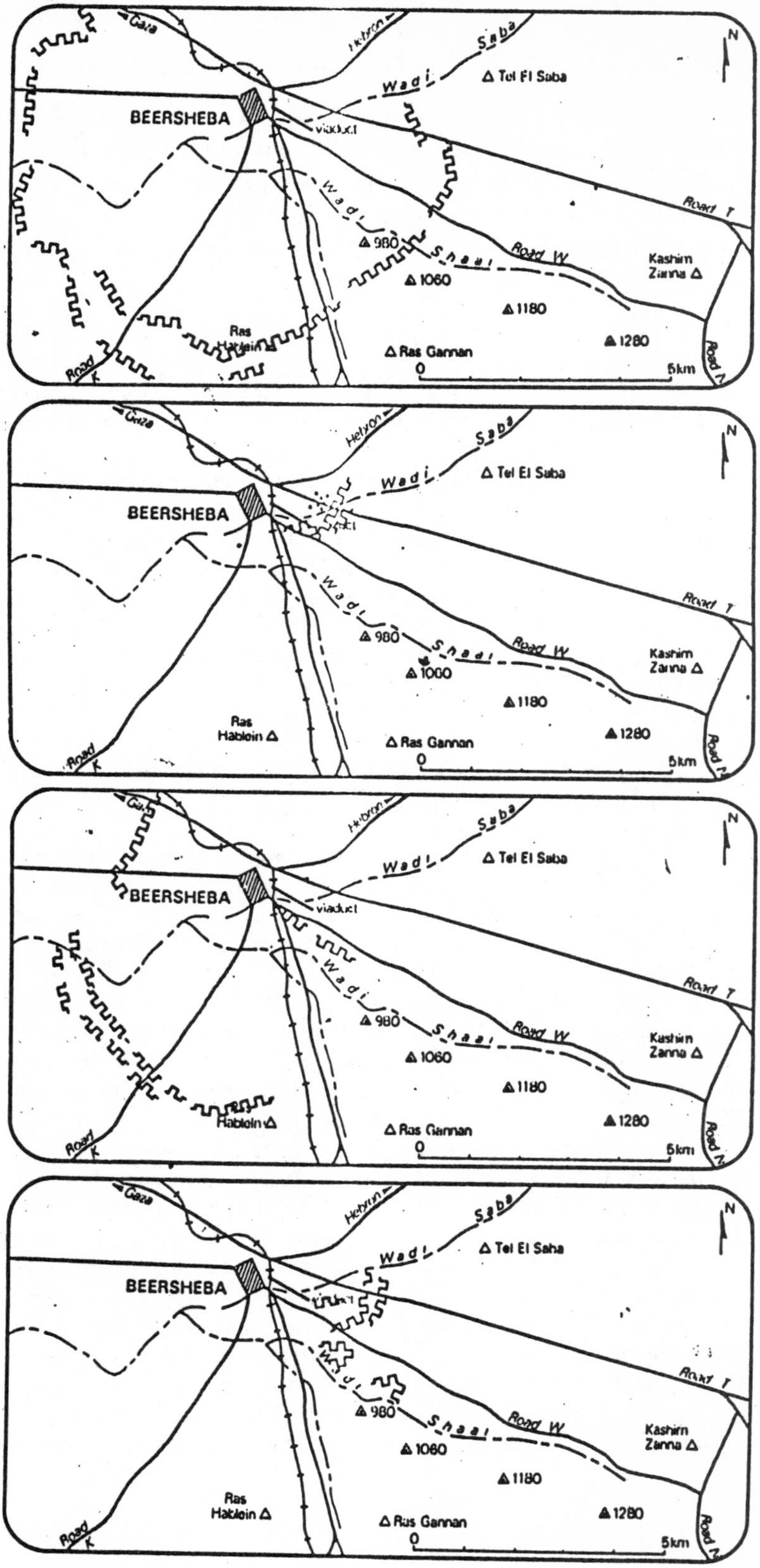
Gaza
Hebron
Wadi Saba
Tel El Saba
N
BEERSHEBA
viaduct
Road T
Wadi Shaal
Road W
980
1060
1180
1280
Kashim Zanna
Ras Hablein
Ras Gannan
Road K
Road N
0
5km

advanced slowly at first, each on a frontage of one squadron in three lines about 350 yards apart, with a space of about four yards between the men. The 4th Regiment started off at a trot until the 12th had completed its deployment and aligned its squadrons with those of the 4th. The complete charge formation now cantered for 0.45 of a mile, only then moving into the gallop. This, according to survivors, means that to reach the forward trench the two regiments galloped for about two and a half minutes over the last one and a quarter miles to the first trench line. Some seventy-five yards ahead of each leading squadron rode two ground scouts, followed by Grant and his brigade-major who soon fell back into the reserve line where the 11th was at last mustered, so as to be in a position (as all good cavalry commanders should) to control subsequent developments. Some way behind, the 5th Mounted Brigade followed the 11th Regiment while well to the left the 7th Mounted Brigade cantered briskly along the Khelasa road. The 4th Regiment, on the right, had all three squadrons in the charge, the leading one led by Major J. ('Porky') Lawson, a Yorkshireman who had emigrated to Rupanyup, Victoria, where he kept an hotel. Major E.M. Hyman, a New South Wales farmer, led the leading squadron of the 12th. The third squadron of that regiment was detached to co-operate with the 4th Machine Gun Squadron (less two sub-sections one of which was riding with each regiment). This had been detailed to protect the left of the advance against the enemy who were in some strength in trenches near Ras Ghannam.* 'A' Battery of the Honourable Artillery Company and the Notts Battery supported the charge at a range of about 2,500 yards. The latter, which had unlimbered at the point of deployment, found the range of the Ras Ghannam trenches with its second shot and quickly drove the Turks from them. Across the shallow, undermanned forward trench the two regiments swept without checking, but at the next group of trenches there was a brief fight. An observer looking down from

*While the charge was getting under way, further to the east, it will be recollected, the 1st and 3rd Brigades were moving in pursuance of earlier orders on a line to the north of Beersheba. The 4th Regiment was in a sense, therefore, the spearhead of two virtually complete mounted divisions 'closing in on the doomed town. Its thrust was swift and deadly.' (Wavell, 123).

'the rising ground on which our guns were in action [saw] a most inspiring sight. It was growing dark and the enemy trenches were outlined in fire by the flashes of their rifles. Beyond and a little above them blazed the bigger, deeper flashes of their field guns and our own shells burst like a row of red stars over the Turkish positions. In front the long lines of cavalry swept forward at racing speed, half obscured in clouds of reddish dust. Amid the deafening noise all round, they seemed to move silently, like some splendid, swift machine.'[10]

Another observer, Trooper Idriess,

'laughed in delight when the shells began bursting behind [the galloping light horsemen], telling that the gunners could not keep their range, then suddenly the men ceased to fall and we knew instinctively that the Turkish infantry, wild with excitement and fear, had forgotten to lower their rifle-sights and the bullets were flying overhead. . . . The last half-mile was a berserk gallop with the squadrons in magnificent line, a heart-throbbing sight as they plunged up the slope, the horses leaping the redoubt trenches – my glasses showed me the Turkish bayonets thrusting up for the bellies of the horses – one regiment [in fact, both regiments less two squadrons of the 12th] flung themselves from the saddle – we heard the mad shouts as the men jumped down into the trenches, a following regiment [the two squadrons of the 12th] thundered over another redoubt, and to a triumphant roar of voices and hooves was galloping down the half-mile slope right into the town. Then came a whirlwind of movements from all over the field, galloping batteries – dense dust from mounting regiments – a rush as troops poured for the opening in the gathering dark – mad, mad excitement – terrific explosions from down in the town.'

Survivors, interviewed in old age, confessed to less joyous sensations. Trooper F. Lindsay Taylor of the 4th, who had been in bushfires back home and who had 'galloped through them to try and help someone', experienced the same feelings now: 'the same – not frightened – strain, as you might say'. He had lain flat on his horse's neck, watching its ears flicking as the bullets zipped past.

Next day, while burying the dead, he noted 'the expression on our men's faces – the strain before. I've seen,' he added, 'plenty of dead men, but not with the expression that was on the Beersheba men – the strain that they'd been through – the extreme tension of what they'd had – told on them and it was very marked on their death.' Trooper Vic Smith, also of the 4th, asked how he had felt, replied: 'Oh, I dunno. Wishing to Christ it was over. Yeah . . . and out of it! Oh! I s'pose there was a certain amount of fear there. But you had to keep going; you couldn't drop out.'[11]

When the leading squadrons had reached about halfway to their objective heavy machine-gun fire was directed on them from a hillock on the left. Fortunately, before it could do much damage the Essex Battery, which was with the 7th Mounted Brigade, put the machine guns out of action with its first few shells, having spotted their flashes in the fading light. At about this stage in the charge the enemy riflemen in the trenches hit and dropped numbers of the horses in the leading line. This made the rest of the light horsemen increase their speed to the maximum and as Lawson's squadron (facing the hottest fire) came within half a mile of the chief earthworks, casualties almost ceased. Over the last few hundred yards the squadrons galloped quite untouched. As Trooper Idriess had guessed, the Turks in their panic had failed to alter their sights, most of which were found after the battle to be set for 800 yards.* The enemy artillery, too, failed to shorten range speedily enough and most of their shells passed overhead. The main trenches over which the majority of the horsemen jumped were thickly lined by Turks. The heaviest casualties were sustained in the next few minutes, in what Chauvel's biographer calls 'a savage hand-to-hand struggle'. Its bitterness was unlike the usual fighting between Australians and Turks. There is evidence that on four occasions Turks who had surrendered killed light horsemen. Retribution was speedy in two cases. 'When one member of a gun crew shot a stretcher-bearer, the entire crew was killed. In another incident, a grenade is said to have been thrown into a group of Turkish prisoners when one of them killed an Australian.' The official Australian historian gives his version of what happened at this time:

*Grant (Osborne: I, 354) says '800 metres' which would mean some 875 yards. Wavell (123) says '800 yards'.

'The excited line, reining up amidst a nest of tents and dugouts, dismounted. Lawson's three leading troops were then joined by a troop from the 12th Regiment; as the horses were led at the gallop to cover, the Australians leaped into the main trench which they had just crossed and went to work with the bayonet, at the same time clearing up the enemy in the dugouts. [Two lieutenants of the 4th were killed at this point.]* The Turks were now so demoralized that they offered only a feeble resistance to the bayonet and any shooting on their part was wild and comparatively harmless. After between thirty and forty had been killed with the steel, the rest threw down their rifles . . . . One of the troopers had galloped on to a reserve trench further ahead. The Turks shot his horse as he jumped and the animal fell into the trench. When the dazed Australian found his feet he was surrounded by five Turks with their hands up.'[12]†

*The only two officers killed in the action. One of them, Lieutenant F.J. Burton, died together with four troopers from one of the leading sections of Lawson's squadron. All five were shot dead as they jumped from their horses into a trench.

One of the mounted stretcher-bearers who always rode with the advanced lines was Trooper A. Cotter, a clerk from Sydney who was a well known international fast bowler. He was shot dead at close range. (Gullett, 401)

†Troopers T. O'Leary and A.E. Healey were the 4th's ground scouts. Both got through unscathed. 'O'Leary,' according to the Australian Official Historian, 'jumped all the trenches and charged alone right on into Beersheba. An hour and a half afterwards he was found by one of the officers of the regiment in a side street, seated on a gun which he had galloped down, with six Turkish gunners and drivers holding his horse by turn. He explained that, after taking the gun, he had made the Turks drive it down the side street, so that it should not be claimed as a trophy by any other regiment.'

During the trench fighting, Armourer Staff-Sergeant A.J. Cox of the 4th, an assayer by profession, noticed a machine gun being dismounted from a mule. He rushed the crew single-handed, bluffed them into surrender and took forty prisoners. Trooper S. Bolton, an engine driver from Geelong, pursued a six-horse, six-man gun which a German officer was galloping out of action. Bolton who had lost his rifle but picked up a revolver fired at the officer, missed him and then used the revolver's butt-end to knock him from the saddle, forcing the Turkish crew to return with the gun. Both Cox and Bolton were awarded Distinguished Conduct Medals. A Military Medal was given to Trooper W. Scott who refused to fall out when his thigh was smashed by shrapnel. He insisted that he was capable of leading horses back and 'galloped five of them out of harm's way before fainting as he was lifted from the saddle.' (Gullett, 400-401)

Charge of the 4th Australian Light Horse Brigade at Beersheba (12th A.L.H. Regiment pictured).

On the left the 12th found that the defences were not continuous and so most of Major Hyman's leading squadron, led by Captain R.K. Robey, rode through a gap straight on into the town. Hyman himself, though, with about a dozen men charged up to the trenches of a small redoubt, dismounted and killed some sixty Turks (numbers shot by Hyman with his revolver) in a bitter little fight before the rest surrendered. Major C.M. Fetherstonhaugh, a thirty-eight-year-old veteran of the Boer War, commanded the second squadron of the 12th. Within thirty yards of the main trenches his charger was hit. He at once put it out of its agony with a revolver shot, jumped into a trench, emptied his weapon into the nearest Turks and then fell, shot through both legs. Most of his squadron followed Robey into the town along the main street, while Robey 'rode hard for the western side of Beersheba, aiming to envelop it by the north.'[13] (see plate 12 and Appendix 2, p. 349)

During the time which had passed since the order for the charge* had been given, the galloping Australians, some time before they

*Blenkinsop (203) says that the actual charge was 'all over in ten minutes' which seems likely.

reached the town, overtook numerous fugitive troops and overran nine guns. Most of the gunners surrendered, but some who resisted were shot or bayoneted. On the 11th Regiment reaching the town, Grant ordered it to push through so as to hold it against attack from the north, west and south-west. It took some 400 prisoners who were retreating from the south-west.

While this remarkable charge was in progress the 7th Mounted Brigade, as soon as it was seen that the trenches at Ras Ghannam had been evacuated, rode into Beersheba arriving in the dark at about 6.30 p.m. Half an hour earlier the Anzac Mounted Division, with the 3rd Brigade attached, had also entered the town.

The overrunning of guns, transport and the columns of retreating men by the audacity of the Australian horsemen brought an abrupt and dramatic change of fortune to the Turks. They had had every reason to rely upon the onset of darkness and the resistance of the still intact defences east and north of the town to enable them to make a more or less orderly withdrawal having completed the destruction of the wells. Now, although a few scattered parties of the enemy continued to set fire to buildings and to blow up railway installations, there was for the vast majority only one panic-induced thought: how to escape from the madly yelling Australians. To the hills north and north-west of the town there was a wild, disordered mob-rush. Ismet Bey, the Turkish corps commander escaped by only a few minutes. In an intercepted wireless message that night he said that his troops had broken because they were 'terrified of the Australian cavalry'.[14]

Even more vitally important to the whole Beersheba plan, the majority of the engineers bolted with the crowd, leaving behind most of the demolition charges which they had laid at the wells and at the more important buildings. Of the seventeen wells only two were completely and another two partially destroyed. Three pumping engines were put out of action.

* * *

As to the quality of the Turkish commanders and soldiers who were manning the trenches, there has been considerable denigration of both by nearly all British historians. This seems to have been generated in large part by Kress von Kressenstein. Having been properly gulled by Meinertzhagen's wily deception and led to believe that only a feint against Beersheba was intended, he laid false blame on Ismet Bey's handling of his men and on the poltroonery of the Arab troops.[15] Unfortunately Preston, without serious enquiry, accepted this version in total.[16] In fact, Ismet Bey, who, according to Liman von Sanders, was 'one of the most capable of the higher Turkish commanders',[17] had by early morning become aware of the light horsemen's advance and by 10 a.m. had estimated that he was about to be attacked by two divisions of mounted men. Two hours later 'he understood that there were two to three divisions in the east and one to one and a half in the south.'[18] By telephone from his headquarters, as he received more and more alarming reports about both the infantry and the mounted advances, he tried frantically to convince Kress that either the town should be evacuated or that he should receive immediate reinforcements. No attention, it appears, was paid to his pleas.[19]

At his disposal were some 4,400 men and twenty-eight field guns. These included his reserves.* Against this puny force were arrayed 58,000 men and 242 guns. The division which chiefly manned the trenches was 'ranked by the Turkish staff as one of the best in the 16th Division'.[20] Most of the men who met the charge directly were from one of the Yilderim regiments raised to recapture Baghdad and consisted of high-class Anatolians. Some of their officers were Germans. There were two other regiments largely made up of Arabs. They were all local farmers who had an interest in defending their properties. There is little evidence that they behaved in a cowardly way. Indeed, Ismet Bey went out of his way to praise their conduct. Something like 1,000 rifles, nine machine guns and three batteries of artillery, supported by two aircraft armed with machine guns and bombs, opposed the actual charge.

* * *

*The 3rd Cavalry Division formed the chief part of his reserves. It was commanded by Colonel Essad Bey, 'the most remarkable cavalry commander in the Empire'. Two lancer regiments were on the eastern flank, commanded by Ferid Bey, who had recently commanded the Turkish Imperial Guard. (Nogales, Rafael *Four Years beneath the Crescent*, 1926, 305).

Seventy officers and 1,458 other ranks were taken that night. The 4th and 12th alone accounted for thirty-eight officers and some 700 other ranks. Nine guns, seven ammunition limbers, four machine guns, a mass of transport vehicles and much other material fell to them. The total casualties of the whole Corps amounted to 197 of whom fifty-three were killed, a large proportion being due to the air attack. The 4th lost, beside Lieutenants Burton and B.P.G. Meredith, nine other ranks killed and four officers and thirteen other ranks wounded.*

From the two regiments there were probably not more than between 400 and 500 men actually in contact with the enemy. Thirty-one killed and thirty-six wounded in a mounted charge against artillery, machine guns and strongly entrenched infantry was no great price to pay for an achievement without which the whole of Allenby's grand plan must have been in serious jeopardy.† Without the success achieved in that one vital hour, all the careful preparations of the summer months would very likely have resulted in another lengthy stalemate. Next day Allenby invested Grant with the Distinguished Service Order. Seldom has that decoration been better deserved.

During the night, while sporadic sniping from the hills to the north of the town caused a few casualties to men who were lit up by the burning buildings, every available party of engineers sought out the vital wells and began to develop the flow of urgently

*In the zone of almost point-blank range through which the light horsemen had to pass,

> 'an entire section of the 4th Regiment was wiped out by a salvo of hand grenades. One man survived. Nine men of the 12th had their horses shot immediately in front of the Turkish trenches and they opened fire on the enemy from the cover of broken ground. Of 'B' Troop, 'A' Squadron of the 4th, only six or seven men got across the first Turkish trench on horseback and fourteen of the twenty-eight horses were killed. A twenty-man troop of the 12th tried to charge through the gap to the left of the redoubt. Only six men made it.'

(Jones, fn 32, 37).

†At Huj (see p. 181), seventy-five casualties out of about 170 men occurred, with some 100 horses lost. In the Wadi Sheria charge (see p. 170) the 11th and 12th Regiments were unable to complete it. The 11th and part of the 12th lost seventy-five men and 100 or so horses (see p. 170), while the only charge which, in low casualties and success, equalled Beersheba was probably that at Kaukab Ridge (see p. 325). But that charge was against demoralized troops.

needed water.* But, in spite of two reservoirs containing 90,000 gallons being left intact, it proved insufficient for the whole Corps and about half of its units two days later had to be withdrawn to Karm to relieve the pressure on the captured wells. Indeed several wells in gardens were not discovered until two and three days later. It took time, too, for new pumping plant brought up on eight caterpillar tractors to arrive, while the dismantled plant used at Asluj did not arrive until the afternoon of 3 November. Meanwhile, had it not been for the pools discovered by the 2nd Brigade and others in the Wadi es Saba after the thunder-storms of 25 October, many more troops would have had to leave the Beersheba area, thereby even further dangerously curtailing the operations of the next few days than in fact happened. Indeed, though these pools just outside the town were rapidly evaporating, large numbers of the horses depended entirely upon them during the night and on the following day.† Thus was Allenby favoured by uncovenanted good fortune. It was also fortunate that the 27th Turkish Division which formed the main part of Beersheba's garrison was a largely Arab formation. Had there been instead Anatolian Turks to contend with, an even stouter fight would have been put up.

> 'It may be urged,' wrote Wavell, 'that the Turks shot badly. But they probably shot no worse than the very great majority of troops would in similar circumstances. It requires extremely

*After removing the fuses, the engineers 'erected five engines and three oil pumps, repaired one engine [for working the pumps] and erected pumping sets brought from Asluj.' Initial attempts were made to ration the horses' water by imposing a time limit for each batch as it came to the troughs. These came to nought since the craving for water of horses that had been without it for some thirty-six hours and more was so great that they got out of control and rushed the troughs. During the night to try to prevent this, a stout rail was hastily erected at every line of troughing. There was no difficulty in limiting the water for camels 'as they invariably drank in two bouts, with an interval of ten minutes between each.' (Blenkinsop, 204). Luckily a number of extensive stone troughs used by the Turkish cavalry were intact (Tallents, 109). At least one ancient and cumbrous *saqqia* was kept in operation and proved as efficient as the oil pumps. 'It drew as much water – about 1,500 gallons an hour – as the well would produce.' (Falls, 92-3).

†Numbers of stray ponies were rounded up in the town and fitted with abandoned Turkish pack saddles and with tins for carrying water. (Tallents, 110)

well-trained and disciplined units to adjust their sights calmly and to produce good fire effect in the face of galloping horsemen. Moreover, in this and in other charges in this campaign, the clouds of dust raised by the leading squadrons formed quite an effective screen to the formation and movements of the units in rear and prevented accurate ranging or shooting by the enemy . . . .

'The moral results of the charge were even greater than the material gains. It set the pace for the whole campaign, inspiring the brigade which carried it out with immense confidence and all the other mounted troops with a spirit of rivalry and emulation. And this demonstration of the power of mounted men to ride home on their infantry undoubtedly shook the nerve of the Turks.'[21]

* * *

The Beersheba charge is especially remarkable in the annals of mounted troops, for the very idea of mounted infantrymen, which the light horsemen really were, indulging in a cavalry charge against entrenched, unbroken infantry, supported by artillery and machine guns, was until then unknown to history. As one chronicler has put it, it was 'against the principles of war'.[22] The one VC, six DSOs, four MCs, four DCMs and eleven MMs awarded to participants, were not, perhaps, excessive rewards for efficient heroism.

17

'The capture of Beersheba with its wells mainly intact and the almost complete destruction of the enemy's 27th Division formed an auspicious opening to General Allenby's operations. But the real crisis lay in the next few days.'

WAVELL in *The Palestine Campaigns*

'Wherever the adventurous light horseman went, he knew that galloping close at hand were the heavily burdened pack-horses of the gunner, ready to come into action a few moments after a halt was ordered.'

The Australian Official Historian[1]

XXI Corps attacks Gaza – Newcombe's detachment surrenders – XX Corps attacks Hareira

The second, most vital, phase of Allenby's great design took longer than he had hoped. Though he kept his plans fluid and flexible for the period after Beersheba had been taken, he had been anxious for XX Corps and Descorps to make a further advance on 3 or 4 November for the main attack, while XXI Corps, so as to occupy the Turks' attention in the interval, started its assault on Gaza. This last, consequently, he ordered for the night of 1 November. The assault was highly successful in so far as it cut deep into the heavily defended enemy positions and by 6.30 a.m. on 2 November, aided by six tanks, it had achieved its furthest objective, Sheikh Hasan on the coast north-west of the town. It had thereby turned the flank of much of the prepared defences of the town. It had also succeeded in preventing any part of the garrison from moving eastwards and had indeed compelled its reinforcement with reserves which would otherwise have been available against Chetwode and Chauvel. The price was high: some 350 killed, another 350 missing and about 2,000 wounded. This was just about double the number of casualties suffered by XX Corps and Desert Mounted Corps in the capture of Beersheba, showing how expensive an attempt to break through on the coast would have been (see p. 141 above). 550 prisoners were taken as well as three guns and thirty machine guns. Over 1,000 of the enemy were buried on the field. There the corps remained until Gaza was evacuated on 7 November (see p. 168).

Meanwhile the initial advance of Chetwode and Chauvel on the right was soon held up by an unexpectedly fierce battle around Khuweilfe where the Turks just beat Chauvel's men to its valuable wells. The waterless area in which it was fought consisted of steep hills, jagged outcrops and stony desert, totally unsuitable for mounted troops and pack animals. Descorps, therefore, fought almost entirely on foot, together with XX Corps, which bore the brunt, on its left. The dominating ridge which if taken would have won the action, was unyieldingly denied to 53 Division. It was eventually captured on 6 November simultaneously with Chetwode's main attack.

The series of complex actions which raged around Khuweilfe added up to a tactically drawn battle. It was so hard fought for two reasons. The morning after Beersheba fell, Descorps had advanced north and east of Beersheba following up the retreating 'Beersheba Group' (numbering now as few as 1,500 rifles), but perhaps more importantly in search of further water supplies which in the event, though there were vague reports of their existence, did not materialize. This movement, combined with the presence of Newcombe's seventy-strong detachment holding the Hebron road north of Dhahriye (see p. 140), so frightened the Turkish command that it started the rapid movement eastward of considerable 'elements of three or four divisions'[2]* from its reserves. Newcombe had hoped that he would be joined by the Arabs of the hills, but though they provided him with guides they did not see fit to fight with him. Nonetheless, expecting Descorps' troops to link up with him, he determined to do what damage he could in the meantime. On 1 November he fended off some hundred Turks, but next morning six battalions of Turkish infantry attacked him from north and south. After losing twenty men, he wisely surrendered, having blocked all communication between Hebron and Dhahriye for forty hours. From his captors he gathered that considerable numbers of further troops were on the move against him from the

*The advance of one of XX Corps' divisions towards Tuweiyil Abu Jerwal, some six miles north of Beersheba, apparently intending to exploit the success of the preceding day, gave further cause for alarm. It had, in fact, been sent there by Chetwode 'to give security to the right flank of his attack towards Sheria'. (Wavell, 130)

north-east.* The alarm caused by his advance, which the Turks supposed to be the spearhead of a major cavalry raid straight on Jerusalem, spread far and wide. This was partially because of the fear of a general Arab uprising. The enemy high command clearly overestimated the capacities and willingness of the local bedouin. At the same time it grossly underestimated 'the requirements of British stomachs as compared with those of their own people'. A cavalry raid on Jerusalem would, of course, have been quite out of the question at this stage. Nonetheless, Turkish critics continued long afterwards to maintain that this would have been the right course for the British to take.[3]

Another reason for the length and the fierceness of the struggle around Khuweifle was the fact that none of the troops of Descorps or XX Corps could remain long in action without having to be withdrawn to water at Beersheba and even at Karm – distances of twelve miles and more.† This chronic shortage of water, occasioned by the drying up of the rain-pools and the time it took to bring the town's supply up to maximum capacity (see p. 166), was aggravated by a particularly intense khamsin which blew for three days from 1 November thereby upsetting the calculations which had been made on the basis of troops working on a reduced water ration. Indeed 'washing and shaving had to be forbidden

*Hebron, said one German officer in that town, 'was like a disturbed ants' nest; the staff of 7th Army headquarters was standing beside its horses saddled for hasty retreat . . . . The Main body of the English was said to be only a few kilometres off . . . . A whimsical touch was given to the affair by an old Arab procuress, who in the midst of the universal terror was calmly preparing to meet the enemy with a band of scantily clad dancing-girls.' (Obergeneralarzt Steuber in *Yilderim*, 114, quoted in Falls, 83)

†Ras el Nagh, a hill near Khuweilfe, was taken by 7 Mounted Brigade, but because of the need for formations to withdraw for water, it had to be subsequently held in turn by 1st, 5th Mounted, New Zealand and the Imperial Camel Corps Brigades. While on 3 November 7 Mounted Brigade was retiring, Wigan, its commander, 'went along the whole column and handed his brandy-flask to those who seemed the most exhausted.' (*Twentieth MG Sqn*, 31)

On 4 November a determined attack was made on the hill while the 5th Mounted Brigade was holding it. The Gloucesters had just been relieved by the Worcesters and had moved back to their horses when they were urgently summoned to return. One squadron mounted and returned at the gallop keeping its footing on ground over which when dismounted it had had difficulty in picking its way. A squadron of the Warwicks also turned up just in time to repel the attack. The brigade's casualties numbered over fifty in the day's fighting.

altogether, so that, in the words of an eye-witness, a conference of field officers looked like a Nihilist meeting.'[4] The horses of 1st, 2nd, 5th and New Zealand Brigades had only one proper drink during 3, 4, 5 and 6 November. Many of them went unwatered for forty-eight hours. Most regiments spent every night looking for water, more often than not without any success, thus depriving the men of sleep and adding greatly to their fatigue. The Medical Officer of the Worcesters says that when 5th Brigade reached Beersheba, while the horses made a rush for the troughs, many of the men, too, 'oblivious of previous orders, could not resist plunging their heads into the horse-troughs'.[5] Not until 5 November did the output of the Beersheba wells reach its maximum, some 390,000 gallons a day. All this, combined with the unexpectedly stiff Turkish resistance, forced Allenby to bow on 4 November to Chetwode's and Chauvel's irrefutable arguments for a postponement of the major attack on Hareira and Sheria in the centre till 6 November. Water had once again proved to be a chief ruling factor.

Though it had taken six days of fighting to prepare for the rolling up of the main enemy line of trenches, by the time that great coup was launched it enjoyed more or less favourable circumstances. As has been shown, the operations to the north and north-east of Beersheba had attracted the whole of the enemy's available reserves, including the 19th Division, one of the finest formations in the VIIth Army.* With the mounted troops acting as right flank guard, XX Corps' main attack, during the 6th made by 10, 60 and 74 Divisions with overwhelmingly superior numbers, was, in Chetwode's words 'completely successful,† all the allotted objectives being attained with comparatively small casualties and the Desert Mounted Corps as a result was enabled to move forward to Sheria in readiness to take up the pursuit.'[6] It was fifteen hours, though, before at 7.30 on 7 November, Chauvel's troops could be assembled to take it up.

*Until 5 November 'there were at least seven infantry regiments between the Hebron road and the left of the defensive line and only two regiments in this line up to the Wadi esh Sheria, a distance of six and a half miles.' (Falls, 95).

†So was 53 Division's final attack on Khuweifle.

The Corps' casualties were approximately 1,300. Over 600 prisoners, twelve guns, numerous machine guns and large quantities of ammunition were taken.

15. Brigadier-General Granville Ryrie (see p.58)

16. Brigadier-General William Grant (see p.127)

17. Major-General Edward Chaytor (see p.60)

18. Major-General Henry Hodgson (see p.93)

19. Chauvel a few days after Romani, August, 1916 (see p.55)

20. Lieutenant-General Sir Philip Chetwode (see p.77)

21. General Liman von Sanders (see p.219)

22. General von Falkenhayn (see p.136)

18

'The Desert Mounted Corps' great moment was come; the moment for exploitation of success, for which the mounted troops had been maintained at a strength far higher in proportion to that of the infantry than in any other theatre of war, was in sight.'

CYRIL FALLS in the *Official History*

'The Commander-in-Chief's foresight and preparations had achieved a great victory and was only denied its full fruits by two factors beyond his control – the resistance at Khuweifle and the absence of water after Beersheba was taken.'

H.S. GULLETT in the *Official Australian History*

'[Allenby] knew, better than anyone else in the force, that cavalry exploitation, to attain its best, demanded perfect timing, maximum numbers and 'that the troops should be fresh. None of these conditions were attained.'

LIEUTENANT-COLONEL REX OSBORNE in *The Cavalry Journal*, 1921

'Launching the depleted Anzac and Australian divisions on Huj and Jemmamah to secure water and cut off the Gaza garrison was not to hurl a thunderbolt into a disorganized rabble – there was neither rabble nor thunderbolt.'

A.J. HILL in *Chauvel of the Light Horse*

'The successful or sustained pursuits of history have been few, the escapes from a lost battle many. The reasons are partly material, but mainly moral. A force retreating falls back on its depots and reinforcements; unless it is overrun, it is growing stronger all the time .... The pursuer soon outruns his normal resources.'

WAVELL in *Allenby: A Study in Greatness*[1]

Barrow's 'special detachment' – XX Corps takes Hareira – XXI Corps enters Gaza – action at Kh. Buteihah – Ameidat taken – pursuit starts – Jemmame taken

At 8.10 p.m on 6 November Chauvel received Allenby's order to collect at once

> 'all available mounted troops except Yeomanry Division [and the New Zealand Brigade] and to push through to Jemmame and Huj. Australian Division is transferred from general reserve [where it had been temporarily resting at Shellal] and has been ordered to send staff officer to report to Descorps to explain situation of division and take instructions. 3rd Light Horse Brigade is being ordered from Shellal at 7 a.m. to-morrow, by Imleih and Irqaiyiq, where Descorps will send instructions to meet them, but the advance will not be delayed to wait for this brigade, which must follow on.
>
> 'Object of Descorps will be to secure the water at Jemmame and Huj with a view to operating rapidly from those places so as to cut off or pursue the Gaza garrison.
>
> 'Chief expects the enemy to be pressed with the utmost vigour.'

Barrow was ordered at the same time to 'take command of the right flank'. For this purpose, the New Zealand Brigade from Anzac Division, the Camel Brigade and the 53rd Division were attached to his own Yeomanry Division.* He was to remain 'on approximately the line he holds at present [near Khuweifle], taking every opportunity that may offer of punishing the enemy.'[2]† 60 Division from XX Corps was given to Chauvel to assist in his advance.

The early morning of 7 November saw Chetwode's XX Corps complete its part in Third Gaza with the final capture of Sheria and Hareira. From 31 October to 7 November the Corps' casualties numbered 932 killed, 4,444 wounded and 108 missing. It had captured 2,177 prisoners, forty-five guns, seven trench mortars and fifty machine guns.

At 9 a.m. on 7 November the Imperial Service Cavalry Brigade at the head of XXI Corps entered Gaza. The Turks had evacuated

*Barrow's 'Special Detachment' also had with it 11 and 12 Light Armoured Car Batteries. These would have been very useful in Chauvel's advance.

†In the event, Allenby allowed Barrow's 'special detachment' to follow up the retreating Turks almost to Hebron, with unfortunate results later on (see p. 184).

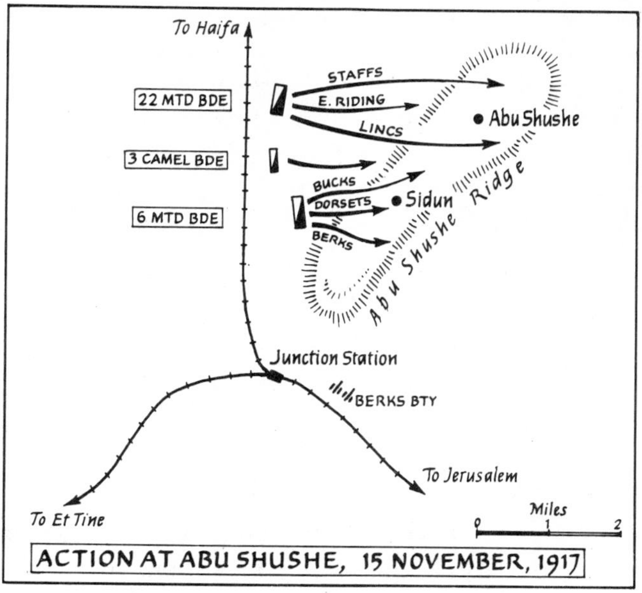

the town and were at last in full retreat from virtually the whole Gaza–Beersheba line – that line, a very extended one for the size of their force, which they had held with such stubborn tenacity for nine long months.* Wavell thought that 'left to themselves, it is likely that the Turkish commanders would have begun a retreat very much earlier, a step which would have gravely increased our difficulties. But their German commanders and staff would have nothing of such suggestions.'[3]

* * *

The prospects for a successful pursuit and interception by Descorps were poor. The lack of water between Sheria and Huj, the fact that

*'A Turkish rearguard, seemingly unperturbed by the presence of British forces behind both its flanks, maintained itself throughout the day in the Tank and Atawine redoubts, the one portion of the front line still intact. [A serious effect of] the presence of this rearguard was the delay it occasioned to the march of the transport columns which were being hurried across from the XX to the XXI Corps to enable the latter to take up the advance along the coastal plain.' (Wavell, 145)

Chauvel could employ out of his eleven brigades the equivalent of rather less than four, all of them intensely weary after a week of fighting and marching, combined with the steady resistance of the enemy's rearguards, meant that progress on 7 and 8 November was distressingly slow. Most of the Turks escaped, marching, as was their wont, under depressing and arduous conditions with rapidity and cohesion. It has been suggested that Allenby should at once have taken the risk of transferring Barrow's Yeodiv and the New Zealand Brigade from Chetwode north of Beersheba in time for the pursuit, but such a course, with the retreating Turks on the British right as yet far from totally broken, was one which he was probably right to eschew.

At one moment on 7 November a troop of 11 Regiment, missing the signal to dismount in the face of overwhelming fire from Turkish trenches near Kh. Buteihah, galloped on. 'The rifle-fire,' according to the Australian official historian, 'was terrific, but the pace saved the Australians, and when they came within a few horse-lengths of the advanced line the Turks raised their hands.' Lieutenant A.R. Brierty, commanding the troop, then galloped over a shallow trench full of riflemen, dismounted his twenty-one men who began to employ their bayonets. As they did so, 'the Turks who had surrendered opened fire at a few yards' range and the other troops all round joined in the shooting. In a few seconds every Australian horse had fallen, eleven men had been killed and the rest of the troop, except one, wounded.' Another troop, under Lieutenant J.S. Bartlett, rushed to Brierty's help with a Hotchkiss and nine riflemen, shooting down a German officer and twenty Turks, losing four killed and three wounded.[4]

Late on 7 November the capture of Ameidat was effected by 1 and 3 Regiments at the gallop. They took the Turks there quite by surprise and were soon rounding up 391 prisoners. Also captured were twenty-seven ammunition wagons, a complete field hospital and masses of stores. Early next morning a squadron of 5 Regiment galloped after two guns which were being hurried towards the Wadi Hesi and captured them with their escort. At this time the 8th, 9th and 10th Regiments of 3 Brigade made a speciality of stalking enemy guns and then charging them 'suddenly from the rear. It became a commonplace,' wrote one witness, 'to find an enemy 5.9-inch howitzer in a hollow in the ground, with the detachment dead around it and the words "captured by the 3rd A.L.H. Brigade" scrawled in chalk on the chase of the gun.'[5]

* * *

At dawn on 8 November 1 and 2 Brigades moved off on a six-mile front over open plain, dotted with prominent hills, nearly every one of which the enemy held on to with tenacity. Chauvel's orders were to close the narrow corridor which still remained open for the Atawine rearguard and to cut off the Turks who were facing 52 Division on the Wadi Hesi. To his assistance came 7 Mounted Brigade from Corps Reserve which was placed under Anzac Div. This brigade was fresher than the others, for it had watered its horses in the Wadi esh Sheria before joining Chaytor's Anzac Division's headquarters at 9 a.m.; but it was only a skeleton brigade, consisting of little more than four squadrons and a machine-gun squadron, together with, of course, the excellent Essex Battery.

Throughout the morning Chauvel's men had before them the tantalizing sight of long columns of Turkish infantry, transport and guns moving across the plain towards Jemmame.* 'It was a sight,' in Cyril Falls' words, 'such as cavalrymen in all ages have longed for, such as would have fired the blood of a Le Marchant (see Vol 1, 53 *et seq.*) or a Scarlett (see Vol 2, 123 *et seq.*). But the day was to prove the vast transformation in warfare made by the machine gun.'[6] The defeated, dispirited, diseased and hungry Turkish foot soldiers were magnificently protected by artillerymen on both flanks and even more effectively by the unwavering machine gunners who time and again prevented both the mounted troops and the Londoners of 60 Division from getting at their quarry. 'Men are remarking,' wrote Trooper Idriess, 'how the Turk fights to the very last charge, until the pounding hooves are upon him, then he drops his rifle and runs screaming; while the Austrian artillerymen and German machine-gun teams often fight their guns until they are bayonetted.'[7]

Though disappointingly small numbers of prisoners were taken, Anzac Div and 7 Brigade managed to reach Jemmame in mid-afternoon. There copious water supplies were fortunately found intact, including the steam pumping plant complete. The engineer who had been left behind to blow it up remained instead to work it 'with great docility'.[8] During the previous night nearly all the

*These were the 26th and 54th Turkish Divisions.

horses of Anzac Div had drunk some water. Numbers of them had then been without a drink for no less than fifty-six hours.

Early in the day the 12th Regiment had ridden twelve miles in an hour and a half to make touch with the Imperial Service Cavalry which was advancing with speed north and north-east of Gaza and to convey orders to its commander to make every exertion to cut across the heads of the retreating columns. The regiment found him near Beit Hanun, but by then the fast-marching Turkish infantry was out of the net.

19

'Machine guns and rifles opened on us the moment we topped the rise behind which we had formed up. I remember thinking that the sound of the crackling bullets was just like a hailstorm on an iron-roofed building, so you may guess what the fusilade was . . . . A whole heap of men and horses went down twenty or thirty yards from the muzzles of the guns. The squadron broke into a few scattered horsemen at the guns and then seemed to melt away completely. For a time I, at any rate, had the impression that I was the only man left alive. I was amazed to discover we were the victors.'

LIEUTENANT W.B. MERCER,
Warwickshire Yeomanry, the only officer in the regiment not hit in the charge at Huj.

'If covering fire could have been given us, our casualties would have undoubtedly been less, but the delay caused would have meant that the infantry would have had many more.

'The prompt action taken gave the guns no chance of getting away and enabled the infantry to advance unchecked for the remainder of the day.

'Undoubtedly the flash of the swords and the great pace which we came at them completely demoralized the gunners, and although a few got away, they would never wait again for a cavalry charge.'

LIEUTENANT A.C. ALAN-WILLIAMS,
Warwickshire Yeomanry

'The enemies [the Yeomanry] charged were no hunted and demoralized fugitives, but unshaken troops – partly German and Austrian – who were holding at arm's length without difficulty the infantry opposed to them.'

WAVELL of the charge at Huj, in
The Campaigns in Palestine.[1]

The charge of the Warwick and Worcester Yeomanry at Huj

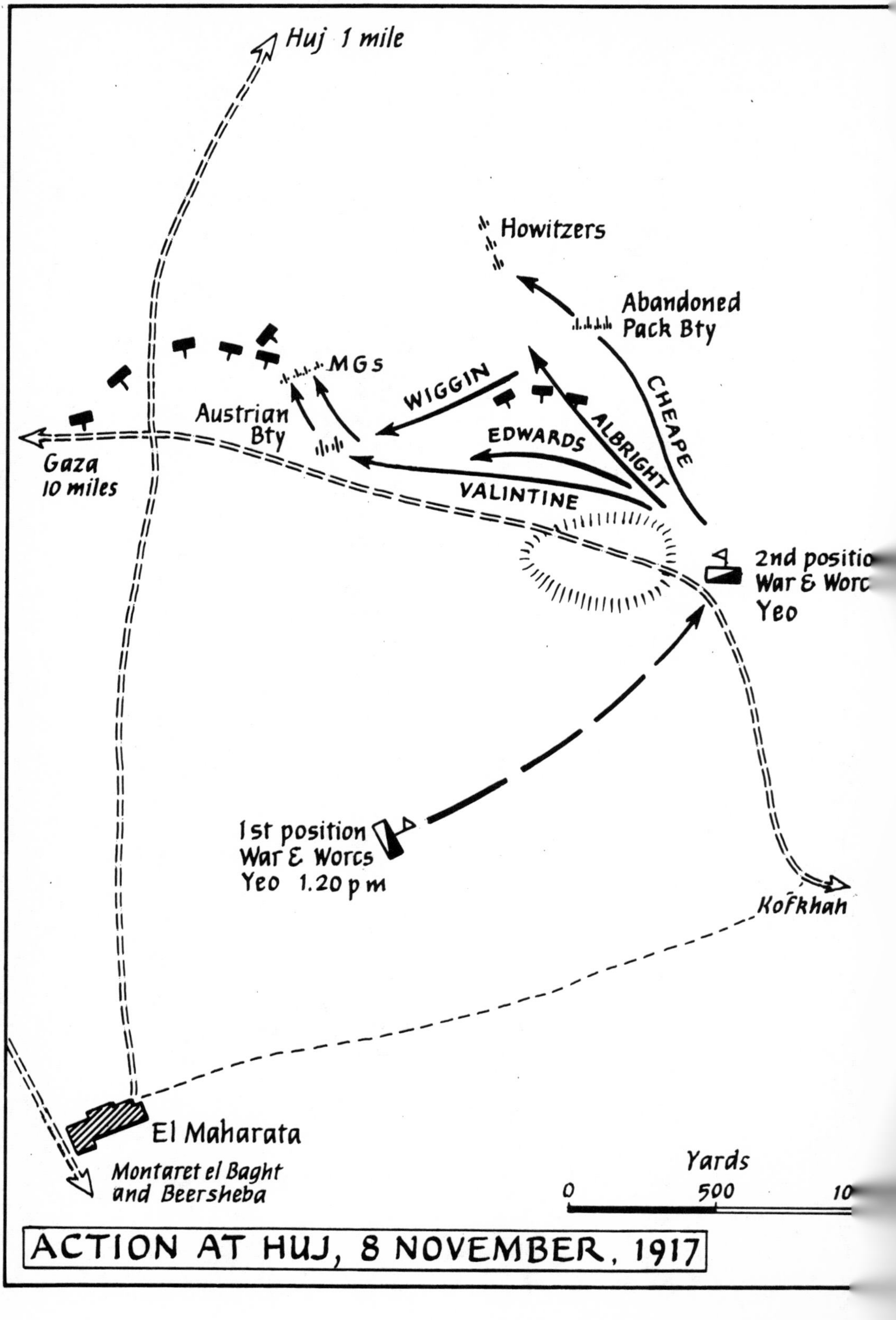

ACTION AT HUJ, 8 NOVEMBER, 1917

In command of 60 Division was Major-General Shea, who had joined the 15th Lancers in India in 1891.* On 9 November he reported to Chauvel (under whose command it will be remembered his division now was) as follows:

> 'Yesterday, after my division had taken the high ground N. of Muntaret-el-Baght, I was scouting ahead in an armoured [Ford] car. I saw many Turks with guns marching N.E. across my front. It was impossible for my infantry to catch them. Some cavalry came up on my right (about a mile away) which turned out to be ten troops of Worcester and Warwick Yeomanry. Judging that immediate action was necessary, I went and ordered Lieut.-Col. [Gray-] Cheape [commanding the Warwicks] to gallop what appeared to me to be the hostile flank guard. I judged the distance at 2,500 yards. This gallop was carried out with the result that some twelve guns and three machine guns were captured, and the gunners (German and Austrian) were all killed or wounded at their guns ....
>
> 'I much regret to hear that the yeomanry lost some twenty-five percent, but they completely broke the hostile resistance and enabled my division to push on to Huj.'[2]

Thus, laconically, the man responsible for launching it, describes the first considerable charge of the campaign made by mounted men employing the sword, a charge which over the years came to exert an important influence on the thinking of cavalrymen.

* * *

Throughout the morning of 8 November the Turkish rearguard, in this area perhaps 3,000 strong, was being slowly pressed towards Huj by 3rd Brigade on the right, 5th Brigade in the centre (the Worcesters acting as its advance guard) and 60 Division on the left. By midday some seven miles had been covered from Muntaret el Baght to Kofkhah, which place fell to 5 Brigade soon afterwards. By now there appeared signs of disintegration among the Turkish infantrymen.

*As a captain he had commanded some Australians during the Boer War (see Vol. 4, 214). He finished his career in 1932 as GOC, Eastern Command, India. He died in 1966 at the age of ninety-seven.

According to Major the Hon. John Cavendish Lyttelton, an officer of the Worcesters,*

'[his regiment and the] Warwicks, two squadrons of each, mounted and galloped forward in hot pursuit. Once for ten minutes they halted for dismounted action, to open a rapid rifle and Hotchkiss fire on the fugitives, then on again towards Huj, until at 1.15 p.m they were held up a thousand yards short of the enemy gun positions covering Huj from the south-east, and the huge depot of ammunition and stores the Turks had established there. Under the shelter of a low ridge the squadrons dismounted to give blown horses a breather and to examine the problem presented by the batteries in front. On the left the 60th Division could be seen striving to advance across the lower ground under a volume of shell-fire from those same batteries and suffering considerable loss in the process. Something had to be done quickly if the whole advance was not to be held up.

'The Colonel [Lieutenant-Colonel Henry John Williams†] decided that a mounted attack was the best solution; but before that could be made it was highly desirable to get the 3rd Australian Light Horse Brigade, which had remained behind at Kofkhah, up to the alignment on the right, both to cover the mounted attack with rifle and machine-gun fire, and to give the necessary protection to the right flank. Accordingly he left the regiment to ride over to the headquarters of the 3rd Australian Brigade.

'No sooner had he gone than General Shea rode up to Colonel Hugh Cheape, who was then in command of the Warwicks, and ordered him to make a turning attack on the artillery position in front, in order to clear the way for the advance of the infantry. Hugh Cheape immediately came across to Major Wiggin, now in temporary charge of the Worcesters, told him of the orders he had received and asked him if he would join in, knowing full well the answer he would receive.'[3]

*He succeeded his father as 9th Viscount Cobham in 1922 and wrote an excellent history of the Worcesters covering the period 1914–1922. (Cobham).

†He had joined the King's Dragoon Guards in 1891 and commanded that regiment after the war.

Gray-Cheape was faced with a daunting task. The Gloucesters, in brigade reserve, were watering far away; 'B' Battery, HAC, attached to the brigade, was far behind and, worst of all, the previous night the machine-gun squadron, with the exception of two guns which followed the leading regiments, and the Hotchkiss riflemen had been ordered to remain behind to find water.* Soon after daybreak they had been given guides to enable them to join the rest of the brigade, but, by the time they were most urgently needed, they had not rejoined.

Sending back a message to Brigadier-General Philip James Vandeleur Kelly (who had only that morning succeeded Fitzgerald in command of the brigade), telling him that he was about to make a mounted attack and urging the guns and the Gloucesters to hasten to his support, Gray-Cheape collected together all the detachments he could find. These amounted to one and a half squadrons each of his own regiment and of the Worcesters on his right. In all he mustered the equivalent of ten troops numbering at most 165 of all ranks, of whom possibly about 120 actually took part in the charge.†

The ten troops had been sheltering behind a boomerang-shaped ridge (see map, p. 174). This appeared to afford a more or less covered approach by which the Turkish left flank might be turned. The Worcesters' medical officer remembered Wiggin giving the order to mount and calling out, 'Now then, boys, for the guns!'[4] The Worcesters led the way: Major M.C. Albright's 'A' Squadron of the Worcesters, followed by two troops of 'C' Squadron under Second-Lieutenant J.W. Edwards. Bringing up the rear was Captain R. Valintine's 'B' Squadron of the Warwicks, followed by two troops of 'C' Squadron.

As they advanced at a brisk trot from the south-west end of the ridge they could not see the hostile guns firing at the infantry of the

*This they failed to do, while ironically the rest of the brigade did. (Osborne: I, 362).

†The Warwicks fielded nine officers and sixty-seven other ranks and the Worcesters seven officers and about eighty other ranks. (Adderley, 135). Cobham, 132, gives nine officers and ninety-six other ranks for the Worcesters, but this includes the Hotchkiss riflemen. It seems that none of the Hotchkiss sections accompanied the charge. Lieutenant Alan-Williams, Second-in-Command of 'B' Squadron of the Warwicks, was ordered by Valintine (see below) 'to fall out the Hotchkiss guns, shoeing smiths and any led horses' before starting.

60th about 1,350 yards away. They then moved north-eastwards under (as they thought) the ridge's protection, in line of troop columns. When they were some 300 yards from the northern tip of the ridge, these hitherto unseen guns – a 75mm Austrian battery – came into sight about 1,000 yards almost due west. Between them and the cavalry it was noticed that the ground, though undulating, was perfectly open.

As the yeomen trotted on they raised clouds of dust. This alerted the Austrian gunners who swung round two of their guns and fired at the horsemen as they came on. Little harm was done for by now the yeomen were gathering speed and it was difficult for the gunners to pick up the range quickly enough. The squadrons halted for a brief moment near to the northern tip of the ridge, but they were instantly subjected to heavy fire from four mountain battery guns and some 200 riflemen, numbers of whom stood up to take aim. These were positioned on a slight ridge to the north-west, some 600 yards distant.

Albright, realizing that the attack on the main target to the left could not go ahead while these guns and infantrymen were in a position to enfilade it, formed his men in column of half squadrons and 'went,' according to Wiggin, 'straight on to attack this lot immediately he realized the position and without waiting for further orders either from me or from Lieutenant-Colonel Cheape.'[5] To avoid the shock, some of the Turkish infantry fired wildly, others wavered, but the majority 'fled down the reverse slope with the victorious horsemen thundering at their heels.'[6] Although many more Turks could have been put to the sword (considerable numbers were) and although the guns of a retreating 5.9 howitzer battery which they were protecting, as well as the mountain guns, were at 'A' Squadron's mercy, Wiggin instantly ordered Albright to break off the pursuit. This wise decision was occasioned by what he saw of the troubled position which Valintine's Warwicks and Edwards's Worcesters were in.

A few moments after Albright had launched his charge, Cheape had ordered Valintine, with Edwards's two troops slightly echeloned to the right, to lead them over the crest of the northern end of the boomerang ridge and to charge the Austrian 75mm guns in flank. Valintine, like Albright before him, formed his men into column of half-squadrons with swords at the 'engage' and the moment they cleared the crest the Austrian gunners opened fire on them with 'an absolute inferno of shells.'[7] Four machine guns behind them and about two companies of riflemen, all protecting the 75mm guns, also opened a fierce fire upon the galloping, shouting yeomen. The distance they had to cover was some 900 yards down a slope and up the other side, 'with the last 100-150 yards very steep indeed'.[8] The Austrians depressed their muzzles to the maximum and set their fuses at zero so that the shells exploded almost as soon as they left the barrels. It was only a matter of moments before Albright, having rallied and reformed his men, joined in the charge in echelon from the right, sweeping down on the gunners' left flank. The Austrians stuck most heroically to their guns. Their final shot, indeed, 'passed through a horse that was almost at the gun's muzzle'.[9] 'Few,' according to Lieutenant Alan-Williams, 'remained standing and, where they did, they were instantly sabred. Others, running away from the guns, threw themselves on the ground on being overtaken and thus saved themselves, for it was found almost impossible to sabre a man lying down at the pace we were travelling.'

Despite terribly high casualties, the yeomen, equally heroically, broke right through the battery, riding down the gunners, sabreing numbers of them, and then hurled themselves, by now perhaps only twenty in number, against the machine guns. These were taken a few seconds later by Albright's Worcesters as they swung to the right. Most of the Turkish foot soldiers, possibly 200 in number, when they saw this second charge bearing down on them, quickly broke and fled, a few stopping to take pot shots at the yeomen who managed to cut down quite a number. They probably thought that the troops opposed to them were far more numerous than they were. The fact that they were not ruled out any question of a pursuit, but at this moment the machine-gun sub-section which had followed the squadrons arrived on the scene and turned its two guns, as well as the four captured ones, on the fleeing Turkish riflemen, mowing down many of them. Some seventy were made prisoner.

While this formidable charge, lasting, from start to finish, according to one authority, about twenty minutes,[10] was in progress, Cheape had led his two remaining troops of the Warwicks off to the right, where he intercepted the 5.9 howitzer battery. This he captured complete, as well as the abandoned camel-pack mountain-gun battery. At this moment Lieutenant-Colonel Williams returned from his mission to bring up the 4th

Australian Brigade which failed to reach the scene of action in time to take up the pursuit. He found a horrible scene of carnage and in its midst the three remaining officers of his regiment arranging the defence of the captured ridge with the few unwounded men who remained. He was helped by the Warwicks' Second-in-Command who brought to the task the few men of his regiment who had been unhorsed or outpaced in the charge. The position was consolidated and the 60th Division, meeting little further opposition, was at once able to establish itself three miles north-west of Huj.*

> 'Suddenly,' noted the Worcesters' Medical Officer as he rode up to the battlefield, 'the terrific din of shrieking and exploding shells ceased and we knew the end had come. A wonderful and terrible sight met our view. . . . The ground was strewn with horses and fallen yeomen, many of whom were lying close to, and some beyond, the batteries. . . . [The guns] were in various positions surrounded by Austrian and German gunners, many of whom were dead or wounded. . . . Our squadrons had not fired a shot and every single casualty we inflicted was caused by our sword-thrusts. Our Second-in-Command had fallen wounded under a gun and was on the point of being dispatched by a gunner with his saw-bayonet when a yeoman from the former's old squadron killed the Austrian . . . .
>
> 'We commenced to dress the wounded at once and found them scattered in all directions. Wounded Turks came crawling in and one could not help contrasting their clean wounds caused by our sword-thrusts with the ghastly wounds sustained by our men from shell fire and saw-bayonet. Part of a Turco-German Field Ambulance, which had been unable to escape, was found in a hollow behind the batteries, and their equipment was invaluable to us, as our dressings soon ran out

*At about this time two squadrons of the 9th Regiment appeared on the right and, according to the historian of that regiment, 'Lieutenant Mueller with a patrol shot down the team of a 4-inch gun and forced the escort to retire, leaving a second gun about 400 yards to the right. A team of bullocks was captured and hitched to the captured gun, but all efforts to shift it failed, probably due to the fact that the Turkish bullocks did not understand the language of the wild and woolly Australians or the terms of endearment that were lavished on them.' (Darley, 103)

> and our Field Ambulance had not yet arrived; the Turkish orderlies were put to work amongst their own men and the intelligent German sergeants proved quite useful.'[11]

It seems that a majority of the grievous casualties sustained by the yeomanry were caused by machine-gun fire. The exact total is difficult to establish but estimates vary between a minimum of seventy and a maximum of ninety, not all of them suffered in the actual charge. All three leaders, Albright, Valintine and Edwards lost their lives, and Wiggin was wounded.* Out of about 170 horses, between 100 and 140 seem to have been actually killed and others were wounded or missing – a horrifying total.

It is hard to say whether or not what Cyril Falls has called 'a monument to extreme resolution and to that spirit of self-sacrifice which is the only beauty redeeming ugly war'[12] was justified by the limited, but successful, object which it achieved. It does appear to be unlikely that Huj could have been occupied that evening had the charge not taken place. It is worth noting that never again in the campaign, except very occasionally against demoralized troops, was a mounted charge made without some measure of fire support or the backing of a second line, or both. Though in public all the commanders were warm in their praise and numbers of decorations were awarded to the survivors, it is difficult to believe that Chauvel, particularly, was not dismayed by the excessively high cost. It is difficult to decide whether the charge could not have

*The Worcesters' casualties were two officers and approximately seventeen men killed, four officers and about thirty-five men wounded. (Cobham, 132). The Warwicks lost three officers wounded – Captain Valintine died of his wounds, but not before he was awarded the Military Cross (which Chauvel now had the authority to award) – about thirteen men killed and some twenty wounded. (Adderley, 130, 136). Some of the men in both regiments later died of their wounds.

Lieutenant Alan-Williams was one of the officers wounded. He spent the night in an infantry field ambulance and next morning travelled 'twelve miles in a cacolet on camel. Never again if I can avoid it,' he wrote. There followed two stages from Sheria in motor ambulance and a twelve-hour rest at Amara, where 'a hospital train was run into the camp which was pitched on a spot on which three weeks previously we had been doing night outpost duty. A wonderful achievement. Kantara was reached next morning and after spending twenty-four hours there we arrived [at 17th General Hospital, Alexandria] after a journey of ninety-six hours – filthy.' (27 November, 1917, Alan-Williams, 5a) achieved its purpose equally well had it been postponed until at least the Gloucesters and the full machine-gun squadron, if not part of the 4th Brigade, could arrive. It seems probable that had Chauvel, not Shea, been in a position to give the order, he would not have done so before some sort of supports became available; but all this is conjecture. What is certain is that the capture of eleven guns, four machine-guns and about seventy prisoners, as well as the killing and wounding of numbers of not easily replaced Austrian and German artillerymen, not to mention numerous Turks, by some 120 horsemen armed with swords was, by any martial standards, an outstanding exploit.

20

'I passed within a few miles of Ascalon and watched a battle from Ashdod!'

CHAUVEL to his wife

'The inevitable delay caused by the necessity of resting our cavalry now gave the enemy the opportunity to collect his scattered forces and organize some sort of line of resistance. Already on the 10th of November, his troops could be seen digging in along the high ground on the right bank of Nahr Sukereir, and aeroplane reports indicated that he was preparing a second line farther north.'

* * *

'One of the chief difficulties of the Corps Commander [Chauvel] at this time, and one which increased daily, lay in the fact that one or another of his brigades was always on the verge of coming to a standstill owing to the exhaustion of its horses. This fact compelled the continual movement of brigades from one part of the line to another, to relieve others unable to carry on the pursuit, thus increasing the fatigue and distress of the horses.'

LIEUTENANT-COLONEL THE HON. R.M.P. PRESTON in *The Desert Mounted Corps*

'The Commander-in-Chief had hoped by 9 November [to reach] both Et Tine and Junction Station, but difficulties in regard to water had baulked the execution of his plan. He was now advancing slowly, and in large measure with infantry, over ground which he had hoped to cover with horsemen at the trot. When the time came to fight for the Holy City, a dear price was to be paid for the inevitable delay.'

GULLETT in the *Australian Official History*[1]

Pursuit continues – Huleikat taken – enemy panics at Et Tine – action at Balin

The Plain of the Philistines was now full of hopelessly confused and despairing Turkish infantrymen, perhaps 15,000 of them, streaming northwards. On the spot to get at them and force them to surrender there were, in the centre, virtually only the rest of the yeomanry from 5 Brigade and the three regiments of 3 Brigade. Neither, though, was sufficiently strong or fresh to brush aside the unwavering staunchness of the German machine gunners. A few successes in the way of captures were made but, as Major L.C. Timperley of 10 Regiment wrote, the enemy 'hung on to their guns and waggons and brought the machine guns into action whenever we threatened them.'[2] Contributing during this afternoon to the demoralization of the Turks were twenty-eight bombers which machine-gunned as well as bombed them. Large numbers of German aeroplanes had just been diverted from Mesopotamia, but they did not arrive in time and most of them were still in their cases. Those which were not bombed on the ground were destroyed by the retreating enemy. After 31 October few German planes were seen in the air. No 1 Australian Flying Squadron did splendid photographic work at this time. They obtained detailed pictures of the enemy's formations and each evening formation headquarters received up-to-date maps based on them.

During the night of 7 November Chauvel had begged Allenby to return to him if not all of Barrow's detachment from the east, at least Yeodiv. To this plea the Commander-in-Chief yielded, but the orders were not sent till the afternoon of 8 November. Since, when they were received, Barrow was still dealing with the Turks near Hebron, it was not until two days later that the division, after an exhausting march, short of water and in a raging khamsin, could take over Descorps's right flank.

An intercepted wireless message received on the morning of 9 November indicated that a counter-attack had been ordered from the Hebron direction, but Allenby, rightly, believed that the enemy troops there were too disorganized to make it. His only action in consequence was to move the Camel Corps to a position from which it could be on the flank of any attack which might develop.

Meldrum's New Zealand Brigade was not ordered to rejoin Descorps till later. Consequently it did not leave Beersheba till 4.30 p.m. on 11 November. With an overnight watering stop it reached Hamame at 11 a.m. next morning. It had covered fifty-two miles in eighteen-and-a-half hours, the Auckland Regiment marching an extra eight miles because it was near Khuweifle when the order was received. The Camel Brigade had rejoined Descorps at Julis in the afternoon of the 11th.

The capture of Huj on 8 November yielded hardly any water supplies. The enemy had destroyed all but two of the wells and these were very deep. The winding gear had been destroyed. For this canvas buckets attached to telephone wire were a poor substitute.* Of Ausdiv, 3 Brigade had not been able to water since the previous day and the yeomanry of 5 Brigade since late on 6 November. Many had to spend the night marching to Jemmame and when they got there to share the limited supplies with Anzacdiv. Because of the crush, many failed to get a drink at all. The men, as well as their mounts, suffered severely from thirst. No part of either Ausdiv or Anzacdiv could therefore be said to be by dawn on 9 November fresh and ready for the pursuit ahead, and, indeed, only Anzacdiv was able to advance at all that morning.

Chaytor's men were exhausted, particularly by sleeplessness, but at least their horses had been watered, yet when they moved forward they did so without rations for themselves or their mounts. These eventually caught them up near Kaukabah, 'where the screen of the 7th [Regiment] galloped a convoy of 110 wagons and took 390 prisoners'.[3] Huleikat, which had been Kress's headquarters, when captured, yielded large quantities of freshly baked brown bread, which the men consumed as they rode onwards. In the evening, 130 Turks with bayonets fixed, 'moved towards two troops of the 7th', about forty men in number, who had just lit their camp fires. Major T.L. Willsallen, who commanded them, 'dashed forward and shouted in Turkish "You are surrounded!"'[4] A short parley followed before all the Turks surrendered, indicating an advanced state of demoralization. Cox's 1 Brigade, meanwhile, on the left had pushed on to a village near Sdud, the ancient Ashdod, where it bivouacked. It had covered sixteen miles as the crow flies, but the main forces of the Turkish VIII Army had escaped the pincers formed by the advance of XXI Corps on the left and Descorps on the right. 'Those captured on [9 November] were either the weaklings . . . or else bodies of transport in which the galled and starving beasts could respond no longer to the flogging of their drivers.'[5]

*Kress and his staff had had to make a speedy exit from his forward headquarters at Huj. Among the mass of equipment captured was their wireless code book. This proved of inestimable worth later on.

Mediterranean
Sea
Philistian Plain
Judaean Hills
Vale of Ajalon
Ayun Kara
Surafend
Ludd
(Lydda)
El Ramle
Kubeibe
Zernuka
Yebna
W. Jamus
Aqir
Abu Shushe
El Mughar
Sidun
Beshshit
Qatra
W. Qatra
Nahr Sukereir
Junction
Station
Burqa
Mesmiye
Sdud
(Ashdod)
Et Tine
El Qastine
To Jerusalem
Tel el Turmus
Balin
Hamame
Julis
El Mejdel
Ashkelon
Kaukabah
Burbera
Beit Jerja
Huleikat
Beit Jibrin
Deir
Sineid
Tram line in 1917
Beit Hanun
Tel el Hesi
Huj
Jemmame
GAZA
To Beersheba
Miles
0
5
Montaret el Baght

PURSUIT AFTER THIRD GAZA, NOVEMBER, 1917

Since the fall of Gaza on 7 November, XXI Corps had not been able to advance very far along the coast because its transport had not yet been returned from XX Corps. On 8 November it encountered stiff resistance and its small body of Corps mounted troops, though they performed well, could not make much impression on the enemy. Like their comrades to the east they suffered from lack of water and, further, the horses were severely tried by excessively soft sand. Early in the morning, the Imperial Service Cavalry Brigade (Hyderabad, Jodhpore and Mysore Lancers) in attempting to carry out its orders to make touch with Ausdiv so as to cut off the enemy remaining in the trenches to the east of Gaza, advanced on Beit Hanun. From there, having captured a 150 mm howitzer and masses of ammunition, it had to withdraw to near Gaza for lack of supplies, especially water. At 3 p.m. it joined up with 12 Regiment. The composite regiment of Corps cavalry, three squadrons strong (one each of the Royal Glasgow, Duke of Lancaster's and Hertfordshire Yeomanry Regiments), also suffered from lack of water, neither men nor horses having had a drop for thirty-six hours.*

Though the pursuit of the mounted troops had fallen very far short of Allenby's and Chauvel's hopes, for reasons which have been made clear, the moral effect of the Australian and yeomanry actions and the threat of further ones was startlingly great.† In the

*Since they were thus immobilized, 'it was an improvised mounted party of nine grooms and signal orderlies from 156th Brigade headquarters . . . which reached the mounds marking the site of [the historic town of Ashkelon], found the place unoccupied, boldly pushed on to Majdal and handed that large village and its stores over to the troops of A. & N.Z. Mounted Division. The Turks had fallen back . . . and patrols reported Beit Jerja, Burbera and Huleikat all clear of the enemy.' (Falls, 138)

†After the fall of Jerusalem, Wavell, acting as liaison officer between the War Office and the EEF, was sent to London with Allenby's report, largely composed by himself. He later wrote:

> 'I had written a few sentences of criticism of the failure of the cavalry to cut off the enemy in the pursuit after the Gaza-Beersheba battle. Although Allenby himself had, I know, been critical of the slowness of the mounted troops, he told me in no unmeasured terms that I had no business to criticise them: "Had I ever commanded mounted troops in a pursuit?" etc. I had the temerity to stick to my guns . . . . I found myself at the end of [the argument] pouring with perspiration though it was not a warm evening. After A. had finished with me he said, "Now come along and have a drink".' (Connell, 133)

afternoon of 9 November a major panic occurred at Et Tine, the VIII Army headquarters. 'Suddenly,' wrote Kress, 'news spread that hostile cavalry . . . was moving against Et Tine. Although this rumour was false and fantastic, yet it caused such agitation that many formations began to retreat without orders . . . . A great number of officers and men could not be stopped till they had reached Jerusalem or Damascus . . . . Not only was all telegraphic and telephonic communication destroyed, but almost all the horses of Army headquarters were stampeded so that it was left unable to send out orders.'[6]

* * *

The failure to cut off the main enemy forces north of Gaza meant that a manoeuvre of interception had been reduced to one of pursuit. This, as has been shown, was due, in the words of Allenby's despatch, to problems 'of supply rather than manoeuvre'. The mass of the mounted troops were now thirty-five miles in advance of railhead at Deir el Belah. The two enemy railways (see map p. 186) were useless as many bridges and culverts had been destroyed. Motor lorries and camels moving along the narrow only partly metalled road between Gaza and Junction Station were virtually the sole means of supply. Though lorries were limited to loads of one ton, they often got stuck in the deep sand.

For fodder, the daily ration was only 9½ lbs of grain a day, invariably consisting of gram (pulse) to which the horses were unaccustomed except in small doses mixed with other corn. There was no hay or other bulk fodder. In all, the horses of Descorps alone required more than 100 tons.* Some of the horses of 2 Brigade and 7 Mounted Brigade had no proper drink for *eighty-four hours*. When Ausdiv arrived at Tel el Hesi early on 10 November, after a night march – the only one made during the

*As soon as the mouth of the Nahr Sukereir had been captured the Royal Navy began landing stores there. The supply situation was thus much improved. Without this help it would have taken much longer than it did for two more infantry divisions to be brought up to the quickly moving front line so that they could take over when the mounted troops forged ahead.

pursuit – it found several large pools of good water and the horses got their fill at last.*

It became clear on 10 November that 'the remainder of the Turkish army which could be induced to fight was making a last effort to arrest our pursuit south of the important Junction Station.' All efforts were therefore directed towards the capture of that place 'as early as possible, thus cutting off the Jerusalem army'. On 10 and 11 November progress was painfully slow, the wells being a chief target of the enemy's artillery. Those which were not destroyed or badly damaged were seldom less than 150 feet and sometimes 250 feet in depth, and it took an unconscionable time to draw the water from them. 'But by the evening of the 11th favourable positions had been reached for a combined attack.'[7] The following day had to be devoted mainly to preparations for the advance on 13 November. Amongst these was the watering of horses, many of which managed to get a second drink a few hours after their first. This always proved especially beneficial when hard work lay ahead. On one occasion a few days later 3rd Brigade made a halt at a productive well. Its horses, which had been without a drink for fifty-six hours, were at first allowed only a small drink, 'but during the remainder of the day they were watered five times and by evening appeared to be quite fit and ready for more hard work.'[8]

On 12 November the Turks made a determined attack upon, chiefly, the 5th Mounted Brigade, which at one point was almost surrounded, its horse battery in danger of capture. A skilful retirement made possible largely by the brigade's machine-gun squadron, whose commander Captain Eludyr Herbert of the Gloucesters was killed, just managed to save both guns and yeomen.[9]

The immediate objective of the advance on 13 November was the line Beit Jibrin, through El Qastine, to Burqa, which was believed to be held by about 13,000 men, most of them high-quality Anatolians. It extended for twelve miles. The final objective for the day was Junction Station. Once this vitally important point

*During the night march 'the advance guard [3rd Brigade] dropped pickets along the route every quarter of a mile, which were picked up by 5 Brigade. This brigade, in turn, dropped pickets to be picked up by the rear guard [4th Brigade]. Signallers with lamps were sent by the two leading brigades on to every prominent hill top during the march, to flash the letters of the divisional signal call intermittently.' (Preston, 61).

had been reached, Allenby meant to strike eastwards and, crossing the Judean Hills, enter Jerusalem. On 11 November he received two telegrams from Whitehall. The first reminded him that at any moment he might be required to send some of his divisions to France, while the second made it clear, notwithstanding, that the War Cabinet did not want to interfere with his plans for securing Jerusalem. He promised in reply not to commit his forces too deeply. Beyond the capture of Junction Station he was not thinking in terms of a firm time-table.

The main attack on 13 November was to be carried out by XXI Corps. Protecting its left flank along the coast, Yeodiv was to be followed by Anzacdiv. Guarding its right flank was to be Ausdiv. Allenby in his despatch portrays the situation in the morning of 13 November:

> 'The country over which the attack took place is open and rolling, dotted with small villages surrounded by mud walls with plantations of trees outside the walls. The most prominent feature is the line of heights on which are the villages of Qatra and El Mughar, standing out above the low flat ground which separates them from the rising ground to the west, on which stands the village of Beshshit, about 2,000 yards distant. This Qatra–El Mughar line forms a very strong position, and it was here that the enemy made his most determined resistance against the turning movement directed against his right flank. The capture of this position by the 52nd (Lowland) Division, assisted by a most dashing charge of mounted troops . . . was a fine feat of arms.'[10]

21

'The action [at El Mughar] must always rank as a notable example of the successful employment of all arms on a small scale. Its conduct was as bold and skilful as its execution was gallant.'

CYRIL FALLS, in the *Official History*

'[The action at Abu Shushe] is considered by some authorities to have been a more brilliant exploit even than the charge at El Mughar, owing to the fact that the rocky hills up which the charge was made appear impassable for horses except at a walk.'

LIEUTENANT-COLONEL REX OSBORNE in the *Cavalry Journal*[1]

El Mughar – Junction Station entered – Abu Shushe – Jaffa entered – end of pursuit over Philistian plain – Turkish VII and VIII Armies divided

The country in the area where the action at El Mughar, which Allenby rightly called this 'fine feat of arms' was executed, is extensively cultivated. The numerous villages are built on the tops of hills where the rock outcrops prevent cultivation. They are usually surrounded by cactus hedges and make excellent fortified positions, especially if occupied by clusters of machine guns. When, at about 8 a.m. on 13 November, 8th Mounted Brigade approached the village of Yebnah, it found that, untypically, it was only lightly held. This was discovered when, as was almost universally the practice during rapid advances, two troops reconnoitred at the gallop to one side and another two to the other side of the village. In this case they were fired on by so few Turks that they could signal back for the brigade to advance.* Pushing through the village they came to two others, Zernuka and Kubeibe, to the extreme right flank of the enemy position. This time they

*'If the village proved to be strongly held the few men in the exploring troops, moving in extended order and at a very fast pace, seldom sustained many casualties, while they nearly always succeeded in gaining a fairly accurate idea of the numbers of the enemy, the location of his machine guns, etc.' before retiring. (Preston, 78).

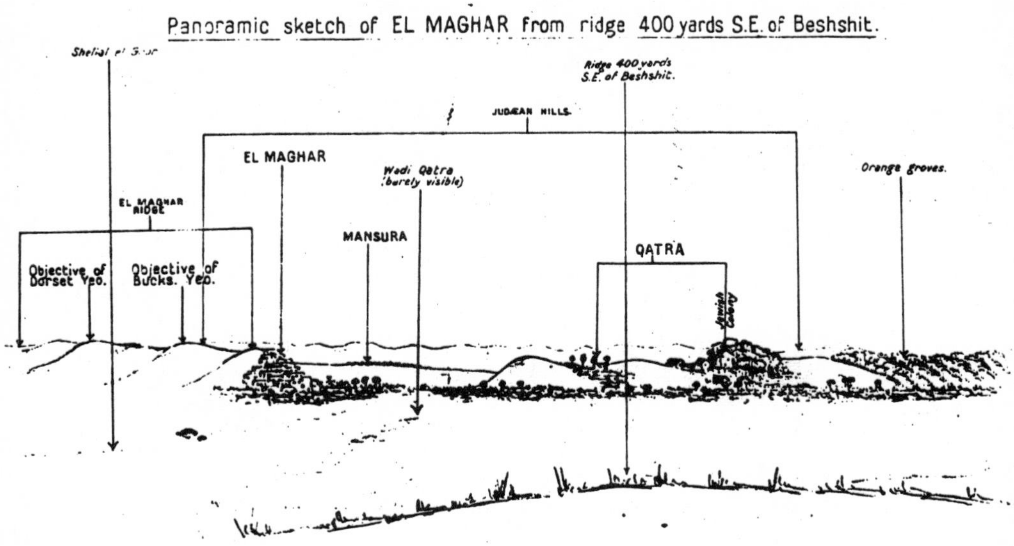

THE QATRA—MAGHAR POSITION.

were greeted with heavy machine-gun fire and were held up. In the meantime Beshshit, Tel el Turmus and El Qastine had been captured by the infantry. But 75th Division soon had a hard fight for Mesmiye, while 52nd Division found its way absolutely barred by heavy fire from the villages of Qatra and El Mughar. These are divided by the Wadi Qatra which separates the ridge on which the villages stand. A maze of allotment gardens surrounded them and south of Qatra there was a small Jewish settlement with large orchards. This was the very first glimpse of western civilization seen since the campaign had started.* Between the two villages and Yebna and Beshshit lay a flat valley. In it the only cover consisted of the beds of the Wadi Jamus and its tributaries.

Since the infantry could make no progress throughout the morning, at about 1 p.m. a quarter of an hour's concentrated

*A party of officers of the 10th Regiment entered 'a sort of village inn' in one of these Jewish villages. Addressing a young girl first in Arabic and then in French brought no response until the girl 'with a merry laugh' said: 'Cut it out. I can speak English all right and what's more I know Perth and Wellington Street and Barrack Street and all the rest of it. Good old W.A. [Western Australia] will "do" me!' (Olden, 209)

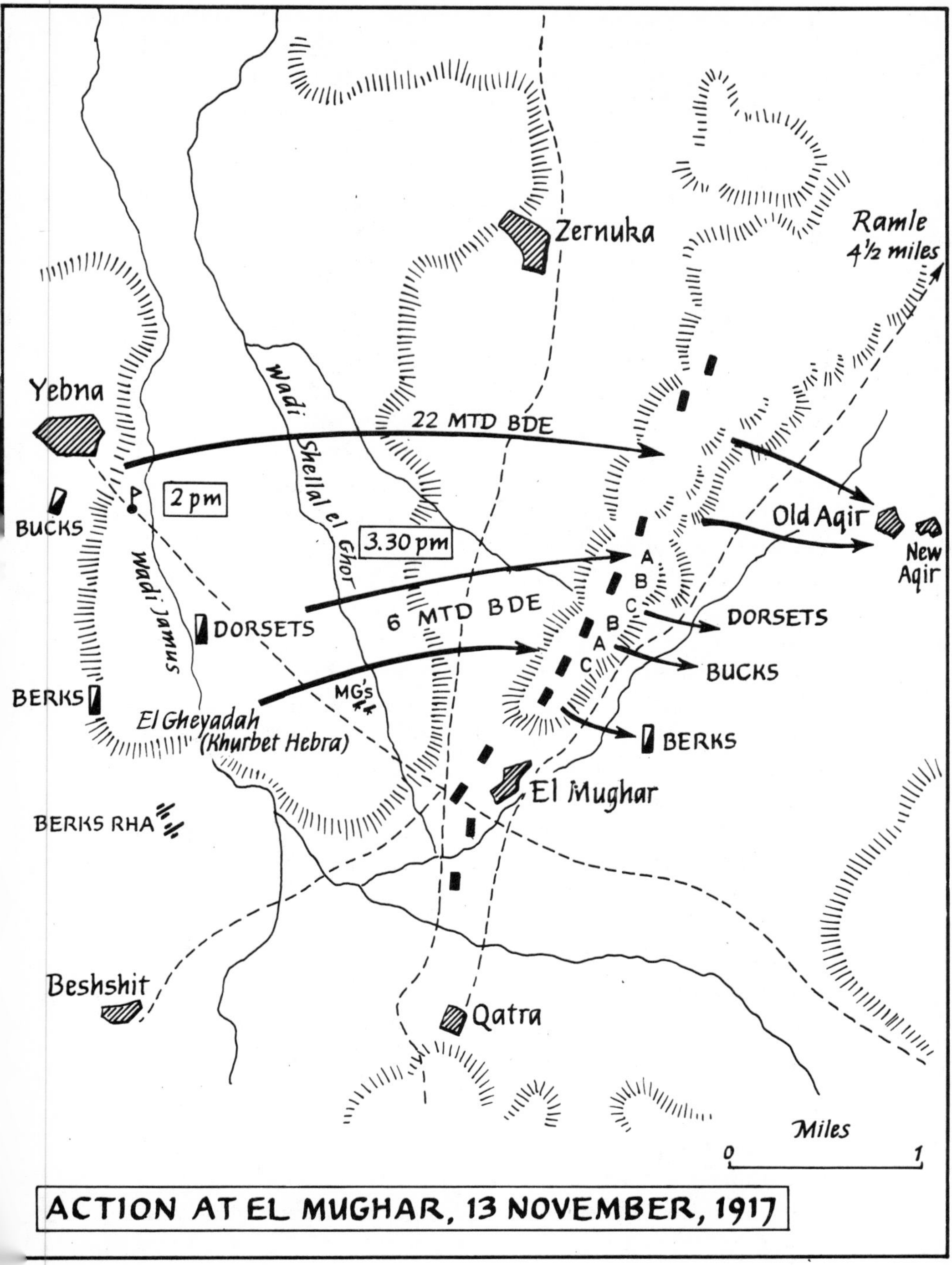

ACTION AT EL MUGHAR, 13 NOVEMBER, 1917

bombardment of the ridge to take place at 3.30 p.m. was decided upon. At about 2.30 p.m. Major-General John Hill, commanding 52nd Division, went over to Yeodiv's headquarters south of Yebna, saw Barrow, its commander, and asked for support on his left against the El Mughar ridge. Barrow at once telephoned to Godwin, commanding 6 Mounted Brigade and ordered him to 'advance and take El Mughar'.[2] Godwin, however, seeing that 52nd Division was held up and foreseeing Barrow's order, had already moved the Bucks Hussars Yeomanry, less one squadron, into the Wadi Jamus which ran almost due south from east of Yebna. This narrow ravine, though difficult to enter and emerge from and broad enough to take a single rank of horses and men only, furnished excellent cover. Godwin moved his own headquarters into it and soon ordered the Dorset Yeomanry 'to dash forward in small bodies and take up a position on the left of the Bucks'.[3] During this 'trickling' operation they came under shell and machine-gun fire. Godwin kept the Berkshire Yeomanry in support some half a mile in the rear. He had also ordered the Berks Battery to unlimber in the cover of a rare clump of trees west of the wadi, from where it could open fire on El Mughar at a range of 3,000 yards.

Meanwhile Lieutenant-Colonel the Hon Frederick Heyworth Cripps, commanding the Bucks, despatched Lieutenant C.H. Perkins to reconnoitre along the line of the Wadi Shellal el Ghor running parallel to the ridge. Perkins 'cantered up and down under a hail of machine-gun fire which, in the words of an eye-witness, "followed him as the spotlight follows a dancer on stage", seeming to bear a charmed life.'[4] Perkins's account of this reconnaissance is slightly in conflict with that of the *Official History*:

> 'Having been taught in my days with the 7th Reserve Cavalry at Tidworth,' wrote Perkins, 'that cavalry charges should not exceed about 600 yards, I had in mind looking for cover for led horses whence a shortish dismounted attack could be made.
>
> 'Although the map showed a depression over a mile away between us and the Turks on the ridge, it wasn't obvious [that it was] deep enough as cover for the led horses.
>
> 'I took my groom with me in case of casualty and went off at a canter in a largish left-handed arc till we could clearly see all the ground up to the ridge. On our immediate front there was no such cover and on my return, somewhat blown, I said

so with some relief, I remember, for judging by the resultant noise [the position] created, especially from El Mughar village on my right, it was obviously defended by an awful lot of Turks and machine guns.

'I thought of course that due to lack of cover for led horses, they might not try it! However, after what seemed an awful long pause, the General [Godwin] said "Then – we'll gallop it".'

Cripps now got in touch with the battalion commander whose men were in the wadi further east. Yeodiv's official report states that Perkins, on returning unscathed from his reconnaissance, reported that there was an excellent covered position near Khurbet Hebra from which the 17th Machine Gun Squadron could support the attack, bringing its fire to bear 'for the longest time possible before being masked by the advancing troops'.[5] Perkins's own account, however, does not mention this.

Godwin now gave verbal orders to his commanding officers. The Bucks were directed to a point about 1,000 yards north of El Mughar, while the Dorsets were to aim for the hill on the Bucks' left, both attacks starting simultaneously at 3 p.m. The two regiments were to advance in column of squadrons extended to about five paces, the squadrons of each regiment following one another at about 200 yards. Moving on the outer flanks, two machine guns on their pack horses were to accompany each regiment. The Hotchkiss sections were also to ride with the squadrons. The Berks were to enter the Wadi Jamus as soon as the attackers had vacated it. These orders were given at 2.30 p.m. This left half an hour for the Berks Battery to register and for the commanding officers to issue their orders.*

Punctually at 3 p.m. the Bucks and Dorsets clambered out of the steep bank of the Wadi Jamus (see illustration on p.193).† The instant that they appeared they drew artillery fire. They had

*While Cripps was giving out his orders a message arrived from 52nd Division 'suggesting that a flanking movement by the cavalry north of El Mughar would enable the infantry to advance. A reply was sent to the effect that the brigade would advance at 3 p.m.' (Yeodiv's official report, Osborne: I, 372)

The 8th Mounted Brigade, which was attacking the village of Zernuka (see p.191), was to afford protection to the left flank of the charge.

†Because of their relative positions in the wadi, 'A' Squadron of the Bucks had to cross the Dorsets' front to rejoin its other two squadrons.

between 3,500 and 4,000 yards to ride without a shred of cover. For the first mile they kept to a trot. When they were about half a mile from the enemy they broke into a canter, swinging up the rocky slope. From now onwards they were subjected to sustained if plunging machine-gun and rifle fire. About 100 yards from the top Captain John Crocker Bulteel,* leading the charge at the head of 'B' Squadron, gave the order to draw swords and charge, 'and the men sat down and rode'.[6]

> 'As the minutes passed,' wrote Perkins, who commanded 3 Troop of the Squadron, 'and things got hotter, we were holding in galloping horses – then swords were drawn and we were breasting the rise with what seemed like pillows hitting our faces – the Turks firing. In seconds there they were, mobs of them, some shooting, some running out of their slit trenches and some sensibly falling flat on their faces.
>
> 'Blown and galloping horses are not exactly "handy", so with a sword in one hand and hindered by clutter of equipment it was nearly impossible to get at a dodging Turk. One missed and missed again until the odd one wasn't quick enough. In just such a case the hours of arms drill paid off for instinctively one leant forward and remained so, to offset the jerk as the sword comes out – in fact, precisely as one had been warned.
>
> 'By then those of us (man and horse) who had got there were among a seething mob – such as a mounted "Bobby" in a football crowd – shouting at them to surrender in the only words we knew – sounding like "Tesleem Olinerz". Years later I learnt the translation in Cyprus. It was "Hands up the lot of you!" All this shouting didn't seem much good – probably because on our bit of the front only about nine blokes had by then arrived.'[7]

Meanwhile Captain the Hon Neil Primrose, leading 'C' Squadron, had come up on 'B' Squadron's right. The leading troops, soon after emerging from the wadi, were amazed to see a fox going away between the two opposing lines. A chorus of view halloas greeted this apparition, 'perhaps,' as Cyril Falls has put it, 'taking some minds back for a fraction of time too small to measure to the pleasant Vale of Aylesbury at home.'[8]

*Who later became Secretary to the Ascot Authority.

Bulteel's squadron tore through the Turks, killing a number with their swords, and then galloped right over the crest before they could pull up. Primrose's squadron was checked, since though many of the Turks threw down their arms as the summit was reached, numbers, seeing how few the attacking horsemen were, soon returned, 'got into the knolls on the flanks'[9] and commenced a fierce fire fight at close quarters. Captain Lawson's 'A' Squadron arrived during this fighting and with his aid the ridge was taken. 'B' Squadron, meanwhile, was playing on the rear of El Mughar with machine gun and Hotchkiss. A little later on some of the machine guns which had been captured were also employed against the village. On the extreme left 'A' Squadron of the Dorsets, commanded by Captain G.M. Dammers, arrived at the base of the ridge with his horses so blown, having had to ride further than the other squadrons, that he decided to dismount, send back his horses and advance to the crest on foot. On Dammers's right, Major R.G.S. Gordon's 'C' Squadron and Major F.J.B. Wingfield-Digby's 'B' Squadron were led forwards by the Dorsets' commanding officer, Sir Randolf Littlehales Baker, Bt. These, like the Dorset's squadrons, charged right up to the crest of the ridge. The six machine guns, which had throughout the charge been giving effective covering fire, now moved to the summit with great speed and 'fired on distant targets'.[10] The Berks Battery expended 200 shells during the charge. Though the mass of the ridge had been cleared of the enemy and consolidated, bodies of the enemy in El Mughar itself continued to hold out. Godwin therefore brought up his support regiment, the Berks Yeomanry, and ordered it to clear the village. At the same time the infantry on the right, which, on seeing that the enemy had turned its guns on the charging yeomanry, showing commendable initiative, had dashed six hundred yards to the allotment gardens and were closing in on them. The Berks met the infantry in the northern outskirts and together they gathered in some 400 prisoners to add to the several hundred taken by the Bucks and Berks in their charge. This was a classic example of fruitful cooperation between horse and foot. The village was not entirely cleared until 5.30 p.m.*

*A sharp controversy raged over whether the infantry (Scottish Borderers) or the yeomanry captured the village. The former claimed that they had done so and that the latter had merely collected the prisoners. (Falls, 172)

As a result of the action at El Mughar, eighteen enemy officers and 1,078 other ranks, together with fourteen machine guns and two field guns were captured. The enemy's casualties probably totalled well over 2,000. 6 Mounted Brigade suffered the death of one officer and fifteen men (eight from the Dorsets, six from the Bucks and one machine gunner). There were 107 wounded, with one man missing, a total of 130 or about 16%. The brigade's losses in horses were 265 killed and wounded, about 33%. It is interesting to note that the Dorset's squadron which dismounted at the foot of the ridge suffered considerably higher casualties, especially among the led horses, than the squadrons which charged mounted. Perkins's comments on the charge were:

> 'One may well ask why Johnny Turk acknowledged to be a tough customer was overcome in perhaps 10-15 minutes by galloping cavalry? Imagine however his position.
>
> 'A line of horsemen appears out of the wadi 2 miles away, but too far to shoot at. Soon they are in range but by then another line of horsemen emerges and apparently your fire and machine guns show little result. In your growing anxiety, your aim is dodgy to say the least and again a third lot of horsemen emerge. It is all very quick and even in those days when most men were used to horses the galloping onrush is frightening – so do you stand pat or run?
>
> 'Foolishly in their indecision, many got out of their slit trenches – shooting or running. Had they sat tight they would have been an almost impossible target for troopers' swords on galloping horses and our success could not have been so sudden.
>
> 'For just these reasons – as a cavalry charge – it was certainly unusual but nevertheless highly successful due to surprise & speed.'[11]

If Katia was the nadir of the yeomanry's performance in the campaign, El Mughar nearly reached the zenith, with Huj, perhaps, half way between. El Mughar was a model of what could be achieved by cavalry charging in extended formation, its objectives pointed out on the ground and supported by both artillery and machine-gun fire.

* * *

23. General Sir Archibald Murray (see p.37)

24. General Sir Edmund Allenby (see p.119)

25. Djemal Pasha (see p.25)

26. Colonel Kress von Kressenstein (see p.25)

27. Major-General Chaytor with Ali Bey Wahaby, Commander, Turkish II Corps, 29 September, 1918

On this eventful 13th November much else was going on. To the south, while the yeomanry's charge was being delivered, another infantry battalion, after a stiff fight, took Qatra. To the north, Brigadier-General Fryer's 22nd Mounted Brigade, the East Riding Yeomanry leading, with the Staffords behind, had been ordered to secure the extreme northern part of the ridge. They were just crossing the Wadi Jamus west of Yebna when they saw the Bucks and Berks attaining the crest. The sight which met them when they, in turn, topped the ridge was an astonishing one. The whole steep eastern slope was thick with the running figures of hundreds of Turks racing for the village of Old Aqir. Major J.F.M. Robinson, who was in command of 'A' Squadron of the East Ridings, conferred urgently with two other squadron leaders (one of them of the Staffords) and decided that he would try to seize the village, while the other two cut off the fleeing enemy to the north and north-east. At his immediate hand was only half his squadron and he had to leave most of that to hold the position on the ridge as he had been ordered. With only fifteen men, he raced for Old Aqir, sabreing a number of fugitives in the process, took the place, dismounted the far side of it and set up a heliograph, but could get no answer from the brigade. To the east of Old Aqir, in New Aqir, a Jewish village not yet included on the maps, was Refet Bey, the commander of the Turkish XXII Corps. Believing that the whole front had broken, he collected his staff, numbering about a company, concealed some of them in a hidden wadi and opened fire on Robinson's gallant band and on the other two squadrons. This brought them to a halt.* Since night was now falling, communication with brigade non-existent and with New Aqir apparently strongly held, the squadron leaders realized that the village could not be taken that night. Shortly afterwards Fryer sent an order for their return. The ridge was to be held during the night.

It seems very likely that, had 22nd Brigade been pushed on to take Old and New Aqir, there would have been little enough to stop it. Unlike 6th Brigade which, having lost so many horses, was not in a position to pursue, the 22nd was more or less fresh. Major Robinson believed that two regiments would have done it with ease. On the enemy's side it was believed that VIII Army was facing

*Robinson's party saw Refet Bey on a white horse, 'riding up and down to organize his defence, and repeatedly fired upon him without success'. (Falls, 171).

'irretrievable disaster'.[12] The two villages were not in the event occupied until 6 a.m. next morning. This gave the Turks flying from Junction Station time to escape northwards to Ramle.*

After hard fighting on 13 November and in the early hours of the 14th, 75th Division entered Junction Station, the first to arrive there being two cars of the 12th Light Armoured Car Battery. These, firing their machine guns, chased a large body of Turks for two miles, killing or wounding more than 200. Beside the great tactical value of the station, it proved to be the very first point since the advance began where water flowed in unlimited quantity. Other inestimable boons were a large store of food and fodder, machine shops and a complete hospital, all in full working order and enormous dumps of petrol, grease and timber. At much the same time Ausdiv entered an evacuated Et Tine and the New Zealanders repulsed with a bayonet charge a vigorous Turkish attack near Surafend. On the 14th, too, the Yeomanry Division moved to a point four miles north-east of Junction Station and made touch with the Composite Cavalry of XXI Corps.†

* * *

The capture of the Mughar ridge made imminent the separation of the two Turkish armies, but a rearguard still held a prominent ridge, a mile and a half in length, near the village of Abu Shushe, which covers the mouth of the Vale of Ajalon, one of the chief passes into the Judaean hills. Further, while it remained in enemy hands this dominating position threatened the right flank of the advance on Ramle. It was a forbidding-looking mass of boulders and scrub. Barrow was ordered late on 14 November to dislodge the enemy from it next morning. The 6th Brigade, riding over ground which afforded some cover, but which from its boulder-strewn nature prevented more than an occasional canter, attacked the southern part of the ridge, while the 22nd Brigade tackled the northern part. The advance was made partly on foot and partly mounted and otherwise, except for the slower speed at which it

*The Middlesex Yeomanry had entered Zernuka at 9 p.m. on 13 November.

†The vet of the Worcesters noted that due to the number of dead horses which had had to be shot due to exhaustion and thirst, 'the shrieks of the jackals made it difficult to sleep that night'. (Teichman, 192)

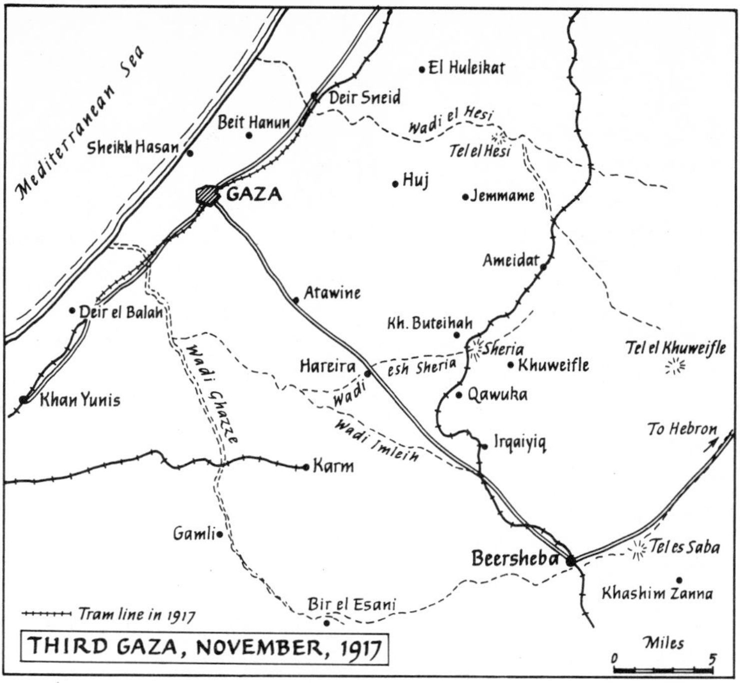

took place, much resembled the action at Mughar. Again the horse batteries (Berks and Leicester in this case) gave vital covering fire; again the machine guns protected the flanks and again when the crest was reached the Hotchkiss sections helped to mow down the flying Turks. An important difference lay in the fact that, although possibly about 4,000 in number,[13] the enemy was a great deal less ready than two days before to resist the assault, many of them deserting their posts before the yeomen could close. It appears that even their machine guns, numbers of them hidden in hillside caves, fired erratically. Certainly the Turkish riflemen did. Some 400 bodies are said to have been counted on the field and 360 prisoners were secured. The losses of the six yeomanry regiments were under fifty. Only one officer met his death: the Hon Neil Primrose, who died of his wounds two days later. At the age of thirty-five he had six months previously left his post as Parliamentary Secretary to the Treasury to serve with the Bucks.*

*Primrose was the younger son of the 5th Earl of Rosebery, Prime Minister in 1894–5, and of his wife, *née* de Rothschild. He had been an MP since 1910 and held two minor government posts before his last one: PUS to the Foreign Office and then to the Ministry of Munitions. His cousin, Major Evelyn de Rothschild, second-in-command of the Bucks, died soon afterwards from the wounds he had received at El Mughar. Thus died two members of the family which had played so great a part in founding the Jewish colonies the area of which Allenby's forces had just entered.

Throughout these closing stages of the aftermath of Third Gaza, the enemy time and again, sometimes fiercely, often half-heartedly, counter-attacked the pursuing mounted troops. For lack of space it is impossible to recount here the numerous occasions on which the Australasians in particular found themselves compelled to throw back these counter-attacks and to drive the Turks from well defended knolls. The New Zealanders, especially, on their way to taking Jaffa, which they entered on 16 November, distinguished themselves on a number of occasions by means of combined mounted and dismounted actions.* As they rode unopposed into the beautiful town of Jaffa, the horsemen plucked oranges from the wayside trees.

That the enemy was able to withdraw considerable numbers of his weary, broken troops to fight another day was due chiefly to the chronic water problem. Night after night the Turks were able to march away and sometimes make a stand next morning because their pursuers were exhausting themselves in the nocturnal search for water. On the night of 15 November, for example, the Worcesters arrived at a small Jewish colony to find a row of canvas troughs in place, but the engine for bringing up the water from 150 feet below had broken down. They moved, therefore, to another place a mile away where water was reported to exist. Here 'only four at a time could get at a single stone trough and the proceedings,' according to the regimental historian, 'were further complicated by the necessity for driving off repeated attempts to rush the well on the part of the native population.' By 6 a.m. six troops had been watered. These horses had gone thirsty for eighty-four hours. Two other troops only found water after *ninety hours*' abstinence.† Brigadier-General Kelly reported on 16 November that in the three regiments of 5th Mounted Brigade there were only 690 men mounted and fit for duty.[14]

It was now clear that unless the horses of Descorps were granted a decent rest, numbers of them would have to be destroyed, while many others would be unable to carry on. By 16 November

*At Ayun Kara, particularly, where on 14 November the attacking powers of the mounted arm against an enemy in position were brought fully into play. (See Powles, 145-50).

†This may be an exaggeration. The longest *continuous* period without water is given by Blenkinsop (208) as eighty-four hours (Lincolnshire Yeomanry).

operations had been continuing practically without intermission for nineteen days and nights. The horses had been watered, on average, only once in every thirty-six hours.* Though Junction Station was sixty miles from Beersheba, Anzacdiv had since 31 October covered 170 miles, Yeodiv 190 and Ausdiv no less than 230. When it is considered that the average weight on the back of each horse was probably about twenty to twenty-one stone, the feat of keeping enough horses, always short of fodder and water, sufficiently fit to carry on the pursuit must be regarded as one of the most wonderful in the history of the mounted arm.

* * *

With an effective wedge driven between the two Turkish armies, Allenby's great battles which had started with the bombardment of Gaza on 24 October had achieved one of their chief objects. Descorps alone had taken 5,720 prisoners, more than sixty guns and some fifty machine guns. In all about 100 guns and 11,000 prisoners had been captured. With some thousands of dead, approximately a quarter of the enemy force had been destroyed. Complete annihilation had certainly been avoided; the enemy had now reached comparative, if only temporary, safety.

* * *

As a postscript to the great pursuit over the Philistian Plain, it is amusing to note what happened to the heavy echelons of the cavalry ammunition columns in the course of it. Last heard of at Sheria on 7 November, 'advancing boldly on the enemy', they turned up at Junction Station on the 19th. For twelve days they had been totally lost, wandering about, 'neglected and forlorn, in the wake of the cavalry'. Receiving no rations they

> 'had been maintained entirely by the predatory genius of the gunner subaltern in command . . . . He brought his command of 600 horses and men into the Station, all fit and well, and

*One of the horse batteries attached to Ausdiv was able to water its horses only thrice in nine days. The intervals between waterings were sixty-eight, seventy-two and seventy-six hours. Remarkably only eight of the battery's horses were lost from exhaustion. (Preston, 94) no questions asked. And if, sometimes,' as Lieutenant-Colonel Preston puts it, 'a battalion waited in vain for its rations; if, now and then, a harried supply officer found that one of his camel convoys had delivered its supplies during the night to some unknown unit, owing to a mistake; if guards on ration dumps are notoriously vulnerable to cigarettes and soft words, one can only reflect that war is a sad, stern business in which "dog eats dog" when opportunity arises.'[15]

22

'*November 16th.* The Turks had now broken into two distinct masses, widely separated by the physical formation of the country. The Eighth Army (XXth and XXIInd Corps), consisting of four damaged and one fresh division, retired into the Plain of Sharon behind the Nahr El Auja, a river about thirty yards broad. This Army was based on the railway running down from the north. The remnants of the Seventh Army (IIIrd and XVth Corps), consisting of six shattered divisions and the 3rd Cavalry Division, were based on Jerusalem and the road running north from there. The Junction Station–Jerusalem railway was of no value to them, as it ended in a cul-de-sac at the last-named place. The only means of effective lateral communications between these two armies was by road to the north, a circuitous route of seventy-five miles. The C-in-C decided to contain the Eighth Turkish Army, to attack the Seventh Turkish Army and occupy Jerusalem.'

LIEUTENANT-COLONEL REX OSBORNE
in the *Cavalry Journal*, 1921

'No operations are to be undertaken within a six-mile radius of Jerusalem.'

XXI Corps Order, No. 14

'A complete rest was promised and we set about putting ourselves in order. Boots and puttees which had not been off for a fortnight were cast, and cramped and dirt-encrusted toes cleaned and stretched out to the glorious sunshine. Little by little and bit by bit off came those crisp curly beards and our mothers would have known us once again. Then determined attacks and onslaughts on the brigades of lice in breeches and shirt. . . . Good water for man and beast and food enough to spare. Life was "bob".'

A yeoman of the Middlesex Yeomanry
after the capture of Jaffa.

'We have had a very bad time. The breakdown of the army, after having to relinquish the good positions in which it had remained for so long, is so complete that I could never have dreamed of such a thing. But for this dissolution we should still be able to make a stand south of Jerusalem even today. But now the 7th Army bolts from every cavalry patrol.'

MAJOR FRANZ VON PAPEN, attached to Turkish General Staff, to Count von Bernstorff, German Ambassador at Constantinople, 21 November, 1917[1]

The battles for and entry into Jerusalem

The decision which Allenby now took to make straight for Jerusalem was an excessively bold one. Warned from London not to commit too many of his troops, faced with 'a country of ambushes, entanglements, surprises, where large armies have no room to fight and the defenders can remain hidden,'[2] with the winter rains about to fall and daunting problems of supply, Allenby's determination might be thought to have verged on the rash. Yet it took only twenty-one days against all odds and the Turkish VIIth Army to reach the great moral and political if not strategic prize of the Holy City. No one was more aware than Allenby, a keen student of military history, that throughout history Assyrians, Romans and Crusaders had recoiled baffled and broken from numerous assaults on the western bulwarks of Judaea.

Three days after Mughar, Allenby had told Barrow that 'he had selected Yeodiv to go into the hills and cut the Turkish line of retreat from Jerusalem along the Jerusalem–Nablus road. He added that it was really an infantry job, but the infantry could not get up in time and he knew "my fellows could do it".'[3]* The 10th Australian Light Horse was attached to 5th Mounted Brigade, replacing the Gloucesters so that the Australian forces should be in on Jerusalem's capture. Allenby intended to make certain that by speedy action the enemy could not recover from his demoralization in time to establish adequate defences. In this he only partially succeeded. On 17 November he gave his orders. 75th Division was to take the only good metalled road running from east to west,

*In 1920 Allenby wrote: 'We used to hear, especially in peace manoeuvres, that such and such a tract of country was suited to cavalry action. The truth is that cavalry can and will fit its tactics to any country. This has been shown repeatedly during the war just ended: in the rocky hills of Judaea . . . and the mountains of Moab.' (Quoted in Osborne, II, 134)

through Amwas, 52nd Division was to advance along the Lydda (Ludd)–Jerusalem road, such as it was, while Barrow was directed on Ramallah and Bire some eight miles due north of Jerusalem along the Nablus road. Chetwode's XX Corps was to remain where it had been ever since Sheria, namely recuperating at Gaza. As he wrote in his despatch, 'before our position in the plain could be considered secure it was essential to obtain hold' of that one good road 'which traverses the Judaean range from north to south, from Nablus to Jerusalem'.[4]

Since virtually all the fighting was now necessarily on foot its details need not be rehearsed here. That it was excessively severe and remarkably unpleasant there can be no doubt. The difficulties of marching through the virtually roadless hills were well nigh overwhelming. The delayed rains added to the appalling transport and supply problems. These, even in ordinary circumstances, would have been intractable enough, especially as railhead was some forty miles behind the forward troops. 'The querulous camel,' wrote Wavell, 'in conditions of cold and wet among unaccustomed hills and rocks that for once at least justified his attitude to life, was the means of transport for the great majority of the troops advancing on Jerusalem.' (see p. 208). Before the operations were over, the condition of many of these long-suffering beasts was, according to an officer of the Camel Corps,

> 'almost desperate. They had cavities on each side of their humps so deep that you could have buried a cricket ball in them. These cavities were, of course, septic and the raw flesh was alive with maggots. . . . Cruel though it may sound we had to saddle and load up the worst cases in order to preserve, if we could, those animals which had relatively minor sores by retaining them as spares.'

A peculiarly virulent form of mange contracted at this time was bestowed by the camels upon their masters, resulting in unbearable irritation. 'A special decontamination centre was established for our benefit. In the meantime,' reported an officer of the Imperial Camel Corps, we were put out of bounds, no other unit being allowed to approach us.'[5] Fortunately in October a farsighted administrative officer had got Allenby's approval for the formation of companies of donkey transport. As early as mid-August twenty-two donkeys had been issued to each regiment. From the first days of December, when some 2,000 became available and after initial acclimatization problems, they proved their worth,* for the camels constantly fell down and broke their legs, while in the plains which became vast spongy quagmires numbers of them sank up to their girths and had to be abandoned. Many were the casualties, too, among their long-suffering Egyptian drivers, 209 of them dying from exposure. On frequent occasions camel transport camps had to be pitched on windswept sites at altitudes of up to 3,000 feet.[6]† There was hardly any water to be found except what fell from the sky. This shortage, fatigue and the increasing cold affected the horses badly, numbers of them foundering. Many of those that survived became veritable skeletons. The mud was so thick in places that they were deprived of essential rest through being unable to lie down.‡ The sudden drop in the temperature also

*Hardy, tireless, indifferent to gunfire and requiring little feeding, the donkeys, known to the troops as 'Allenby's white mice', (Vernon, 129), proved invaluable for the rest of the campaign. By 31 December, 12,790 had been bought. They were particularly free from lameness and vice. There were complaints, though, that they were too slow to keep up with mounted troops.

The officers' batmen were issued with one each, but in the 5th Regiment at least 'they acquired horses and disdained to ride on donkeys. We accordingly handed our donkeys over to the Brigade Band.' (Wilson, 118)

The batmen of the yeomanry, being perhaps more observant of the rules than the Australians, kept to their donkeys. A batman of Jacob's Horse seeing 'an extraordinary procession of Yeoboys riding white donkeys at the tail of their column, shouted to a colleague: "'Ere, come and look at the Scots Greys marchin' out."' (Maunsell, 191)

Their shoes were made by civilian farriers in Egypt. 'Each donkey was shod up before leaving Cairo and was equipped with a spare set of shoes and nails.' The shoes proved unsuited for work in the hills and shoes of a horseshoe type had to be substituted. Recurrent conjunctivitis is common in Egyptian donkeys. Eye fringes were therefore issued. (Blenkinsop, 215, 231)

When the sowar batmen of the 34th Poona Horse were issued with donkeys 'at first all classes of the men and more particularly the Rajputs, strongly objected to riding these animals, which in India are ridden only by men of low caste. Before the campaign came to an end, however, the donkeys had become very popular, especially among the Rahtore Indian officers.' (Wylly, 141)

†2,090 camels died of exposure between 1 November and 31 January and 2,620 had to be evacuated to the four Camel hospitals at Zeitoun, Kantara, El Arish and Rafa. (Blenkinsop, 213)

‡The heavy rains ruined much of the forage. Over 900 tons of crushed gram was lost through fermentation. The barley began to sprout and became mouldy, while the tibben became sodden and musty. (Blenkinsop, 207)

imposed severe hardships on the men for they were for the most part still dressed in light, khaki drill and they had no blankets, greatcoats or, of course, tents.

At first the horses of Yeodiv came in useful for helping to bring up supplies, but otherwise they soon proved to be a tiresome encumbrance. The valleys soon became 'beds of viscid black mud',[7] while the ravines were raging torrents. Neither could be crossed by horses. As early as 20 November commanding officers represented that the horses were lessening rather than increasing mobility. Three days later Bulfin allowed Barrow to send back to Ramle and Lydda all his horses, except for officers' chargers, a few for carrying machine guns and for pack transport. All his horse artillery guns had to be left behind. The six screw-guns of the Hong Kong and Shanghai Battery were his only artillery support. By 20 November the fighting strength of the whole division was down to 1,200 rifles, since at least a quarter of the division's strength was employed in looking after its horses. 'Some squadrons were reduced to twenty rifles and four Hotchkiss guns in the firing line.'[8]

* * *

On 24 November there was a general pause in the fighting. The objectives set on 17 November for the first attempt on Jerusalem had not been achieved. The reasons for the pause were not the strong enemy resistance and spirited counter-attacks but even more the 'cold, wet, lack of sleep, hunger and thirst', which had 'added to the fatigue of fighting'. Indeed 'marching across this harassing ground,' as Barrow put it,' 'had brought men and animals nearly to the point of exhaustion.' During one sixty-hour period the troops had received only one half-day's ration and the horses but a few handfuls of tibben.[9]*

Vast exertions had now so improved the lines of communication that XX Corps could be brought up to relieve XXI Corps which

*On 28 November a freakish phenomenon occurred. There was virtually no darkness that night. 'The rocks shone fiery red in the light of the sinking sun, yet threw deep shadows on the same side from that of the rising moon. The signalling lamps of the 22nd Mounted Brigade had been destroyed by shell fire, but the East Riding Yeomanry was able to keep touch with its brigade headquarters by helio on the moon. It seemed that once again the sun stood still in the Vale of Ajalon.' (Falls, 227)

returned to the plain north of Jaffa. Proper rations and fodder could also once again reach the troops. By the first week in December all was set for a second attempt on Jerusalem. With Chetwode and his fresh infantry now in charge, the exhausted Yeodiv could take an immensely well deserved rest.* Its fighting strength had been further reduced to a total of 800 rifles: it had suffered 499 casualties, or over 41%, and it had lost much material. The 17th Machine Gun Squadron, for example, had only two guns left. When on 1 December Allenby came to thank the division he told the men that had they not held the field during the past critical fortnight, 'the whole army would have been compelled to give up the hold it had secured on the mountain passes, and that if this had occurred it would have taken three months' hard fighting and thousands of casualties before we should have been able to capture Jerusalem.'[10] This was no exaggeration.

Chetwode's attack began on 8 December. The next day, in the words of the Official Historian, 'at a moment when a resumption of fighting and, indeed, a stouter resistance than that of the previous day were to be expected, it was found that the enemy had vanished.'[11] Allenby, whose orders that the Holy City was under no circumstances to be fought over had been religiously carried out, made his famous entry on foot on 12 December – through the Jaffa Gate which was traditionally opened only to a conqueror. Robertson was responsible for suggesting this as a contrast with the Kaiser's flamboyant entry on horseback through a breach in the walls nineteen years before. When on the same day Lloyd George announced Jerusalem's capture, the news had an electrifying effect on morale. It followed the German counter-offensive at Cambrai, the elimination of Romania from the war, the Italian reverse at Caporetto and above all the collapse of Russia.† The fact that the city had been spared any of the horrors of war did high honour to both Allenby and the army he commanded. To his wife the Commander-in-Chief wrote: 'It was a great feat; and our losses were light.'[12]

*'During the withdrawal regimental transport took twenty hours to travel twelve miles and horses and men had the greatest difficulty in keeping their footing. . . . The black, soft soil seemed bottomless. Many animals collapsed from starvation and exhaustion and had to be abandoned.' (Blenkinsop, 207)

†The armistice on the Russian front was signed on 5 December.

* * *

After the fall of Jerusalem, to make his position truly secure, Allenby needed more elbow room. On 20 December, therefore, the infantry forced a crossing of the Auja, driving the Turks well away from Jaffa which was to be used as a port for the landing of supplies. A week later, just as a further advance in the hills was about to begin, Falkenhayn launched a counter-attack in an effort to recapture Jerusalem. Though pressed with great gallantry it was soon checked and the enemy was forced back to Ramallah, a good ten miles from the city. As Wavell put it, 'the objectives of the campaign that culminated in the capture of Jerusalem had been to frustrate the Turko-German expedition against Baghdad, to engage Turkey's last reserves of men and to invigorate the nation's resolution at a time when the apparently unbreakable deadlock in the main theatre was tending to discouragement. All these purposes had been accomplished before the end of 1917.'[13]

* * *

Between 31 October and the end of the year casualties in horses, mules* and donkeys amounted to about 10,000 or 11.5% of the total strength.

About 5,000 had died, been killed or were missing. The rest were evacuated to the six vet hospitals on the lines of communication. In January, 1918, a further 2,684 became 'dead' casualties, while 5,677 were transferred to vet units. After Gaza fell railhead for horses was for some time sixty miles in the rear. It is very remarkable that of the 3,000 invalids undertaking this journey up to 1 December only forty-five were lost in transit.

The yeomanry's horses on their return from the Jerusalem area were 'exhausted and debilitated to a dangerous extent. All horseshoes and nails had been used up or left behind to relieve transport, with the result that many animals,' as the official veterinary history puts it, 'were lame from broken and bruised feet. . . . Diarrhoea was very prevalent in all mounted formations,

*At about this time light South American draught mules had mostly replaced draught horses. Immediately after Beersheba the mules, in anticipation of the rough country towards Jerusalem, were shod for the first time. (Blackwell, 98, 106)

as a result of the animals being fed almost exclusively on gram [pulse], which was the only grain available.'[14]

Of the average strength of horses, mules and donkeys during the period July to December, 1917, just under eight per cent, 6,597, died, went missing or were sold. The figure for camels was 12.8%. 2,840 horses and mules were killed or destroyed in the field and 358 died in hospital. Of digestive diseases, chiefly sand colic, there were 17,212 cases during the period. 984 died in the field and 119 in hospital – 1.33%. In veterinary circles 'colic is a disease of work' was a well known saying. Its truth was borne out by comparing the preceding six months' losses which amounted to only 296, 0.4%.[15]

"COO-EE"

Here's a "Coo-ee", Sister Billjim,
From a Billjim over-seas,
Where there aint no scented Wattle,
And there aint no Blue-Gum trees.

We're among the wavin' date-palms,
Makin' Jacko Turkey-trot,
And send sincerest Christmas Greetings,
From this Gawd-forsaken spot.

Copyright.

A PALMY XMAS 1917.

23

'The real conflict of opinion was on the impregnability or otherwise of the Allied line in the West. If the Turks were to see the last effort of Germany definitely held, they would no doubt accept terms as soon as they suffered a fresh defeat. But should their German allies be at or near the gates of Paris, neither the occupation of Damascus nor of Aleppo would be likely to compel their surrender.'

WAVELL in *The Palestine Campaigns*

'General Smuts's mission had the merit of bringing the discussions down from the somewhat fanciful realms in which they had hitherto revolved to the solid earth of reality.'

* * *

'The Desert Mounted Corps was a more formidable weapon after the reorganization than before it.'

CYRIL FALLS in the *Official History*

'Well do I remember galloping past a Turk Battery of big guns with the gun crews standing with their hands up, and some just standing. There seemed to be no one guarding them as we swept past. They would have been taken care of by other units following.'

CORPORAL HENRY P. BOSTOCK,
a scout of the 10th Australian
Light Horse Regiment,
of the dash for Es Salt, 30 April, 1918

'To us fresh from the climate and the flesh pots of France the contrast in conditions was immense.'

An officer of the 18th King George's
Own Lancers[1]

Westerners v. Easterners – Supreme War Council, January 1918 – Smuts's mission to Near East – Ghoraniye bridgehead across Jordan established – Jericho entered – the Amman Raid – re-organization of EEF on withdrawal of troops to France – Second trans-Jordan operations

Lloyd George's Christmas gift of Jerusalem – by courtesy of Allenby and the EEF – encouraged him in his anti-Western views. In these he was strongly supported by Henry Wilson, who was on the point of succeeding Robertson as CIGS. Their increasing revulsion at the carnage in Flanders led them to press ever harder for a solution in the Near East. They believed that a strictly defensive stance on the Western Front, awaiting the arrival of the Americans, would allow Allenby to knock Turkey out of the war. Though the reinforcing of the Palestine theatre would be wasteful of shipping, it was argued that the offensive on the Western Front was wasteful of lives. It was also suggested that a Turkish surrender would lead to Bulgaria's defection, thus opening a path to Austria and Germany.

Robertson, on the other hand, led the Westerners, with the fervent and anxious support of Haig (and, incidentally, at a much lower level, of Wavell) in reiterating their view that every available man at the time of a manpower crisis must be employed on the Western Front. Their case was vastly strengthened by the alarming fact that the German divisions released by Russia's defeat were even now pouring into France in preparation for a spring offensive of unequalled violence which would be delivered before the American divisions could be brought into play.* In spite of this, the Prime Minister and Henry Wilson prevailed upon the Supreme War Council which met at Versailles on 21 January, 1918, to agree to stand on the defensive in the West and in the Balkans and to 'undertake a decisive offensive against Turkey with a view to the annihilation of the Turkish armies and the collapse of Turkish resistance.'[2] The French insisted on the proviso that no white troops should be taken from France while the security of the Western Front was still to take priority over everything else.†

*Russia's withdrawal from the war extinguished the benefit of free passage through the Dardanelles which had hitherto been a chief reason for crushing Turkey.

†The War Office in early January calculated that some 60,000 to 80,000 enemy combatants might be assembled to oppose the EEF before the summer. Against these Allenby, whose long-term plan was to occupy the Tiberias–Acre line, could pit 69,000. On 3 January he told London that he 'could not deploy more troops even if I had them until my railway is doubled.' (Quoted in Gullett, 652)

28. A regiment of Australian light horse on the march up to Jerusalem (see p.209)

29. A wire-netting road across the desert

30. Water express on desert railway drawn by London and South-West Railway engines

31. Dummy horses in the Jordan Valley (see p.256)

32. Watering horses in the Jordan Valley

33. The camp of 'A' Squadron, 9th Australian Light Horse, near Jericho, August, 1918

To discuss the strategy of a major Near Eastern offensive, Lloyd George despatched to Egypt in February Smuts, now a member of the War Cabinet. He was to discuss the problem with Allenby and a representative of the Mesopotamian force. Allenby's attitude to the emissary of the 'frocks' was far from cordial. According to Chetwode, 'he was bored to death with his being there.'[3] But Smuts was highly impressed by the excellence and fitness of the troops and by their confidence in Allenby. His consultations resulted in a plan for going over to the defensive in Mesopotamia and attacking in Palestine. The War Cabinet approved the plan. It included moving two Indian infantry divisions from Mesopotamia and at least one and possibly two Indian cavalry divisions from France* as well as additional heavy artillery and aeroplanes.

* * *

In the meantime it was clear that in the near future any further major advance northwards was out of the question. The rainy season put an end to any hopes of the speedy construction of roads, the necessary accumulation of supplies and ammunition or, most important of all, the extension of the railway to a point nearer the front. A prolonged rest period was essential, too, for the horses of Descorps. The shortage of remounts was acute, due chiefly to the shipping situation. The only way to bring the regiments anything like up to establishment with fit horses was to nurse the exhausted and the invalids back to robustness with ample rations and dry lines. The majority of the corps was therefore withdrawn on 29 December to Deir el Balah where supplies could be received by rail and water with relative ease.

* * *

Allenby, with his capacity for planning ahead, had already made up his mind that when he was ready he would advance up the great coastal plain where his superiority in mounted troops could be employed to the full. Before he could achieve this it was essential to

*In the event eleven regiments of Indian cavalry were transported across the Mediterranean without a hitch at a time when torpedoing was almost at its zenith.

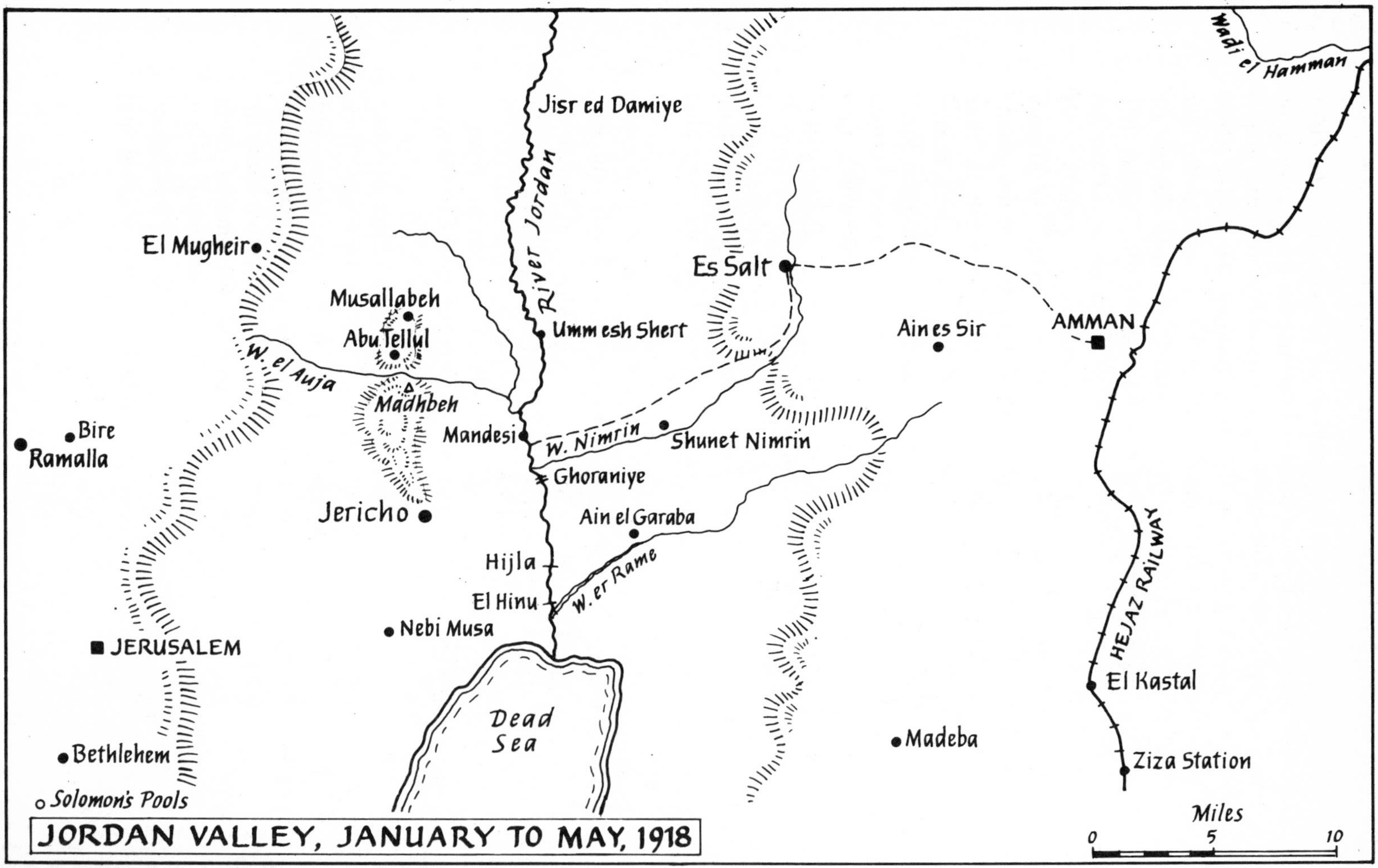

JORDAN VALLEY, JANUARY TO MAY, 1918

drive the Turks across the Jordan so as to make his right flank secure. This was necessary, too, as a preliminary to any expedition against the Hejaz railway. Further, it would rob the enemy of the copious supplies which they brought across the Dead Sea from the grain-growing districts at Kerak and Maan. The 60th Division, starting on 19 February, advanced directly on Jericho, while Anzacdiv was ordered to turn the Turkish left, entering the Jordan Valley near Nebi Musa so as to cut off the enemy's retreat from Jericho. Because of an endless series of ridges and the 'tumbled mass of hills' encountered, the advance was slow. 'The chief obstacle lay in the difficulties of the ground rather,' as Allenby's despatch of September 1918 put it, 'than any opposition the enemy might offer. . . . The chief feature of the enemy's resistance was the volume of machine-gun fire.'[4] This the enemy used effectively to check the Anzacs' turning movement till the night of 20 February, during which they withdrew across the river. They left only a triangular bridgehead about two miles in length and one and a half miles deep at Ghoraniye (inevitably known to the troops as 'Gonorrhoea'), on its west bank. On 21 February, Ausdiv, advancing across the plain, entered Jericho. The Turks had withdrawn from the town that night.

The next stage in improving Allenby's position took place between 8 and 12 March. Employed were both XX and XXI Corps. The mounted troops took virtually no part. The fighting was stiff, the steeply terraced hills giving the defence every advantage. XX Corps eventually secured the water supply of the Wadi el Auja and commanded the Beisan to Jericho road, a principal route into the Jordan Valley from the north. To the west XXI Corps conformed.

All was now set for a raid on Amman, one object of which was to help the operations of Faisal's Arabs in the Hejaz. These had been of considerable help to the EEF, ever since, with British encouragement, gold and munitions, they had started extending their revolt against Turkish rule northwards along the Medina–Damascus railway. Their ceaseless wrecking of trains and dynamiting of rails under the leadership of Lawrence had contributed to safeguarding Allenby's right flank and had caused a serious drain on the enemy's limited quantity of men and materials.* The essence of Lawrence's tactics was the avoidance of pitched battles, but exceptionally on 26 January, 1918, his Arabs inflicted a crushing defeat on a Turkish force which was armed with mountain and machine guns. This had tried to recapture the recently taken town of Tafile. Now, in March, it had been reoccupied by a large column which included a German battalion.

Allenby's projected raid on Amman, which was the most vulnerable point on the railway, was partly intended to force the recall of the Turkish Tafile expedition. 'He also,' in Wavell's words, 'hoped, by the destruction of the tunnel and viaduct near Amman, to isolate the Turks south of that place for some considerable time.'[5] Of these it was calculated that there were some 20,000. Allenby's long-term object, though, was to make the enemy think that he meant to attack by way of Amman the railway junction of Deraa, the most vital point of all the Turks' communications, some fifty miles north of Amman, and thus force the strengthening of the enemy's defences east of Jordan. This would weaken the Turkish line in the west where Allenby intended eventually to launch his main attack.

For the Amman raid Chetwode formed 'Shea's Force' consisting chiefly of 60th Division, Anzacdiv and 1st Imperial Camel Brigade. Shea's orders were to occupy Es Salt and to destroy 'the railway at and south of Amman to a sufficient extent to cause a complete cessation of railway traffic'. 60th Division was to effect the former and the mounted troops were to effect the latter. 'You can expect,' Shea was told, 'no direct co-operation from the Arabs.'[6] Nor was there any.

*The Arab Revolt had started in June, 1916. Husein, Grand Sherif of Mecca, was the prime mover. He soon disposed of the garrisons in his area – with the exception of Medina. At the end of the year, Lawrence, a member of the staff of the Arab Bureau in Cairo, got in touch with Faisal, a son of Hussein's. Instead of besieging Medina Lawrence persuaded Faisal that the best way to achieve his ends was to counter the reinforcements, which were being sent from Damascus to reestablish the Mecca garrisons, by marching northwards to Wejh. From there he could threaten the communications with Medina from the north. By February, 1917, this move had proved successful in halting the Turkish advance on Mecca. The enemy was compelled to scatter his troops along the railway.

The next stage was the occupation of Akaba. This Lawrence achieved by attacking from the land side, having enlisted a force of tribesmen from as far north as the outskirts of Damascus. Early in July Akaba was occupied and Faisal's men were then moved round by the Royal Navy to establish their base there.

Since there were virtually no mounted actions in the hilly terrain over which this first trans-Jordan raid took place, there is no need here to detail its progress.* It was a disaster, the first that Allenby's forces had experienced. At its end Shea reported that 'adverse weather conditions [persistently atrocious] and the opposition encountered prevented [its] objects from being completely attained.'[7]† Nevertheless it had fulfilled, as Wavell puts it, 'a part of its purposes. The Turkish expedition to Tafile had been promptly recalled and a part of the Maan garrison had been moved north, thus facilitating the operations of Faisal's Arabs. And the Turks, thoroughly alarmed for the safety of their communications east of Jordan, had been compelled to make a permanent increase to their forces in this area.' By the middle of April these numbered some 8,000 – about double their strength before the raid.[8]

For Liman von Sanders, the Turks' new commander, and for the valiant Turkish troops who defended Amman and its railway installations, the boost to morale was greatly inspiriting. Liman, as soon as Shea's Force had been withdrawn on 2 April, speedily

*On 23 March the Aucklands cleared the country opposite Ghoraniye by a charge during which they fired at the gallop from their saddles, capturing sixty-eight prisoners and four machine guns.

On 28 March the 5th Regiment 'made a mounted dash down a large wadi between the Camel Brigade and the infantry. They rapidly covered half-a-mile; but, after dismounting, their attempt was halted by machine-gun fire.' Gullett, 570).

These seem to have been the sole occasions during the eleven-day raid when mounted action occurred. A squadron of the Wellingtons, however, on 27 March, 'working round Amman from the south, perceived a train coming up [with reinforcements] and galloped to cut it off, but became bogged within 300 yards of the line. With a whistle of triumph the train ran on into Amman.' (Falls, 339)

†About 350 Turks were killed and 900 wounded. Twenty officers and 595 other ranks were taken prisoner. Anzacdiv and the Camel Brigade suffered 671 casualties and the infantry 447, a rare example of the mounted troops' losses being greater than those of the infantry.

There was much discontent among the Australians at this time with respect to the recruiting methods being employed in Australia. One officer wrote, referring to the last days of March, of

> 'horseholders, sole survivors in many sections, leading the saddled horses of their missing mates, silently echoing the "empty saddle" appeal of Sydney recruiting picnics. No women here leading around saddled-up horses though, and no recruits heartened by Dutch courage, bribed by insurance policies or conscripted by white feathers, to fill empty saddles.' (Berrie, 128)

developed a strong, if somewhat isolated position at Shunet Nimrin, astride the road up to Es Salt and Amman, from where observation of the Ghoraniye bridgehead on the east bank as well as the plain on either side of the river could be observed.

Liman, who incidentally, like Allenby, was a cavalryman, understood the Turks' strengths and weaknesses far better than von Falkenhayn ever had. He at once replaced most of his German staff by Turkish officers and soon began to show, as he had done at Gallipoli, his supreme ability as a defensive general. But on 11 April, in two abortive actions designed to prevent another raid across the Jordan, his troops lost four times as many men as Cox's group defending the Ghoraniye area. The bridgehead there was never again assaulted.

* * *

On 21 March, while the Amman raid was in progress, Ludendorff's great offensive on the Western Front was launched. This at once put paid to the offensive in Palestine as agreed after Smuts's visit. On 26 March Allenby was instructed to send nine yeomanry regiments to France.* The following day he was ordered to fall back upon a policy of active defence. Taken from him were twenty-three well-tried battalions which were replaced by Indian battalions from India. These, which arrived in driblets, had no fighting experience, hardly any training for modern war and were chronically ill-equipped. In all, the EEF was deprived of about 60,000 front-line troops.

Allenby at once set about reorganizing his force. By mid-August and after numerous alterations, Descorps came to consist of four divisions: Anzacdiv, the only division to remain in every respect as it was, Ausdiv, unchanged except for the substitution of 5th Light

*'Out of the blue,' says a member of the Lincolnshire Yeomanry, 'we were told that our horses were to be handed over to the Indian cavalry and that we were to be sent to France and retrained as machine-gunners. A ceremony was held at Gaza to mark the occasion.

'Items of saddlery and spurs, etc. were buried and a wooden memorial erected bearing the inscription (much abbreviated): "Stranger pause and shed a tear – A regiment's heart lies buried here."' (Seely, Frank, letter to *The Times*, 1 May, 1990).

Allenby also lost five and a half heavy batteries and five machine-gun squadrons. According to Falls (421) he also 'sent to India one garrison battalion and three cavalry regiments, made up by withdrawing squadrons from ten different regiments of Indian cavalry.'

Horse Brigade* for 5th Mounted Brigade, and 4th and 5th Cavalry Divisions. The 4th, commanded by Barrow, consisted of the 10th, 11th and 12th Brigades. The 5th, a new division, commanded by fifty-two-year-old Major-General Henry John Milnes Macandrew, 5th (Indian) Cavalry, who had commanded 5th Cavalry Division in France,† consisted of the 13th, 14th and 15th (Imperial Service) Brigades.‡

*This brigade was at first commanded by Brigadier-General Gregory (see p. 223 below), who after ten weeks took command of 11th Cavalry Brigade, and from late August by Brigadier-General George Macleay Macarthur-Onslow, who had commanded 7th Regiment since 1915. It consisted of two new Australian Regiments: the 14th and 15th Light Horse, made up from the Australian battalions of the Camel Corps, which, as such, ceased to exist, though the British battalion kept its camels and was soon sent to operate against the Hejaz railway south-east of the Dead Sea (see p. 249). The New Zealand Company of the Corps was formed into the 2nd New Zealand Machine Gun Squadron for use by 5th Brigade. The brigade's third regiment was one of French cavalry known as the 'Regiment Mixte de Cavalerie' consisting of two Algerian Spahi squadrons and two of Chasseurs d'Afrique. It was fully equipped with automatic weapons and mounted on Moorish barbs, small in size but great stayers.

†Macandrew, three and a half years earlier, had reported Allenby to Haig as not being 'cavalry minded'. Now, ironically, he was to lead the last great mounted action in history under that commander. He had joined the 5th Bengal Cavalry in 1889 and held a number of staff appointments during the South African War. In *Who's Who*, he gave as his recreation 'gentleman rider across country and on the flat'. He had commanded the 18th Lancers on the Western Front. He died at Aleppo in 1919 having been burned as a result of an accidental ignition of petrol.

‡*10th Cavalry Brigade* (Howard-Vyse): Dorset Yeomanry, 2nd Lancers, 38th Central India Horse. *11th Cavalry Brigade* (Brigadier-General Charles Levinge Gregory of the Indian Army): County of London (Middlesex) Yeomanry, 29th Lancers, 36th Jacob's Horse. *12th Cavalry Brigade* (Wigan): Staffordshire Yeomanry, 6th Cavalry, 19th Lancers. (Berks, Hants and Leicester Batteries, RHA – 13-pounder guns). *13th Cavalry Brigade* (Brigadier-General Kelly, 3rd Hussars. From October, 1918, Bt-Col George Weir, 3rd Dragoon Guards): Gloucester Yeomanry, 9th Hodson's Horse, 18th Lancers. *14th Cavalry Brigade*, (late 7th Mounted) (Brigadier-General Goland Vanhalt Clarke, 18th Hussars): Sherwood Rangers, 20th Deccan Horse, 34th Poona Horse. *15th (Imperial Service) Cavalry Brigade* (Brigadier-General Cyril Rodney Harbord): Jodhpore Lancers, Mysore Lancers, 1st Hyderabad Lancers. (Essex Battery, RHA – 13-pounder guns). The 15th Machine-Gun Squadron belonging to this brigade was on a special establishment. While all other machine-gun squadrons consisted of British personnel, this one was made up of Imperial Service personnel from the three

(Continued over)

Chauvel's command by the time of the opening of the great September offensive included thirty-six regiments, of which fourteen were Australian, thirteen Indian, five British, three New Zealand and one French. His artillery was still all British except for three Indian mountain batteries.*

To some European officers in the Indian regiments

> 'Egypt was a step backwards into a secondary theatre of war, and, what was worse, to one closer to India. It must be remembered,' wrote an officer of Jacob's Horse, 'that at this period India was regarded by most Indian Army officers serving out of it as some dreadful antediluvian spot, to be avoided at all costs. . . .
>
> 'The Indian ranks were extremely sniffy about Egypt, and Kantara especially. They were particularly contemptuous over the Kantara cinema which they considered, strange to say, greatly inferior to those of "Marsellaise" as they called Marseilles.'[9]

Lieutenant Roland Dening of the 18th Lancers found

> 'everything very amateur and run with much less efficiency and organization than in France. No doubt about it being a side show. With the Turks opposite you can do things which would be criminal in France. Large bodies of mounted men can wander about in no man's land in the dark and so on. . . . People here have very little idea of *real* barbed wire. . . . The war is on the more pleasant old fashioned style where second rate arrangements are good enough for a victory.'[10]

States represented in the brigade.

XX Corps troops included the Worcesters. XXI Corps troops included a composite regiment of one squadron, Duke of Lancaster's Yeomanry and two squadrons of Hertfordshire Yeomanry. Corps Troops also included two Light Armoured Motor Machine Gun Batteries and two Light Car Patrols.

The Indian regiments, unlike the yeomanry and Australasian regiments, which consisted of three four-troop squadrons, were made up of four three-troop squadrons each.

*Godwin was his Brigadier-General, General Staff; his Deputy Adjutant and Quartermaster-General was Brigadier-General Edward Fynmore Trew, late of the Royal Marines and his GOC, Royal Artillery was Brigadier-General Algernon D'Aguilar King of the Royal Horse Artillery.

Before Beersheba Hodgson, it will be remembered, had begged to be allowed to equip his two Australian brigades with the sword, but permission had not been granted. (see p. 137). Now in the early summer he once again pressed upon Chauvel the desirability of this step. This time, since there was a period of comparative inaction ahead which could be devoted to training in the weapon, Chauvel agreed. He was influenced also by the fact that intensive training in shock tactics with bayonets used as if they were swords had been going on in a number of regiments since March.[11]*

Chauvel got hold of Brigadier-General Gregory, a famed expert from the 19th Lancers of the Indian Army, to educate 5th Brigade especially in the use of the sword. After some weeks of concentrated training he brought even the ex-camel men, who had not had the advantage of the light horsemen's bayonet tuition and some of whom had never in their lives ridden a horse, up to a high standard of efficiency. Something like half the cameliers had been light horsemen originally, but the rest had been drawn from the infantry. Much instruction was also given in cavalry drill which was of course very different from that of the light horse. The advent of the sword meant that the rifle could no longer be slung. Consequently rifle buckets had to be issued, and the quantity of ammunition carried on the man was reduced to ninety rounds, the balance being hung round the horses' necks.

Chaytor's Anzacdiv elected to remain as mounted rifles.† Thus nine out of the twelve brigades were armed as cavalry proper. The lancer regiments, of course, retained their lances. The 34th Poona Horse and the 20th Deccan Horse, before they were sent to the

*He was swayed, too, by the splendid work done by the Indian lancers in patrol encounters across the Jordan in July (see p. 237). An account of their exploits and the lessons to be drawn from them were circulated throughout the corps. Hodgson replied 'that if his men were armed with an equally suitable weapon, they would do equally well, but as such a weapon had been denied them he must respectfully request that accounts of the undeniably brilliant exploits of the Indian Cavalry with the lance should not be sent him in future for issue to his Division because they could only result in dissatisfying them with their own comparatively inferior armament.' (Osborne, II, 140).

†Soon after the division had been formed when some form of *arme blanche* was mooted, the only weapon available had been 'a form of lance', of which some 300 were issued. After trials with these they were 'found unsuitable'. ([Anon] 'Operations of the Mounted Troops of the EEF', *Cav. Jnl*, X, 1920, 387).

Jordan Valley, were equipped with lances at their own request, 'because,' as the brigade-major of 14th Brigade said,

> 'the division commander [Macandrew] insisted on patrols charging at sight any enemy parties they saw. . . . It was so successful that deserters from the Turks said more men would desert to us if regiments without lances were given the patrolling, as they were so afraid of lances – the idea, I suppose, being that would-be deserters were afraid to leave their lines for fear of being stuck with the lance before they could explain!'

In the second week of April the mounted troops received their first shrapnel helmets, otherwise known as steel helmets or tin hats. They were almost the last troops in the course of the war to be thus equipped. 'Most of us,' wrote an officer of yeomanry, 'found them very hot in the heat of the day.' So much was this so that in June they were withdrawn from some units and not, it seems, issued again.[12]

* * *

Allenby had hoped that most of his major re-organization, which added up to a considerable increase in his mounted troops, could be ready by mid-May. At that time, anxious as ever to imbue the enemy's minds with the belief that his next big thrust would be aimed at the railway junction at Deraa, he had planned to launch a second trans-Jordan attack. He now intended to occupy Es Salt and the surrounding country permanently, thus sparing his troops 'the ordeal of a summer in the Jordan Valley'.[13] A subsidiary object was to deny the Turks the fruits of the barley and wheat harvest which were soon due to be reaped in the hills of Moab.*

*On 18 April Allenby ordered Chetwode to demonstrate with three brigades of Anzacdiv against Shunet Nimrin 'with a view to stimulating the idea that further operations against Amman are contemplated by us' (*Chauvel*, 145). Falls, Preston and Wavell: *Pal.* do not mention this demonstration. A.J. Hill in *Chauvel*, 145, and Gullett, 596, suggest that it was a mistake which unnecessarily alerted the enemy to the May operations, thereby making them much harder. Both authorities hint at the possibility that Allenby had not at that date decided to proceed with the May attack. To the present author it appears more likely that the demonstration was merely in Allenby's mind another small part of his plan to concentrate the enemy's thoughts and forces on their left flank, and that it did not in any way compromise the May attack.

On 20 April Allenby put Chauvel in charge of the operation. Three days later an envoy from the Beni Sakhr tribe arrived from Madeba, only nineteen miles south-east of Ghoraniye, to tell Allenby of their willingness to cooperate with him in an attack towards Amman, provided this took place before 4 May by which date their grazing would be exhausted and they would be forced to retire southwards. Foolishly, without consulting either Lawrence or Captain Hubert Young, his liaison officer with the Beni Sakhr, Allenby accepted the proffered help, which if it were to materialize, would fit well into his plans. As will be shown it did not materialize.* In the belief that it would, the Commander-in-Chief decided to advance the operation to 30 April, although by then his reorganization was far from complete. In particular Barrow's new division was not yet ready. There is evidence that Chauvel was far from happy with the task which was set him. Since, except in two instances, most of the operation took place dismounted, the temptation to recount its intricacies must be resisted. In brief, the plan was to assault frontally the strong Shunet Nimrin position with two brigades of 60 Division and the New Zealanders, while sending Ausdiv, 4th Brigade leading, northwards up the valley to block at Jisr ed Damiye the main route by which the enemy could march reinforcements from west to east of the river. 3rd Brigade, following the 4th, was to branch off so as to capture Es Salt, while 5th Brigade was to strike at the town by the Umm esh Shert track. From Es Salt Ausdiv was to intercept the hoped-for retreat of the Turks from Shunet Nimrin. However, successive attacks over the following days failed to dislodge them from that strong position.

Meanwhile Ausdiv rode up the plain between the Jordan and the escarpment to its east at great speed. Unlike the frontal attack, this flank attack took the Turks by surprise, but they were in sufficient strength at Jisr ed Damiye to prevent the vital bridge there coming

*In fact the envoy could speak for less than 400 men. Young found the leader of the Madeba Arabs 'perplexed and frightened by GHQ's reaction'. (*Chauvel*, 145; see also Young, H., *The Independent Arab*, 177-8 and Mousa, S. *T.E. Lawrence: An Arab View*, 166-7). Lawrence, when he reached GHQ as the operations were coming to an end, expressed shock at the gullibility displayed. It has been generally supposed that Major-General Louis Jean Bols, the Chief of Staff, unlike his predecessor, Guy Dawnay, was 'too inexperienced in local matters to smell a rat'. (Bullock, D.L. *Allenby's War*, 1988, 112). This was also Lawrence's view (Lawrence, T.E. *Seven Pillars of Wisdom*, 525) under fire from all three of the division's horse artillery batteries. Wilson's 3rd Brigade, on the other hand, by a very dashing and skilful exploit was able to enter Es Salt that evening. The final advance was made at the gallop by 8th Regiment, leading the brigade. Commanding the first troop to enter was Lieutenant Charles Foulkes-Taylor, only recently commissioned and temporarily detached from the 10th Regiment. As he entered the town he found an English-speaking German staff officer trying to organize resistance. Foulkes-Taylor raced up to him and forced his surrender. He then put fourteen rounds from his own automatic and two clips from the German's pistol into the hesitating enemy and smashed the pistol over the head of one of them. Brigadier Wilson reported that 'the men of Lieut. Taylor's Troop were using their bayonets as swords. One sergeant got two on the point (sword in line). The general opinion was that they were not good for mêlée fighting – too blunt. They used them for striking. Swords would have been invaluable here. The men with revolvers, Hotchkiss gunners, were using them freely.' Foulkes-Taylor's troop pursued for two miles, taking some 200 prisoners. The commander of the Turkish IV Army and his staff escaped in motors with one minute to spare.

All resistance was soon overcome and much transport and twenty-eight new machine guns still in their packing cases were captured. On the next day the brigade received medical effects dropped by aeroplane.[14] The taking of Es Salt was the sole tactical success of the operation, which otherwise proved to be the biggest reverse of the whole of Allenby's campaign. This was because the information upon which the plan had been based was perilously defective and in some respects erroneous. The fact that the Ain es Sir track had been vastly improved since the Amman raid was not detected by aerial reconnaisance. The Arabs, who had apparently promised to block this track, made absolutely no attempt to do so. In consequence there was no bar to the easy flow of reinforcements for Shunet Nimrin from Amman. The air observers had also failed to pick up a concealed pontoon bridge five miles south of Jisr ed Damiye. Even more important was the failure of intelligence to discover the fact that considerable reinforcements had been assembled for an intended attack on Allenby's position on 2 May. Beyond these it turned out that there were some 2,000 more Turkish and German troops opposed to him than GHQ had estimated. It was the ignorance of the presence of two reinforced

Tommy. "WHERE DID YOU GET THAT BUNCH?" *Australian.* "OH, *I* DIDN'T GET 'EM—THE DAWG BROUGHT 'EM IN."

divisions on Chauvel's left flank which sealed the fate of his plan. These fell upon Grant's 4th Brigade on 1 May in overwhelming strength, forcing a precipitate retreat in which he lost, beside much other material, nine of his twelve guns.* These were the only guns lost during the whole campaign, but the gunners managed to remove the breech-blocks and the sights and to escape capture. They were immediately equipped with new guns and were ready for action again in less than two days.

Though Chauvel tried during the next two days to continue the operations, Allenby, who came to Chauvel's headquarters in the afternoon of 3 May, agreed that there was no alternative to withdrawal. This was very skilfully executed with surprisingly few casualties. In the whole operation out of the five mounted brigades employed, fifty were killed, 310 wounded and thirty-seven reported missing. The infantry's casualties were much higher: 1,116 of all ranks. The enemy's losses were probably greater: about 1,500 were killed and wounded, while 981 prisoners were taken, 666 of them by 3rd Brigade.

Allenby criticized Grant, but in fact when he had asked Chauvel

*Those of 'A' battery, HAC, and the Notts Battery.

for reinforcements they had been denied to him. It seems clear that the chief blame for the fiasco was Allenby's and his staff's. Chetwode's demonstration on 18 April and Allenby's easy duping by the Arabs, thus depriving himself of Barrow's aid, were sufficient reasons for the lack of success. Chetwode had tried to argue Allenby out of the operation but had failed to convince him.[15] Two decades later he wrote to Wavell: 'These two expeditions of Allenby's across the Jordan were the stupidest things he ever did, I always thought, and very risky.'[16] When Chauvel told Allenby how sorry he was for the failure, the Commander-in-Chief barked back at him: 'Failure be damned! It has been a great success!' adding that he would explain later.[17] Whether he did so or not is unknown, but he probably meant that what really mattered was that Liman's attention had yet more firmly been concentrated on the inland flank. From now onwards more than one third of the total Turkish forces was stationed east of Jordan, a most important pre-condition for Allenby's intended major stroke in the west. Further, he now held two bridgeheads across the Jordan, that at Ghoraniye and another near Mandesi.

24

'Literally an oven; no trees, no water, nothing but rock and dust, the only protection a single piece of canvas between one and the pitiless sun! Gasping for breath, one reaches for the water-bottle, but it is quite warm. Still, a warm drink brings perspiration and *that* is cooling to a certain extent – in its after-effects. . . . Here we are in the place of sweltering sun, sand-spouts, scorpions, snakes, spiders and septic sores, of scorching wind and shadowless waste, that hellish place – the Jordan Valley!'

An officer of 20th Machine Gun Squadron

'Though the actual heat may not have equalled that of a Punjab hot weather, it was more than counterbalanced by the feeling of depression caused by the heavy atmosphere.'

An officer of the 19th King George's Own Lancers

'Well, I reckon God made the Jordan Valley and when He seen what He done, He threw stones at it.'

An Australian trooper.[1]

Summer stalemate – Descorps stationed in Jordan Valley – horrors of life there – Abu Tullul – El Hinu

Static trench positions along a seventy-mile front were now speedily established, demanding ceaseless digging and wiring. There were many gaps in the line owing to the impassability of the country in many places. XXI Corps had its left near Arsuf on the coast. XX Corps' right rested on the cliffs looking down into the Jordan Valley near Wadi el Auja. Neither side possessed sufficient numbers to hold its lines securely, but neither side was strong enough to launch from it any major attack. Thus for four weary oppressively hot months a state of virtual stalemate faced the opposing forces.

In a perfect world the mounted troops would have been sitting behind the line in reserve undergoing the intense training which the newly formed divisions required before being able to undertake major operations. Allenby, though, felt that to counter the continual menace posed to his inland flank by the Ghoraniye to Amman road it was essential that the Jordan Valley should be occupied throughout the summer. It was equally important thus to sustain his own threat against the Hejaz Railway. Further, to have kept infantry there rather than the mounted troops would have entailed much larger numbers of men because mounted troops could be more quickly moved to any threatened point. Most vital of all, it was known from intercepts of Turkish messages[2] that wherever Descorps was stationed would be assumed by the enemy to be where Allenby's next blow would fall. However, he gave Chauvel

> 'the option of withdrawing from the actual valley if,' as the Corps commander later wrote, 'I thought it better, but he told me that, if I did so, I would have to retake the bridgeheads over the Jordan before the autumn advance. I considered that I would lose more lives retaking the Valley than I would through sickness in holding it and, furthermore, there was neither room nor water for large bodies of cavalry in the jumble of hills overlooking the Southern end of the Valley. . . . I told him I considered it better to hold the Valley. He agreed and I was instructed to do so.'[3]*

The official military handbook for Palestine stated that 'nothing is known of the climate of the lower Jordan Valley in summer time, since no civilized human being has yet been found to spend a summer there'.[4] Even the local Arabs, except for a small tribe of negroid origin, invariably left the district during the summer. Each day from mid-May onwards the temperature rose until during July it reached as much as 130°F in the shade and on occasions a degree or two higher. In that month the maximum daily temperature in 1918 on top of a hill near Jericho was 113.2°F in the shade. At its foot it was 3° higher. For an eight-week period the thermometer seldom registered less than 100° day or night. All wheels had to be covered over during the day to prevent the heat and humidity shrinking the wood.

*The only infantry formations which took their turn in the valley were the 20th (late Imperial Service) Indian Infantry Brigade and two battalions of the British West Indian Regiment. In mid-August a Jewish battalion (38th Royal Fusiliers) took up a position on Wadi Mellaha.

34. Sand sleighs with wounded (see p.54)

35. Machine-gunners of the Australian light horse

36. Megiddo. Transport moving up across the Turkish lines. On right, wounded coming back borne on camels, 19 September, 1918

37. Megiddo. 5th Cavalry Divisional transport crossing one of the two pontoon bridges over the River Auja, 19 September, 1918.

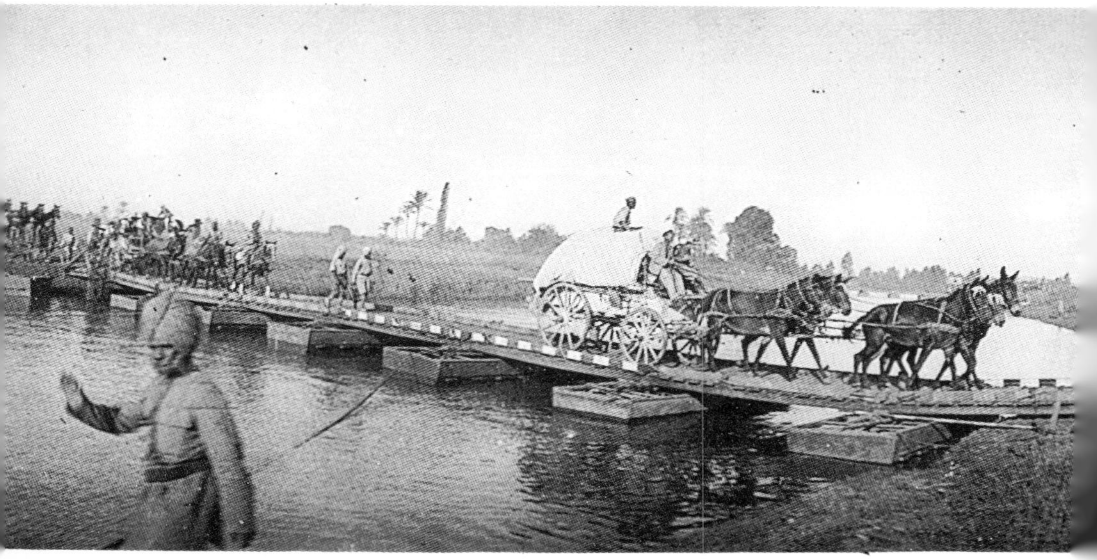

The southern end of the valley, in places fifteen miles wide, lies 1,290 feet below sea-level, but the mountains which enclose it are some 4,000 feet high. The humidity is excessive. The Jordan discharges about six million tons of water a day into the Dead Sea. All of this is lost by evaporation during the summer months. The appalling degree of moisture and the abnormal heaviness of the atmosphere induced in everyone a ghastly feeling of physical oppression and mental hopelessness. An officer of 6th Regiment described what it was like to return from the cool joys of the high ground to the purgatory of the great gorge:

> 'The clear air of the uplands had completely gone. The heat became more stifling with every mile of the descent and its sudden upward rush at quick intervals made body and brain shudder. Unrelieved, save by two touches of colour, the green which enveloped the hideous squalor of Jericho and the distant lifeless blue of the Dead Sea, the Jordan Valley, steeped in the breathless heat of the forenoon, lay in all its dreadful reality. The vast expanse of white dust extended from foothill to foothill. . . . As if in mockery of the morning stillness, dust rose in funnel-like whirlwinds into the very heavens and on every side desolation, disease and death reigned supreme.'[5]

The valley bottom is covered with a layer of white marl impregnated with salt, several feet deep. In winter a little thin grass sprouts. In summer the fierceness of the sun scorches this into brittle dust. Constantly disturbed by feet and hooves this was pulverized into a powder of the consistency of fine flour. It was often so dense that a rider could not see his horse's head. An officer of the Poona Horse found that in the dark 'dust hanging above marching columns had a curious refractive effect, turning stars into luminous blobs that seemed to float only a few feet above one's head.'[6]

Invariably the night was breathless. In early morning a strong, burning northerly wind arose to sweep the dust down the valley in thick, suffocating clouds. Suddenly in mid-morning it died down as abruptly as it had arisen. A good half hour of utter stillness and torturing heat was followed by another violent draught of air, blowing now from the south and continuing till about eight o'clock. This regular climatic régime was repeatedly punctuated by 'dust devils' of inordinate height, capable of lifting a tent high into the air. To add to the joys of life in the valley, every evening at sunset there arose 'a sickening sulphurous stink'.[7]

The chief inhabitants of this infernal region were, beside numberless flies, sand-flies, lice and malaria-bearing mosquitoes, elephantine black spiders, centipedes six inches long with pincers that could inflict an almost fatal injury, many varieties of snake and stinging scorpions.* In the swamps there were hares, gazelles, partridges, wood pigeons, bustards and wild pigs which frequently gave sentries unwarranted frights.

Enormous efforts were made by the Medical Corps and Field Hygiene Sections to lessen the incidence of malaria. Wherever possible, stagnant water was drained and where this was impossible it was oiled. But there were gaps in the line where there was little that could be done to make the mosquitoes harmless. Mosquito nets were, of course, provided – one for every 'bivvy tent'. Veils and 'mosquito gloves' also made their appearance. In swampy areas a repellent paste for smearing on face and arms was issued. The troops were kept away as much as possible from the scrubby swamps, while the courses of streams 'were contained within lines of carefully placed stones; the watering of horses, except at canvas troughs to which water was lifted by pumping, was forbidden, because of the holes made by hoofs along the margins.'[8]

The Turks took, and seemed hardly to need to take, any steps to prevent the disease, so that when the wind blew from the direction of their lines, the fatal insects descended on the luckless troops in their millions. Large numbers of men had to be evacuated in consequence. For instance more men had to go to hospital in the 5th Regiment, chiefly from malaria, but also from blood disorders such as 'sand-fly fever', 'five-day fever' and stomach complaints, than during the previous eighteen months. While 20th Machine Gun Squadron sojourned in the valley, it was deprived through illness of 116 of all ranks. It had to have its numbers made up by recently dismounted yeomanry.[9] By the end of the first week of August, the net monthly loss from Australian units alone, owing to

*Scorpion stings were treated by cutting the skin where they had penetrated. This promoted bleeding. The poison was then squeezed out and permanganate of potash (Condy's Crystals) rubbed in.

sickness, was 600.* Deaths and evacuations averaged about 1% of the Corps' total strength per day. This would just about entail the replacement of the whole of it in three months. Curiously enough the Indian native troops suffered even more than did the white troops. Turkish aeroplanes dropped leaflets with the encouraging message: 'Flies die in July, men in August and we shall come and bury you in September.' There is no doubt that by September, in spite of everything that could be done, the men of Descorps were much weakened by disease.

The horses tolerated the conditions a great deal better than the men. But they, too, often became tired and listless in spite of having very little work to do. This was probably because their forage, though sufficient in quantity, was lacking in nutritive value. An average ration was six pounds of barley, four of gram and twelve of tibben. The tibben formed too great a proportion of it, while the more nutritious foods could not be easily procured. Further, the horses lacked exercise, there being too few men to give it to them. By the time the daily sick men had been evacuated and others had been found for outposts, patrols and anti-malarial measures, there was on average only one man to look after six or seven horses. On occasions in some regiments there was only one for every fifteen horses. Lieutenant Wilson of the Gloucesters says that three or four blankets had to be kept over his horses' backs to prevent heat stroke to their spines. Because of the dust raised, which attracted long-range fire, the horses were mostly watered after dark.[10]

* * *

The frontage maintained by Descorps was twenty-five miles long. It stretched northwards from the Dead Sea up to Wadi Mellaha and thence westwards through Musallabeh and from there it was thrown back by Abu Tullul to Wadi Auja. The enemy opposed to the corps both east and west of the river consisted of some 8,000 rifles, 2,000 sabres and ninety guns.

*Captain Maunsell of Jacob's Horse was a lone voice when he wrote that 'the horrors of life in the Valley have been greatly exaggerated. The flies and dust were the trying things, but in some bivouacs, notably in the Ghoraniye bridgehead, we had little of either. The mosquitoes were both deadly and annoying in certain portions, notably close to the varying wadis where there was water. In most other places they did not worry one.' (Maunsell, 194)

With some variations, the principle adopted was that two of Descorps' divisions took their turn in the valley, while two recuperated in the hills, reliefs taking place every month. The unfortunate horse gunners, however, because of the shortage of artillery, had to spend much longer in the valley of desolation. The front, divided into a northern and a southern sector, was held by a minimum of dismounted men and a maximum of machine guns. These were dug in defended localities, each small but strong. The majority of each division was kept well in rear with its horses, ready to counter-attack mounted at short notice. The wisdom of this method of defence was confirmed by the success of the action in Abu Tullul (see p. 236).

The camps and horse lines were exposed to continual shelling from long-range guns sited on both sides of the river as well as to aerial bombing. Against this neither reply nor much shelter were possible. The casualties were not generally great,* but consequently all the work, such as road building, camp construction and swamp draining, which would normally have been carried out by the Egyptian Labour Corps, fell instead to the troops. One of their chores was to camouflage their tents by applying to them 'red soil and water'. Digging, wiring and standing to arms before dawn, interrupted by numerous patrol encounters, day observation and night outpost duties, comprised the weary daily round.[11] 'The difference in climate between Solomon's Pools [near Bethlehem] and the Valley,' wrote an officer of 6th Regiment, 'was almost unimaginable. Horses had to be rugged, extra blankets and overcoats were needed.' The disparity in temperature was as much as 60°F.[12]

Strenuous efforts were made during this period of inactivity to keep up morale. Leave to Egypt was granted as freely as possible. Camps for leave parties were formed at Alexandria as well as at Cairo. When relieved from the line units were entertained by

*Nevertheless on 4 June 3rd Brigade and the Indians were badly bombed, 103 horses being killed or so badly wounded as warranting destruction. One man was killed and eleven were wounded. (Bostock, 159) Again on 16 July the horse lines of 1st Regiment were so badly shelled by guns firing across the river by direct observation that it lost in a few minutes fifty-eight horses killed and twenty-seven wounded (Gullett, 674). This was 'a great loss to the regiment, so many of the best horses being killed'. The remainder were at once moved 'about 400 yards, where good cover was available and they remained there during the day.' (Vernon, 163)

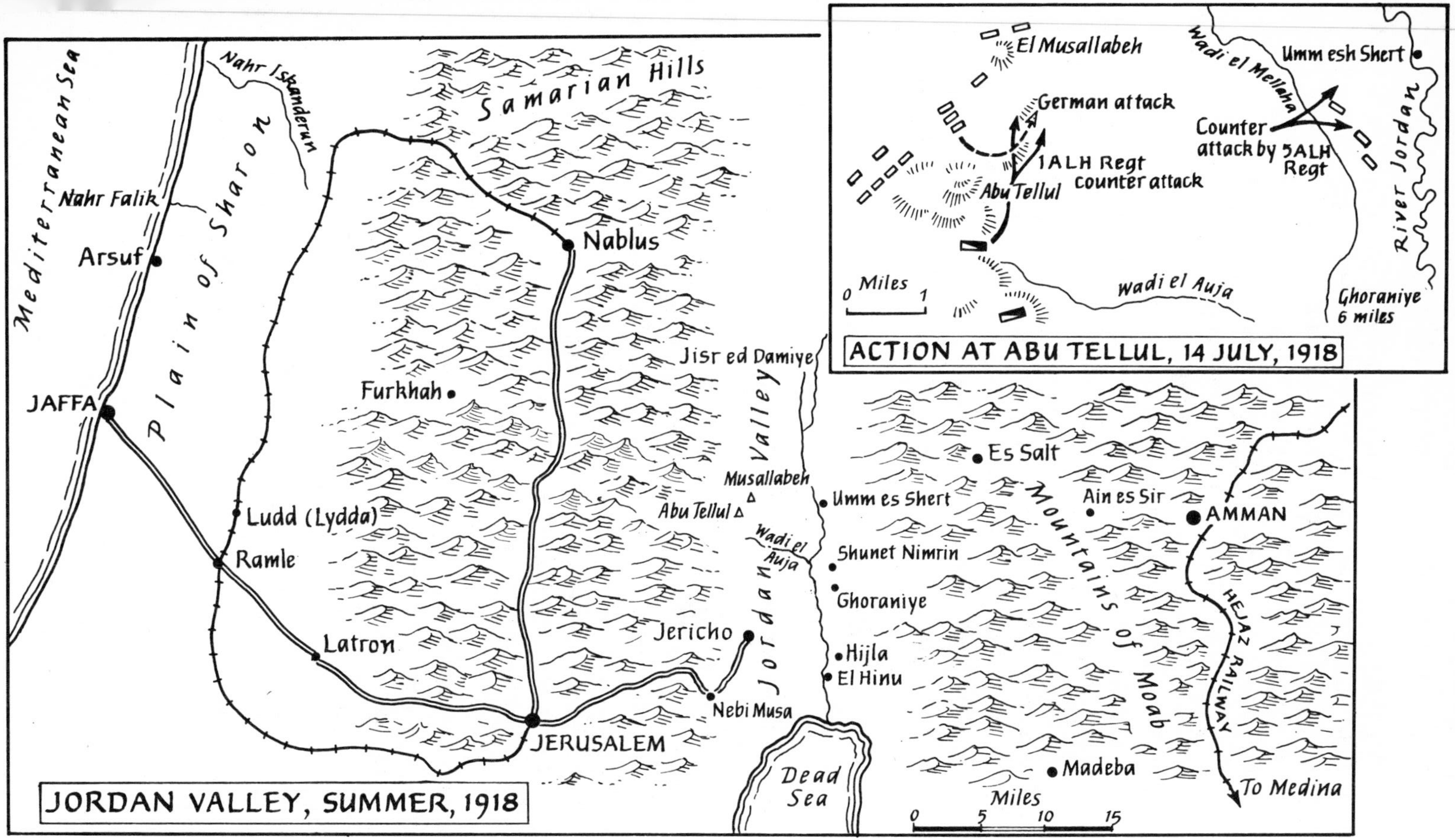

ACTION AT ABU TELLUL, 14 JULY, 1918

JORDAN VALLEY, SUMMER, 1918

concert parties and cinema shows. Every possible sort of sport, including numerous race meetings, was organized. The Canteen Board provided canteens within a few miles of the front. The whole force in both Egypt and Palestine spent £4,500,000 in the various canteens during 1918. Regimental funds were generously maintained out of a rebate of 8% on all cash sales.[13] Rations, except for those in the valley outposts, were adequate and always supplemented by the finest oranges the world produces.

The provision of supplies was undertaken from a depot at Jerusalem. At one time as many as 150 lorries a day were working between Latron and that city, the drivers turning out at 05.00 hours and returning to camp at about 18.30. Between Jerusalem and Jericho the drop of about 2,000 feet seemed almost impossible for lorries to negotiate. It was done nevertheless day after day, each run taking a good ten hours, the drivers becoming 'so encrusted with dust and sweat that it was impossible to recognize them'.[14]

* * *

During this stalemate period the enemy launched only one deliberate offensive. It was the very last that it attempted in the whole campaign. At 3.30 a.m. on 14 July the most vulnerable part of the line was attacked. This was the Musallabeh salient, held because without it the essential water supply from the Wadi Auja would be unavailable. Two Turkish divisions led by two and a half crack German battalions broke in between the advanced posts, but, just as had been planned in case such an attack should be made, Anzac Division counter-attacked between the posts at 4.30 with complete success. As soon as Cox, 1st Brigade's commander, realized that the enemy had penetrated between the posts held by 2nd and 3rd Regiments, he ordered Lieutenant-Colonel C.H. Granville, commanding 1st Regiment, waiting with its horses where they had been moved to avoid shell fire, to 'Get to them, Granny!' This he did not hesitate to do. After a very stiff six and a half hour fight the original line was restored. Chauvel was delighted by this crushing little success, writing that it would improve the image of the Australians 'in the minds of their detractors, who are many'. Well over 100 of the enemy were killed and 425 prisoners, of whom 358 were Germans,*

*These figures are given by Gullett, 669. Falls, 433, says 448 and 377 respectively.

were taken. The division's casualties numbered 108 of all ranks and 101 animals. 19,000 rifle, 30,000 machine-gun and 20,000 Hotchkiss rounds were fired. The Abu Tullul engagement was the only one in the campaign in which German infantry was used as storm troops. Together with the action at El Hinu (see below) Abu Tullul was intended as the preliminary to a scheme to overwhelm the entire Descorps position, but it failed because the four Turkish regiments which were meant to support and follow up the Germans' spearhead failed, according to German sources, to play their part. Liman later lamented that nothing showed 'so clearly the decline in the fighting capacity of the Turkish troops as the events of the 14th July'.[15]

* * *

While this action was in progress a considerable body of Turks came into contact with the outposts of the Imperial Sevice cavalry and the Sherwood Rangers in the scrubby plain between Ghoraniye and the Dead Sea. A squadron each of the Jodhpore and the Mysore Lancers crossed the river by small swing-bridges at El Hinu and Hijla respectively. At about 6.45 a.m. these speedily established that an enemy force consisting chiefly of cavalry about 1,200 in strength,* on a two-mile frontage, lay with its right flank north of the Wadi er Rame, a mile and a half east of Hijla. Harbord, the brigade commander, suggested to Macandrew, the divisional commander, that the rest of the Jodhpores with two machine guns should cross the Jordan at El Hinu, ride unobserved, under cover of the low hills, some 4,600 yards down the east bank, to attack the enemy's left flank from the south. He proposed that at the same time the rest of the Mysores, with another two machine guns, should cross at Hijla and attack the right flank from the west, while the Sherwood Rangers cooperated with them from the north. They were to start as soon as they saw the Jodhpores charge. This rather audacious plan Macandrew approved.

When the Jodhpores,† who started off soon after 10.30 a.m.,

*The 7th, 9th and 11th Regiments of the Caucasus Cavalry Brigade.

†They were led by Major P.F. Gell, the regiment's Special Service Officer. Each of the Imperial Service regiments, though commanded by its own Indian officers, had attached to it several British officers. The senior of these, always an experienced Indian cavalry officer, took his orders direct from the brigadier.

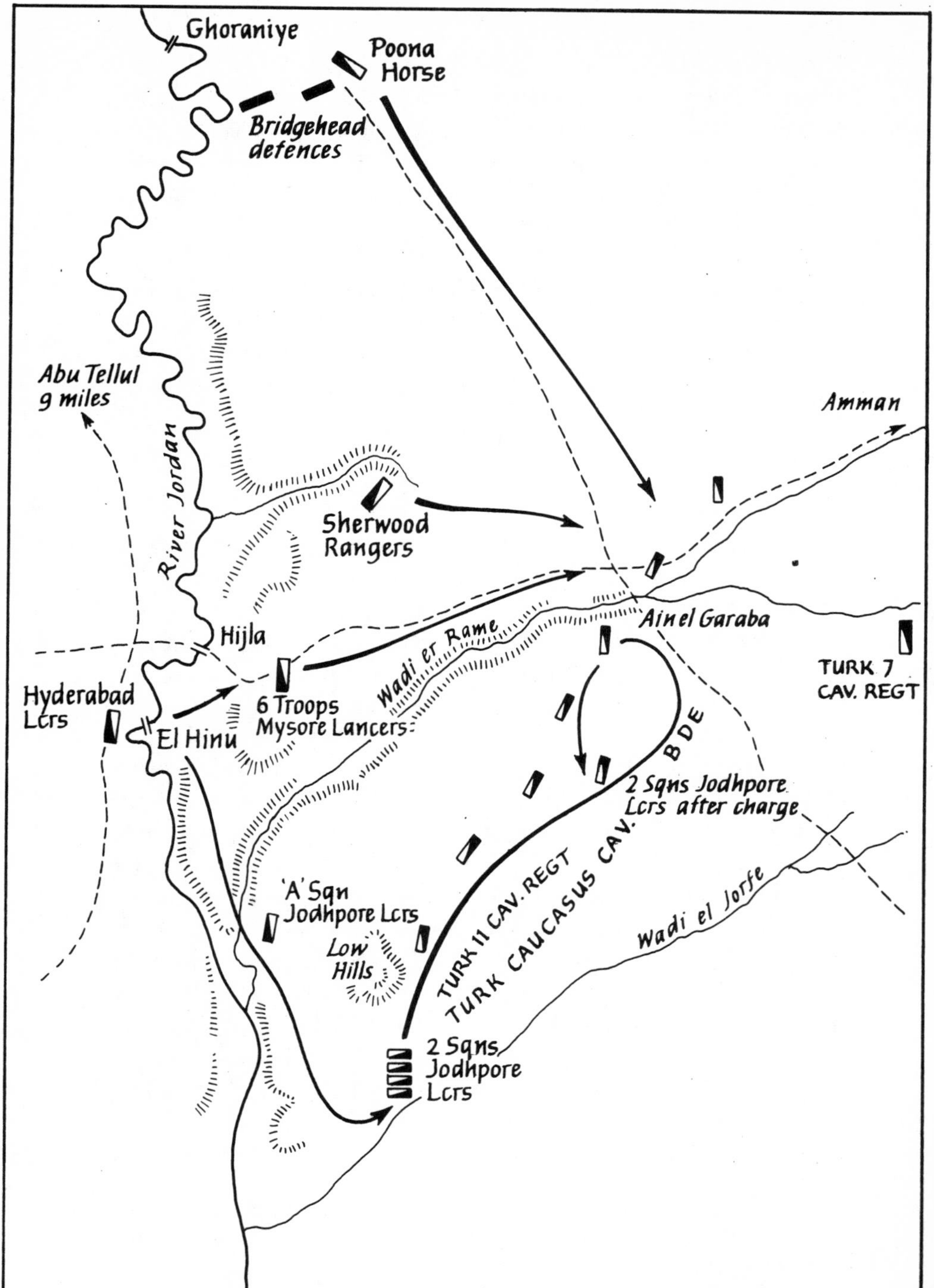
Ghoraniye
Poona
Horse
Bridgehead
defences
Abu Tellul
9 miles
River Jordan
Sherwood
Rangers
Amman
Hijla
Ain el Garaba
TURK 7
CAV. REGT
Hyderabad
Lcrs
El Hinu
6 Troops
Mysore Lancers
Wadi er Rame
BDE
2 Sqns Jodhpore
Lcrs after charge
TURK CAUCASUS CAV.
TURK 11 CAV. REGT
'A' Sqn
Jodhpore Lcrs
Low
Hills
Wadi el Jorfe
2 Sqns
Jodhpore
Lcrs
Miles
0
1
2
Dead Sea

ACTION AT EL HINU, 14 JULY, 1918

reached a point due south of the Turkish flank, they came under fire, and at once extended to an interval of two horses' length, swung to their left in column of troops and galloped due north. They were covered on their right by the two machine guns. As they charged towards their given objective (the ford at Ain el Garaba, two and a half miles east of Hijla) they dashed into a large body of Turks, speared numbers of them with their lances and took more than fifty prisoners.* When, at the ford, they came under machine-gun fire, they swung round in a large circle to rally a mile to the south.

Meanwhile the Mysores, supported by their machine guns, charged along the north bank of the Wadi er Rame and speared about thirty Turks before they could reach the shelter of some impenetrable scrub. Just before this, the Jodhpores, having rallied, had fallen back, pretty well exhausted by the midday heat. They took with them their wounded,† their prisoners and a large number of captured horses. The Mysores, seeing the Jodhpores' retirement, conformed, falling back to cover the crossing at Hijla.

Earlier, three squadrons of the 34th Poona Horse had been ordered to march on the ford from Ghoraniye.

> 'Moving off, with the vanguard troop commanded by Jemadar Pem Singh, the regiment,' as its historian records, 'came at once under enemy shell fire. That there were practically no casualties from this was due to the fact that the Turkish shrapnel was bursting high. Crossing the pontoon bridge at a walk with 300–400 yards' interval between troops, artillery formation was adopted on reaching open ground. . . . No sign of any enemy was seen until about 3 p.m., when just as the vanguard-troop was getting into the Wadi er Rame, light automatics opened on the squadron from the left flank. . . .
>
> 'Expecting that the rest of the regiment would shortly come up and offer a tempting target to these automatics, the

*An example of the offensive spirit of the junior Indian officers was the action of Risaldar Shaitan Singh. He outpaced his troop and galloped alone into a troop of enemy cavalry, shot two with his revolver and unhorsed three 'with a loaded stick' before his men caught up with him.

Allenby next day rewarded Major Gell with a DSO. He also awarded fifteen other decorations to members of the regiment. (Falls, 437).

†They suffered twenty-eight casualties out of the 125 men who rode in the charge.

O.C. "A" Squadron ordered his left troop to extend and charge these Turks, who were firing from some high grass and bushes about 800 yards away on the left, the O.C. squadron deciding that he himself would follow his vanguard into the wadi. The left troop extended to orders and moved off at a gallop led by Lieutenant [A.F.] Dickson.

'This troop was some 300 yards away – quite beyond possibility of recall and the other troops were trotting quietly forward to the shelter of the wadi when a message was received from Major [G.W.C.] Lucas [commanding the regiment] stating that . . . the Imperial Service cavalry had withdrawn, that both banks of the wadi were strongly held by enemy machine guns and riflemen and that without help from the south any advance in the direction of the ford was impossible. Major Lucas stated that in consequence of this information he was withdrawing the regiment to the shelter of a nullah in the direction of Hijla and that "A" Squadron was to join him there at once. The squadron commander, however, judged it best to remain where he was until the result of Lieutenant Dickson's charge was seen.'[16]

Dickson's charge proved to be against a formidable entrenched position designed to protect the ford. His troop jumped a thorn-bush abattis which covered it and 'inflicted many casualties on the defenders. In the first contact Lieutenant Dickson and Ressaisar Zalim Singh were killed, together with seven Indian other ranks'.[17] 'Bloody swords and lances on the trampled ground [bore] witness to the desperate bravery with which they sold their lives.'[18] The rest of the troop rallied and managed to rejoin the squadron.

The Turks did not wait for a second attack on their strong point, for when Lucas sent two squadrons, together with some of the Sherwoods, to deal with the entrenchment, it was speedily evacuated. In all, the Poona Horse lost nine men killed, ten wounded and seventeen horses killed. In the two actions at Abu Tullul and El Hinu the total casualty list was 189, while a possible 1,000 of the enemy were killed and 540 prisoners were taken. Between them the three Indian regiments speared about ninety Turks. Their captures numbered ninety-one, including six officers and four machine guns.[19]

Sergeant. "Off again, are you? You shouldn't 'ave no 'orse. What you want is a cart with a pig-net over the top of you."

The El Hinu action, though small-scale, has been gone into in detail because it was a good example of the numerous spirited performances given by the Indian cavalry in the valley and not untypical of them. Except for Dickson's tragic mishap, the daring plan was excellently carried out, especially by the Jodhpores.[20]*

*Though not present at El Hinu, General Sir Pertab Singji, over seventy years old, accompanied the Jodhpores into the valley and remained there as long as they did. (Falls, 423)

25

'The defeat of the VIIth and VIIIth Turkish Armies west of the Jordan would enable me to control [the Jisr ed Damieh] crossing. Moreover the destruction of these armies, which appeared to be within the bounds of possibility, would leave the IVth Army isolated if it continued to occupy the country south and west of Amman. I determined therefore to strike my blow west of the Jordan.'

* * *

'The main difficulties lay in concealing the withdrawal of two cavalry divisions from the Jordan Valley, and in concentrating secretly a large force in the coastal plain.'

* * *

'That the enemy expected an offensive on my part about this date [September] is probable. That he remained in ignorance of my intention to attack in the coastal plain with overwhelming numbers is certain.'

ALLENBY in his despatch of 31 October, 1918

'In the year 1918, when the troops of every combatant were suffering grievously from war-weariness . . . the British Army [in Palestine] was well fed, was supplied with every material necessity, enjoyed a reasonable amount of repose, had its health guarded by an admirable sanitary service.'

CYRIL FALLS in the *Official History*.[1]

Railway improvements – Turkish armies' situation – Allenby decides to attack in late summer – his strategy – EEF's orders and objectives – supply and communications arrangements.

One of the advantages of the lengthy respite from major operations was the time given for the improvement of railway communications between the Egyptian bases and the front line. There existed by the beginning of September a network of lines as efficient as those behind the front in France. A number of the narrow gauge lines were relaid in standard gauge and for a period both were in use at the same time. Probably unique in railway history was the running of trains made up of the two sets of rolling stock! Taking the line to Jerusalem was a major engineering feat, involving much blasting. Behind the front line over thirty miles of metalled roads and many minor ones were constructed for other wheeled traffic. New roads were often camouflaged with dried grass and stable refuse.

The question of supply was of greater importance in this theatre of war than in any other, for the proportion of animals was much greater. Further, Britain ceased in effect, because of the submarine menace, to be for the EEF 'even an *entrepôt* of supply so far as food and forage were concerned'.* There were restrictions, too, on the export of coal from the United Kingdom which meant that the use of oil in its place had to be urgently improvised, even for the railways.[2]

* * *

In the late summer of 1918 the situation of the enemy was very different from that of the EEF. Enver Pasha, now virtual dictator of Turkey, believing that the great offensive in France would lead to German victory and, keen to achieve an easy triumph in the Caucasus, was quite prepared to starve the Palestine front.† Further and almost unbelievably, the Germans withdrew from it certain of their units. Most of Liman's most experienced senior officers, German and Turkish, were also taken, amongst other

*The opening of the port of Jaffa was of considerable assistance.

In 1918 the Egyptian Government alone collected for the EEF 30,000 tons of wheat, 30,000 of barley, 12,000 tons of beans and 6,000 tons of lentils, as well as 275,000 tons of tibben and 25,000 of millet.

The monthly consumption of grain was 27,000 tons. (Falls, 440-1)

†The signing on 3 March of the Treaty of Brest-Litovsk had assured Turkey of the Russian evacuation of the Anatolian provinces and the Trans-Caucasian territory which had been lost in 1878.

'[The Turks] squandered the bulk of [their men and resources] in sending armies to make the facile conquest of Batum, Kars and Tiflis and to penetrate Azerbaijan towards Baku. These activities could have little effect on the issue of the war, but they alarmed the British General Staff into organizing a special force [Dunsterforce] to limit the Turkish encroachments in the Caucasus. It arrived too late and had to be withdrawn.' (Wavell: *Pal.*, 193) things for the German occupation of Georgia. Mustafa Kemal Pasha, the best of the Turkish generals and to be the nation's post-war leader, who was sent to take over VII Army, was shocked to find the state of affairs so hopeless. Though himself a pan-Turk, he was acute enough to realize that while the war lasted Palestine had to be held if at all possible.

Throughout the three armies forming Army Group F, both resources and communications* were increasingly inadequate and morale was alarmingly low. Each could muster little more than 10,000 effective rifles. Desertions were so numerous that behind the lines the roads were patrolled by machine-gun parties with orders to discourage potential deserters. In the fifteen weeks ending 19 September there were 645. In August 116 Turks gave themselves up to Descorps alone. Outnumbered, hungry and without hope it is no wonder that these poor men lost heart and discipline.

* * *

On 12 July, 1918, Allenby informed the War Office that he hoped to 'resume active operations that year'. In this he persisted against Henry Wilson's advice.[3] Since the early rains usually come in Palestine in the beginning of November, he had long since decided to attack in mid-September. The shape of his plans for the great offensive had been germinating in his brain from before the second trans-Jordan operation. On 1 August he issued secret orders to his three corps commanders giving them an embryonic scheme to work upon. On 22 August he delighted and astonished them by informing them that he had decided to extend the scope of his operations and thereby cast the net much wider than at first envisaged, so wide indeed that if the plan succeeded it would encompass nothing less than the annihilation of Army Group F. Chauvel's immediate response was 'I can do it'. It was an amazingly bold and imaginative plan. Its success depended to an unusual degree on surprise and speed and it took great risks with the supply situation. One who was present when, two days before the operation, Allenby disclosed his plan to brigade and regimental

*Largely owing to a fuel shortage, the supplies reaching Deraa in August were only slightly more than a third of those carried in May. Due to lack of railway capacity, troops had mostly to march. The Turks' transport animals were receiving only two pounds of grain a day and were dying in great numbers.

commanders commented on the expression of astonishment on their faces at the scope of the role allotted to them.[4]

Allenby based his ideas for this *attaque brusquée* upon both his superiority in total numbers and his especial superiority in fast-moving mounted troops. These numbered about 12,000 in the fighting line, while his infantry numbered 57,000. He could deploy 540 guns against the enemy's 370. He kept no GHQ reserve. Even the Corps cavalry, pioneer battalions and details from reinforcement camps were to be put into the assault. The enemy's fighting strength was estimated at a total of about 29,000 men including 3,000 mounted troops with a general reserve of some 3,000 men, and including another 6,000 on the Maan and Hejaz railway.* This disparity, however, was misleading, for the Turkish returns included only riflemen. Some 800 machine gunners per division were excluded. In fact to the west of Jordan there were some 600 heavy machine guns opposed to 350 in Allenby's force, and about 450 light machine guns. The British estimate of the actual rifle strength of a Turkish division in September was 1,646. This low figure was probably very nearly accurate. Whether or not it included the machine gunners, it meant that the divisional capacity for defence was much greater than the bare figure suggests. The ration strength, as opposed to the fighting strength, of the three Turkish armies on 8 September was 103,500. The ration strength of Allenby's three corps was 140,000.†

Lack of manpower was less important to the Turks than the failure of their organization, but they had the advantage always accorded a retreating force, namely that of bringing into action a

*In mounted troops Allenby had a superiority of about three to one, in infantry seven to four and in artillery eleven to eight. (Osborne, III, 352)

†A captured Turkish return of 8 September showed a *total* ration strength, including native workmen, of 247,000, while the *total* British ration strength at the same date covering the whole of Egypt and Palestine and including 80,000 natives was 450,000. There were some 62,000 horses, 44,000 mules, 36,000 camels and 12,500 donkeys.

'One ought, perhaps, to deduct a small percentage from the Turkish figures to allow for faked returns concealing the peculations of supply officers.' (Falls, 454)

The Turkish VIII Army had five infantry divisions and three German battalions in the coastal sector. The VII Army had four divisions between Furkah and the Jordan Valley. The IV Army had three infantry and one cavalry divisions and one German regiment lying in the Jordan Valley and on the hills of Moab.

percentage of its total strength greater than that of an enemy pressing on its heels. When it came to the entrenched defences through which the British infantry would have to break, the Turks suffered from a chronic shortage of barbed wire. The importance of this conspicuous lack is hard to exaggerate. By it, throughout the campaign, the effectiveness of the Turkish defensive positions was gravely impaired.

* * *

The basis of Allenby's strategy is best appreciated by a glance at the map. This shows the Turkish lines of communication. VII and VIII Armies drew their sustenance from the Deraa–Beisan–Afule railway, running roughly parallel to their front line. The chief roads also centred on these three places. Their occupation would close the lines of retreat of all three Turkish armies. Deraa was clearly beyond the reach of a rapid, uninterrupted mounted dash from the coastal area, but Allenby 'could and did', as Wavell put it, 'use the apprehension of a raid on Deraa, which he had so effectively instilled into the mind of his adversaries, to immobilize a large portion of the Turkish forces. And he proposed to use his Arab allies to foster this apprehension and to dislocate the working of the communications round Deraa, though he did not ask them to seize and hold the place.'[5] Afule and Beisan were a different matter. They were forty-five and sixty miles respectively from the British front line. Since the country to be covered was open and good going for most of the way they were within the range of a continuous mounted ride. Once wrested from the enemy, only a single narrow and difficult escape route would remain for VII and VIII Armies. There was, however, one potentially serious obstacle in Chauvel's way. The hills of Samaria, about seven miles wide, divide the Plain of Sharon from the Plain of Esdraelon and the two-mile-wide Valley of Jezreel. The two routes across them, though not physically difficult for mounted troops, would not be easy to force against determined opposition.

To Bulfin's XXI Corps, holding a twenty-one-mile front (eleven miles of open plain and ten of lower slopes), was given the task of breaking through the enemy defences, practically continuous, partially wired and some 2,500 yards deep, between the sea and the railway. On its actual attacking front, the Corps had an infantry superiority of about four and a half to one and in artillery

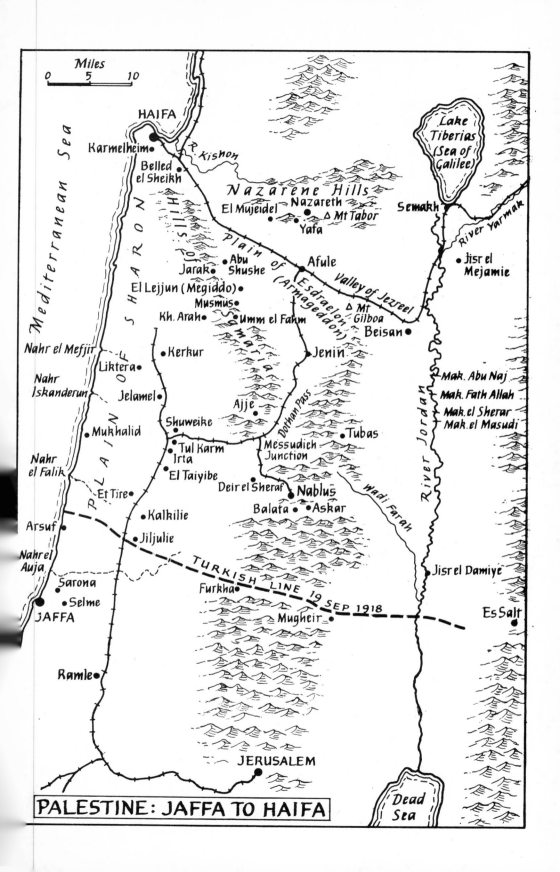

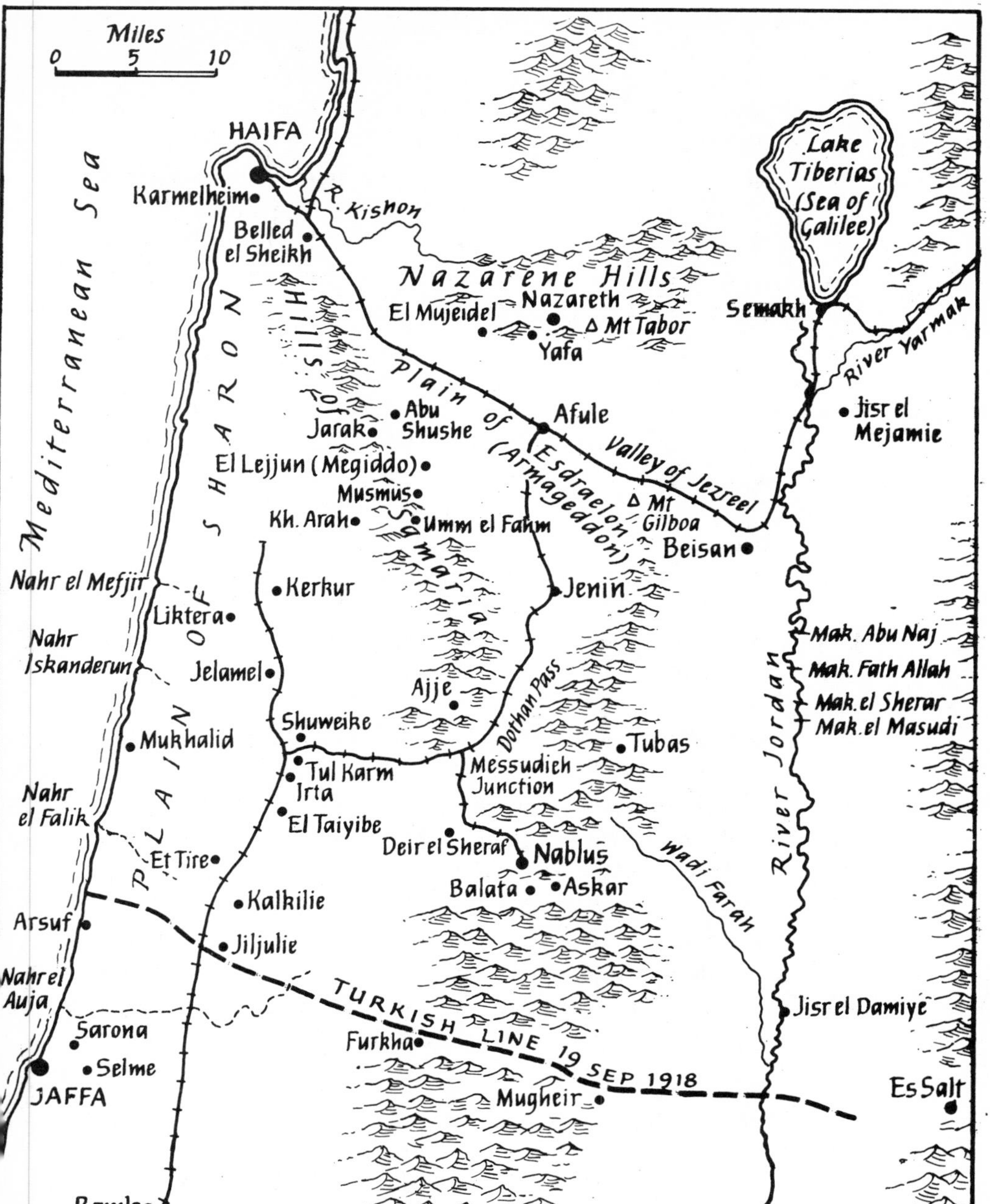

PALESTINE: JAFFA TO HAIFA

of three to one. After a short but intense bombardment, it was to assault at dawn on Z-day. Chauvel's mounted troops were to burst through the gap thus punched in the Turkish line. Beside his own four infantry divisions, Bulfin was given one from Chetwode's XX Corps* as well as 5th Australian Light Horse Brigade and nearly all the heavy and medium artillery. Seven horse batteries of Descorps were to take part in the initial bombardment, before returning to their own divisions.†

Chauvel's mounted force was to consist of Ausdiv (less 5th Brigade), and 4th and 5th Cavalry Divisions. The two remaining infantry divisions of XX Corps (10th and 53rd Infantry Divisions) were to be extended westwards so that XXI Corps could concentrate more closely. When Descorps had been secretly assembled behind the left of XXI Corps, there would remain in the Jordan Valley Anzac Division and eight infantry battalions including two West Indian and two Jewish units, the latter nicknamed the 'Jordan Highlanders'. These were to be commanded by Chaytor and known as 'Chaytor's Force'. Here Allenby took a certain amount of risk, for opposed to Chaytor were some 23,000 men as against his 22,000.

Once XXI Corps had broken the enemy line, it was to swing north-east, drive the Turks towards Messudieh and thence down the road to Jenin where the mounted troops, having earlier reached Afule, would be ready for them. 5th Brigade was to cover the Corps' left flank and make for Tul Karm where Djevad Pasha, who had succeeded Kress in command of VIII Army, had his headquarters. Chauvel's orders after erupting through the breach pierced by Bulfin, were 'to advance to Afule–Beisan, cut the enemy's railway communications at their most vital point and get in a position to strike the enemy's columns if they endeavour to

*3rd (Lahore), 7th (Meerut), 54th, 75th and 60th Infantry Divisions. The French Brigade also came under XXI Corps.

†About three-quarters of the total of 258 guns and 126 howitzers were on the seven-mile front between the Jaffa–Tul Karm road and the sea. There were some 113 enemy artillery pieces on the XXI Corps front. XX Corps had 130 pieces and Chaytor's Force thirty-eight.

450 rounds for each horse artillery 13-pounder had to be carried by units, with 150 rounds ready loaded in lorries and a further 500 rounds dumped at suitable points.

Attached to 5th Brigade was the New Zealand 2nd Machine Gun Squadron (Powles, 238).

escape in a northerly or north-easterly direction.'[6] (For the other important parts of the operation order and for the special instructions in amplification of it, see Appendix, p. 345). Any enemy force which was not directly barring the way forward was to be totally disregarded. When the horsemen reached the line of the Nahr el Mefjir, where a Turkish reserve position, partially dug, unwired and unoccupied, was believed to be, they were to pass into the Esdraelon Plain and despatch a detachment to seize Liman's headquarters at Nazareth. Another detachment was to make for Haifa.

Depending on the progress of XXI Corps, Chetwode was to employ XX Corps to 'block the routes leading from the hills to the Jisr ed Damiye crossing over the Jordan as early as possible'.[7] It was also to advance on Nablus, the headquarters of Mustafa Kemal's VII Army. Between the two corps there was a seven-mile gap. This was to be watched by a temporary formation known as 'Watson's Force', consisting of the Worcesters (Corps Cavalry Regiment), two pioneer battalions and men from the corps reinforcement camp.

Chaytor, in the Jordan Valley, covering fifteen miles of river with three bridgeheads, having concealed the departure of the last of the mounted troops for the coast, was to demonstrate in such a way as to make the Turks believe that another attack on Amman (the headquarters of Djemal the Lesser's IV Army) was imminent and thus prevent as many Turkish units as possible from moving westwards. This he achieved by particularly active patrolling on the nights of 17 and 18 September.*

*Further to the east Faisal's Arabs were given an important part to play. Lawrence and other British officers had been organizing this final Arab effort. Faisal's regulars numbered some 8,000. From these, forming the core of his legions, a mobile force had been formed mounted on 2,000 camels which Allenby had given them when the Imperial Camel Corps was disbanded. Small numbers of British armoured cars, Indian machine gunners and Algerian artillerymen had also been sent to strengthen the force. Starting three days before the British, from its base fifty miles east of Amman, it was to cut the communications north and west of Deraa. (Wavell: *Pal.*, 200). This by some spectacular demolition work it achieved. By Z-Day all railway traffic moving south towards the three Turkish armies had been suspended. The RAF cooperated by bombing Deraa. To these attacks Liman reacted by sending a part of his reserve, including some Germans, from Haifa and Damascus towards Deraa.

In his life of Allenby, Wavell, who, it will be recalled, was Brigadier-General, General Staff to Bulfin, wrote:

> 'It was a daring plan, even against an enemy so inferior in numbers and morale. It would involve a continuous ride of over fifty miles for the majority of the horsemen, and over sixty for some, in the course of which they would have to cross a range of hills in the enemy's possession, passable only by two difficult tracks. There is no parallel in military history to so deep an adventure by such a mass of cavalry against a yet unbroken enemy. But Allenby had not made up his mind lightly, and there was no shaking it by suggestion of difficulties. He left it to his staff and to his Corps commanders to work out the details of the design, but of the main framework there was to be no alteration. The long Turkish domination of Syria and Palestine, and the military power on which it was founded, were to be given the death-blow in the grand manner.'[8]

To deliver the death-blow extraordinarily complicated and arduous arrangements for supply and communication had to be worked out. They were hard enough for the mass of infantry,* but for Descorps they were as intricate and difficult as any in the whole war. However many supplies could be brought up on the cavalry's heels, it would still be necessary for it to live off the country for a period. Because wheeled transport would be temporarily left behind 'every offside horse was issued with a complete set of pack saddlery'. Numbers of pumps and engines for developing local water supplies had to be carried with the advanced troops, together with portable troughs of rubberized canvas with brackets and pegs. 'Explosives for demolition purposes, hand and rifle grenades, parachute smoke signals and Popham pannels for signalling from the ground to aeroplanes were also issued, as was an additional supply of entrenching tools.' Beside the first day's ration, two days' emergency rations and the iron ration, all on the mobile scale, had to be carried on the horses and in the limbered wagons. Twenty-one pounds of corn were to be carried on the horse in nose-bags

*'For example, in XXI Corps two special water columns, each consisting of two companies of the Camel Transport Corps with 2,200 camels carrying about 44,000 gallons, had to be filled up and the camels themselves watered, all within a few hours.' (Falls, 457).

and in sandbags fastened across the saddle. Two blankets were to be carried under the saddle. 'Greatcoats, line gear, tin helmets, all clothing not absolutely indispensable were left behind. The average weight on the horse,' according to Barrow, 'was seventeen stone'. Officers were to be allowed one pack horse each. A fourth day's ration was to be carried by more wagons. A fifth was to be carried by a camel convoy organized by GHQ, which in the event fell hopelessly behind and was diverted to serve an infantry formation. 'By these means', according to the Official Historian, 'it was hoped that an ample margin of time had been given for the corps to exchange the greater part of its camel transport for lorries with XXI Corps.'[9] Between them Descorps and XXI Corps had 390 lorries.*

Ten bridges had to be thrown across the Auja in preparation for the two corps' advance. From mid-July several bridges had been thrown across the river as part of bridging training, only to be dismantled within a few days. When therefore new ones were built they aroused no suspicion.†

*The whole force had 590 lorries, thirty-two tractors and sixty-four other vehicles (trucks). The Ford cars and vans were the only vehicles which had worked well in the desert. They continued to be the most useful of the light vehicles. Of the heavy lorries used in the supply columns now that roads were more abundant, the four-ton American Peerless and the British three-ton Albion were used exclusively. Signal units and anti-aircraft batteries employed three-ton and thirty-hundredweight lorries. For hauling guns, in addition to ten-ton Holt tractors and caterpillar trucks, Peerless lorries proved the best. For ambulances, Talbots, Sunbeams, Siddeley-Deasys, Napiers, Studebakers and Fords were chiefly used. The motor-cycles were nearly all four-horsepower Triumphs. (Badcock, 247–50)

Descorps transport was divided into three echelons. 'A' Echelon, with water, ammunition, medical carts, tools and explosives in 'the fighting vehicles', and 'B' Echelon, carrying a proportion of the supplies for men and horses, were to start with the formations. The divisional trains were to follow and the camels to move in rear of the trains. 'B2' Echelon with the heavy baggage was of course left behind to follow on later.

Over the three weeks from 19 September the mileage covered by transport lorries totalled 720,000, a daily average of seventy miles per lorry. (Massey: *Triumph*, 14)

†Seven were made of heavy piles and could bear the weight of 60-pounders, two were pontoon bridges carrying 18-pounders and one trestle bridge which could take tractors. Another complete trestle bridge and more pontoons and trestles were also carried by the engineers so as to span the Nahr Iskanderun when it was reached.

Timber, netting and bill-hooks had to be carried for constructing causeways over swampy places. (Falls, 459)

Elaborate communications arrangements had to be made. They ranged from wireless to pigeons, from motor cyclists to contact aeroplanes. One troop of cavalry was to form a despatch riders' relay system. Posts were to be established every two miles. The speed required in delivering messages was to be marked on each envelope: 'trot', 'canter' or 'gallop'. Each corps had attached to it a Royal Air Force squadron, though one flight of that allocated to Descorps was under Chaytor's control. Beyond these there were four further squadrons at Allenby's command.

26

'Very significant was the contrast between day and night on the Jaffa–Jerusalem road and the country immediately north and south, as the fateful day of September 19th approached. By day the maritime plain was to the casual observer as it had been through the summer. The traffic on the roads and tracks was normal; there was no outward indication of a vast concentration, although a ride through the olive and orange groves discovered endless lines of horses and mile after mile of resting camels, wide areas of horse-transport, and great parks of motor-lorries. . . . At nightfall the still countryside awakened miraculously into intense but orderly activity. . . . So vast and widespread was the movement that all men feared it must become known to the enemy only a few miles away.'

H.S. GULLETT in the Australian Official History

'Some re-grouping of cavalry units is apparently in progress on the enemy's left flank, otherwise nothing unusual to report.'

A German airman's reconnaissance report, 15 September, 1918

'Even when the divisions were concentrated in the Selme–Sarona area, many officers of the 5th Cavalry Division were under the impression that the 4th Cavalry Division had only left the Jordan Valley two days, instead of nearly three weeks before!'

The regimental historian, 19th Lancers

'On the morning of 17 September in the mess tent of [Ausdiv] headquarters, a historic gathering assembled to meet the Commander-in-Chief. . . . General Allenby, after greeting each officer present, said: "I have come, gentlemen, to wish you 'good luck' and to tell you that my impression is that you are on the eve of a great victory. Everything depends – well perhaps not everything, but nearly everything – on the secrecy, rapidity and accuracy of the cavalry movement."'

* * *

> 'The Turks were in absolute ignorance of the fact that two nights before the attack there was one of the biggest bodies of cavalry ever assembled in war in one small area, under their very noses, and within easy shelling distance of their long-range guns.'
>
> W.T. MASSEY in *Allenby's Final Triumph*[1]

Elaborate means of deceiving the enemy

The day before the great assault was to take place the Turkish Intelligence Service issued a map of the British positions. It showed three mounted divisions still stationed in the Jordan Valley area on 17 September and 60th Division still in XX Corps' right sector.* This was remarkably accurate as to the situation *before* the final concentration had taken place. It shows that this massive operation had been concealed with total success. The means by which this unparalleled deception were effected were numerous, varied and ingenious. None of them would have been of much use without the temporary mastery of the air won by the Palestine Brigade of the Royal Air Force. During the concentration period only four German aeroplanes crossed the lines, whereas in one week of June there had been more than 100. Men were detailed to act as 'special police' and given field glasses with which to observe the approach of enemy aircraft. When one was seen they blew four whistle blasts whereupon every man had to remain absolutely stationary.[2] If it was essential to water horses by day the RAF guaranteed to keep enemy aircraft at a distance between midday and 2 p.m. The extra forward artillery and machine-gun emplacements needed for the

*One authority says that a paper of this date found at Liman's headquarters stated that 'far from there being any diminution in the cavalry in the Jordan Valley, there are evidences of twenty-three more squadrons.' (Wilson, 143)

A havildar from 7th Division was captured on 17 September and is supposed to have told of the coming assault, but Liman apparently believed him to be a plant. VIII Army commander and that of XXII Corps are said to have begged the Marshal to allow them to withdraw to the Et Tire line so that the bombardment which the Indian had predicted 'might uselessly exhaust itself in battering empty trenches'. He refused. (Statement from Historical Section of Turkish General Staff in *Army, Navy and Air Force Gazette*, 18 June, 1927, quoted in Falls, 468). Had he not done so Allenby's task would have been considerably more difficult to carry out.

initial blow had been begun in mid-July. Each was temporarily occupied so that it could be photographed from the air. If in spite of its camouflage it could be seen it was either made less visible or moved.

Fortunately there were thick orange groves and olive woods which afforded excellent cover for any number of troops south of the Auja. Their irrigation channels provided water for the horses, which could thus for most of the time be kept concealed. In the barer ground north of the river, camps had been much increased in size two months earlier. This meant that many more troops could now be accommodated in open bivouacs than were normally there. On the shore below the cliffs numerous well hidden bivouacs were able to secrete large numbers of men. For these the Meerut Division had dug eighty wells. In the 'open' camps to which the enemy was accustomed, there was no restriction on movement about them unless they were increased in size. In the officially classified 'concealed' camps no movement of any kind was allowed between 4.30 a.m. and 6.30 p.m. No fires were permitted in them by day or by night. Cooking had to be done with smokeless methylated spirit blocks, specially issued for the purpose.

The movement of the cavalry from east to west was all done at night. Barrow's 4th Division, for instance, had to make five consecutive night marches each averaging fourteen miles on bad and dusty roads. During the daytime great heat and hosts of flies deprived the men of their missed sleep. They arrived at their assembly points behind XXI Corps only two days before Z-Day. 'The march, coming,' as Barrow puts it, 'at the end of the long summer months in the Jordan Valley, caused [the division] to be a tired formation when it went through the gap on 19 September.'[3]

As always throughout the campaign, enemy agents, chiefly in the guise of local Arabs, were able to carry news across the lines with ease and rapidity. Especially was this so in the present case as there were numerous gaps in the chain of posts which constituted much of the front line. Spies of both armies could and did negotiate these in the dark unseen. Allenby and his staff decided to feed the enemy spies, as had been done on a small scale before Third Gaza, with a mass of false information and clues. A grand race meeting, one of a number which took place in the coastal plain, was announced for 19 September. Fast's Hotel in Jerusalem was suddenly evacuated, sentry boxes were placed at its entrances and the rumour was started that it was to become Allenby's advanced headquarters.* Numerous new billets were also commandeered in the city. Further to the east, Lawrence's agents began buying up large quantities of horsefeed, dropping hints that they were required for the British cavalry. They also wrote provisional contracts for sheep and marked out landing grounds with Arabs hired to watch them.

The most famous of the hoodwinking ruses was the construction of dummy horses. 15,000 of them were erected in the Jordan Valley area. They had been used there in smaller numbers earlier in the summer, purely so as to attract enemy aircraft from the real horse lines. They were made of canvas stuffed with straw, bamboo and wooden poles, many of them 'with real horse rugs and real nose-bags upon their dummy heads.'[4] (See illustration no. 31). In fact our pilots reported that sacks stuffed with brushwood cast in the strong sunlight just as deep a shadow and were effective enough to cozen from the air, so these were used as well.[5] The valley camps were left standing when the mounted troops left them. New camps made up of old, unserviceable tents were pitched, occupied by a few men to show signs of movement and to make tracks to and from them. For a week or two before the attack, one or two battalions, chiefly of the unfortunate West Indians, known to the Australasians as alternatively 'the Golliwogs' and 'the Black Anzacs',[6] were marched down the Jerusalem to Jericho road arriving soon after dawn each day to occupy these sham camps. During the night they were brought back in returning empty lorries, ready to resume their unpleasant daily promenade next day. In the Jericho area mule-drawn brushwood harrows raised huge dust clouds. Across the Jordan several dummy bridges were built with wide roads leading to them,[7] while some of the real bridges had been camouflaged.

An officer of horse artillery remembered his divisional commander telling him to do nothing 'to discourage the idea that the

*Before the war this hotel, later renamed the 'Allenby', had been a German residence. The British Canteens Board had taken it over as an officers' hotel. (Gullett, 685).

'Staff officers busied themselves installing office furniture and telephone equipment and painting the names of a multitude of departments on the doors of the rooms.' (Preston, 196–7).

Signal offices and wireless stations were left *in-situ* after formations had left them. An extra (and bogus) signal office was established in Jerusalem for GHQ, which was in fact near Ramle. A trunk telephone route was built to Nablus to mislead as to the axis of the advance. (Nalder, 185).

cavalry would once again find themselves in the Valley of Desolation'. He also remembered 'vividly the lurid language that arose on all sides when this report spread about the camps!'[8] To an exceptional degree all important orders were given verbally except at the highest formation level.* Officers who visited Chauvel's new headquarters in the Sarona area on the day after his arrival on 16 September 'received only headshakes and even emphatic denials'.[9] Wireless traffic was continued from its previous position near Jericho long after it had moved. Until the actual eve of the assault exact timings of the attacks were withheld from even the divisional and brigade commanders.

By 17 September Descorps' two leading divisions, the 4th and 5th, were safely settled into their orange groves near Sarona, ten and eight miles respectively behind the front line. Ausdiv, which was destined to follow them, was ensconced in olive groves near Ludd (Lydda), seventeen miles behind the line.

*'Secret conferences were called in turn at the various divisional headquarters, when the scheme was explained to staffs and commanders of brigades, each of whom then prepared his scheme and submitted it verbally to his immediate superior.' (Preston, 197).

27

'The battle was practically won before a shot was fired.'

WAVELL in *The Campaigns in Palestine*

'If thou hast run with the footmen, and they have wearied thee, then how canst thou contend with horses?'

JEREMIAH[1]

Battle of Sharon – arrangements for cavalry to enter gap – barrage – infantry assault

At a GHQ headquarters conference on 11 September Bulfin had asked Allenby for a ruling that the troops of Descorps should pass neither north nor east of XXI's artillery wagon line until the artillery had moved forward, and the infantry wheeled to the right, totally clearing the original front. This, even as a simple manoeuvre with no opposition, would take some two hours. Chauvel might just have accepted it had his leading divisions both been going to use the shore route, but, quite rightly, for greater speed and to lessen the possibility of confusion, he wanted clearly separated axes: 5th Division on the beach and 4th Division on a track two and a half miles inland. According to Barrow, Bulfin said to him, 'I know you damned cavalry fellows; you'll go getting in the way of my people, obstructing their movements, masking their fire, delaying them – in short, making a damned nuisance of yourselves if I let you go before we are all clear.'[2] Barrow tried to persuade Bulfin that a formation could be adopted which would not impede a single rifleman or machine gun, but the Corps Commander was adamant. Barrow, not unnaturally, was much perturbed at his division being assembled over two miles from the front line. He, like all contemporary cavalrymen, had in the forefront of his mind what he saw as the distressing delays which marred the battle of Cambrai.

There was a case, too, for believing that at Sheria on 7 November, 1917, time lost in getting the mounted troops through had enabled the enemy to form rearguards. (See p. 170). Barrow's very real concern was to avoid missing the fleeting opportunity which was nearly always all that the successful employment of mounted troops ever enjoyed. Bulfin, on the other hand, was fearful that should even a half-hearted counter-attack be launched, dreadful confusion including the masking of his guns, might be caused by the cavalry following his infantry too closely.

> 'It was essential,' as Barrow put it, 'that we should not become entangled in a fight with the Turks which would delay our march. It was inadvisable therefore to follow immediately on the heels of the infantry, while it was of the utmost importance not to lose one "unforgiving minute" when once the way was open. Consequently, a compromise in time and space had to be made which complicated the staff problem considerably. The object to be aimed at was to bring the head of the column up to the gap at the moment the enemy troops would be cleared from the front. The exact time when this would occur could only be guessed at. . . . Major-General [Sir Vere] Fane, whose division (the 7th Meerut Division) was to make the gap through which we were to pass, was an old acquaintance. I went to him with my difficulty and without hesitation he . . . undertook to give us the word himself directly the enemy had been cleared out of our way, wherever his own troops might be at the time.'[3]

The divisional orders (as in the case of all divisions, not issued till the day before Z-Day) stipulated that the division would 'be assembled . . . in a position of readiness at 08.00 in column of brigade masses . . . . The order [to move forward] will come from the division.' A 'brown line' was drawn on all maps. No mounted man was to pass north of it without permission from Divisional headquarters.[4] As Barrow puts it,

> 'The brown line was some 6,000 yards behind our own trenches. In the course of its march up to the gap the division would have to negotiate two narrow river crossings, water horses and make its way through the complicated routes of our own back areas. If, therefore, it waited at the brown line until I got the word "Go!" from General Fane, it could not arrive at the gap until two hours or more after the infantry had cleared the way. I therefore told [Lieutenant-] Colonel Foster, GSO1, not to await the message from me but to lead forward the division directly it was assembled at the position

> of readiness. The divisional order was written with one eye on the XXI Corps, which would receive a copy in the ordinary course of business. But it was not possible to suffer the loss of those two priceless hours if there was any way of avoiding it. The proceeding was slightly irregular.'[5]

Both Barrow and Macandrew (5th Division) slept the night at the two infantry divisional headquarters, so that they themselves could be with Fane and Shea (60th Division) at the vital starting time.[6] It was fortunate that all four generals were on good terms, having served with each other in the Indian cavalry. At dusk dismounted pioneer parties were assembled well forward. Their horses were led as close behind them as possible, liaison with their brigades being maintained by gallopers. Since the movement was to start before dawn, removing obstacles, ramping of trenches and flagging tracks through the tangle of dug-outs and 'the hotch-potch of the area lying directly behind the trench system' was vital. 4th Division's 'advance report centre' was established 2,500 yards behind the trenches at 10.30 p.m. on 18 September.[7]

During the night of 18 September XX Corps swung forward its right on the east of the Bire–Nablus road. After heavy hand-to-hand fighting its limited objectives had been taken, together with over 400 prisoners. Not long after midnight the headquarters of VII and VIII Armies were heavily bombed so as to disorganize signal communications. At the same time XXI Corps' infantry deployed along taped lines on no-man's-land, which averaged about 1,000 yards in width. A Handley-Page bomber, which had been flown from England three weeks earlier, also dropped 1,200 lbs of bombs on Afule aerodrome, railway station and telephone exchange. It did not take long to learn that the main exchange at Afule had been destroyed, while the exchanges at all three army headquarters, as well as VIII Army's wireless station, had also been put out of action. Over Jenin aerodrome, where there were stationed three German two-seaters and eight scouts, a standing patrol prevented any reconnaissance throughout 19 September.*

*The eighteen machines, mostly SE5s, which effected this, 'sat up above' the aerodrome in pairs for an hour and a half at a time throughout the day. Each carried four twenty-pound bombs 'and directly any movement was noticeable down came a bomb. The aerodrome became as silent as the grave.' The

(Continued over)

In the coastal plain at 4.30 a.m. on Z Day, 19 September, thirty-five minutes before dawn, 385 guns, trench mortars and machine guns opened an intense bombardment lasting fifteen minutes. There was a gun approximately every fifty yards, the weightiest concentration of artillery of the whole campaign.* Two destroyers assisted by firing on the coastal road to the north. During this first violent barrage (over 1,000 shells a minute) the leading infantry waves advanced to the enemy's front line, lit up by hundreds of Turkish signal rockets. The guns which replied fired on the now vacated British front line, thus passing over the advancing troops' heads. The enemy's feeble barbed wire was either cut by the leading troops or crossed by planks. While it was yet dark the infantry swarmed into the enemy's trenches. In Wavell's words, 'There is little to be written of the infantry assault. It was completely and overwhelmingly successful. Only in a few places was it even temporarily held up.' The stoutest resistance was on the right, opposite XX Corps, but before long that was overcome, all objectives were reached and a secure pivot for the wheel of XXI Corps was formed. Of that corps 3rd (Lahore) Division soon turned right and carried Jiljulie and Kalkilie. 75th Division had severe fighting at Et Tire, having advanced 11,000 yards in six and a half hours, while 7th (Meerut) Division made for El Taiyibe. Soon after 11.00 hours all the defence systems in the Corps front had been taken and the Turks were in full retreat. 60th Division broke through on the coast at great speed, covering 7,000 yards in two and a half hours. By 7 a.m. its two leading battalions had crossed the Nahr Falik ready to cover the debouching of 5th Cavalry Division.[8]

---

squadron set up a record for a day's flying, being in the air 104½ hours, and next day only one of the machines was unable to continue.

Two days earlier at the request of Descorps' medical services it had dropped tyre tubes filled with food from 2,000 feet as an experiment. It proved a success. (*Chauvel*, 165).

Three balloon sections were employed in keeping formations in touch with each other by telephone and heliograph. The packing up and moving of balloons to advanced positions as the battle proceeded took place at great speed. (Massey: *Triumph*, 143–6)

*In France for Fourth Army's attack on 8 August, the proportion had been about one gun to every ten yards.

28

'During our three years in France and during the many times we went up to the line in the hope of going through, I, personally, never had the slightest feeling that we should, whereas on this occasion I felt perfectly certain that we should.'

An officer of the 18th Lancers,
second week of September, 1918

'If Intelligence was correct in its belief that the enemy had no reserves of importance south of Nazareth, the mounted divisions had little to do but ride straight and hard on their objectives. . . . Once clear of the gap they were to speed for El Afule, Nazareth and Beisan and rely upon the infantry to drive the enemy back to them.'

H.S. GULLETT in the *Australian Official History*

'The two hours' work of the infantry on the morning of the attack was all that was required to enable the mounted men to finish the war with Turkey.'

W.T. MASSEY in *Allenby's Final Triumph*

'As we cleared the Turkish trenches and rode unopposed through the debris of defeat, we all felt that the "G" in "GAP" for which we had waited patiently [in France] for years had at last been reached.'

A squadron commander of 19th Lancers

'From 10.00 hours onwards [on 19 September] a hostile aeroplane observer, if one had been available, flying over the plain of Sharon would have seen a remarkable sight – ninety-four squadrons, disposed in great breadth and in great depth, hurrying forward relentlessly on a decisive mission – a mission of which all cavalry soldiers have dreamed, but in which few have been privileged to partake.'

LIEUTENANT-COLONEL REX OSBORNE
in *The Cavalry Journal*, 1923

38. The Inverness Battery, RHA, attached to the 3rd Brigade, going into action

39. A section of the Auckland Mounted Rifles Regiment. The leather buckets hanging round the horses' necks had to be put on their noses to prevent them sucking the sand for its salt

40. The Royal New South Wales Lancers watering their horses at Esdud, 1918

41. Indian lancers at Jisr Benat Yakub on the advance to Damascus

Megiddo: Sharon, 19 September, 1918

'Saul would have wondered, Jehu would have glowed had they seen our cavalry sweep down this grand old battlefield.'

SERGEANT M. KIRKPATRICK,
2nd New Zealand Machine-Gun Squadron,
attached to 5th Brigade.[1]

Battle of Sharon: 5th Cavalry Brigade advances – 4th and 5th Cavalry Divisions go through gap made by XXI Corps – Nahr Falik

It will be recollected that Onslow's 5th Brigade had been temporarily attached to XXI Corps 'for local and immediate exploitation'.[2] (See p. 248). It had orders to move up behind 7th Division, advance directly on Tul Karm, capture the town (with VII Army headquarters) if possible and hand it over to 60th Division when it arrived there. However, Onslow had been told by Shea of 60th Division not to concern himself with the town if he met serious resistance, but to advance with all speed to cut the road to Nablus. Mostly disregarding the Turkish infantry, which, reeling from the infantry breakthrough, was streaming across the foothills south of the town, and keeping clear of the infantry battle, the brigade arrived soon after midday on the plain west of it. 'A' squadron of 15th Regiment pursued a column moving north from Irta, while the rest of the brigade galloped to the north and into the foothills to the north-east of Tul Karm. About 2,000 yards north of the town a squadron of the French *Régiment Mixte* galloped down and captured an Austrian battery in action. Since the town was defended by numerous machine guns, Onslow, as ordered, left it alone and concentrated instead upon the thousands of fugitives pouring eastwards towards Nablus. They were an easy prey. As the British and Australian aeroplanes bombed and machine-gunned the head of the fleeing swarms, ridden down or challenged from the adjoining hills with dismounted fire, they surrendered in droves. The greatest havoc was caused and in several places the road was blocked by overturned vehicles. By 6 p.m., an hour after Tul Karm had been taken by 60th Division (which against opposition and over sand had marched sixteen and a half miles in twelve and a half hours), the brigade had captured about 2,000 prisoners, fifteen guns and enormous quantities of material. Part of the French regiment was detached to pursue a body of Turks retreating towards Shuweike. 'The chase was a long one, but very successful, and the Frenchmen returned on the following morning with some hundreds of prisoners.'[3]

While 5th Brigade and 60th Division were thus engaged, the centre of XXI Corps also wheeled eastwards and pressed forward in pursuit into the hills. The infantry had now completed the first part of its task of clearing the way for the mounted troops.

* * *

Macandrew's 5th Cavalry Division had arrived by 04.00 hours[4] in rear of 60th Division, disposed in depth with its head about three miles behind the front line. 'It was an inspiring sight,' wrote one of the Gloucesters, 'to see such a mass of horses in column of troops stretching for over three and a half miles along the beach.'[5] By 08.30 hours Barrow's 4th Cavalry Division, having left its bivouacs at 04.15 and watered its horses, also disposed in depth, was lined up, off-saddled, about two miles east of 5th Division. Both divisions were here rejoined by their horse artillery.* After Barrow's horses had been fed, his division 'moved forward and halted with its head [11th Brigade] 100 yards from the gap.† The pioneer parties,' he wrote, 'had reached our own wire at 7 a.m. and, covered by the infantry, had cut the wire, marked the track with red and blue flags and cleared the way through the enemy's defences.' At 08.40 Fane, commanding 7th Division, told Barrow that the way was clear for him to go forward.

> 'I hastened to the head of the division,' wrote Barrow, 'which had arrived only a few minutes earlier, and at 8.58 a.m. the leading unit [Jacob's Horse] passed through the gap. It did not take us long to cross no man's land and pick our way through the Turkish trench system. Dead and wounded Turks, litter of camp equipment, stores and all the paraphernalia of trench warfare lay scattered around as the result of our bombardment. We were soon in the open and delighted to be free at last from the oppression of trench warfare. . . .

*'B' Battery HAC had been detached from Ausdiv and allotted to 5th Division.

†12th Brigade came next, followed by 10th, which went through the gap at 10.30 in the following order:– Dorset Yeomanry, 2nd Lancers, 17th MG Squadron, Berks Battery, 38th Central India Horse (less two squadrons), Brigade Field Ambulance, one squadron of the CIH. The other squadron was acting as escort to the divisional first line transport. (Whitworth, 132–3)

'The country we were about to traverse is favourable to the movement of large mounted formations. The undulations, the low hills 200 to 300 feet high, the patches of cultivation favoured cover and concealment without impeding the rate of march. The only obstacles are the wadis, which break up formations and are often not seen until one reaches the brink, but they are easily crossed. The "going" on the whole is good, except for some wide patches of uneven sand that were very trying for the guns, ammunition columns, field ambulances and wheel transport. The division went rapidly forward.'[6]

It was now in three brigade columns in echelon from the left. When it came to the enemy's rudimentary third line of defence, it was fired on. Two squadrons of Jacob's Horse charged and took 115 prisoners.[7] A little further on, 6th Cavalry, leading 12th Brigade on the right, charged and captured a small rearguard. One of the officer prisoners had on him a marked map which showed that it was intended to cover both of the routes over the Samarian hills.[8]

* * *

On the seashore, meanwhile, 5th Cavalry Division had begun to pass through the Turkish entrenchments at about 06.00 hours. Unlike the 4th, it had marched to its position of readiness after dusk the previous evening. Thus it had less ground to cover before entering the enemy's lines. At 7 a.m. Shea told Macandrew that there was 'no longer any shelling on the beach south of the Nahr el Falik and that in his opinion the cavalry could now move up to the river'.[9] With great rapidity the division now raced along the narrow strip of sand on the seashore hidden from view and from fire by the cliffs. Macandrew watched the start of the leading brigade, the 13th, 'and when he saw how fast it was going galloped after it in hopes of steadying it, but was unable to catch it.'[10] The naval gunfire 'effectively dissuaded the enemy from any activity on the beach and as a result, the division made its exit in column of troops through the trench system and battle front down what was in effect a deep and broad communication trench – an ideal covered route.'[11]* Nevertheless the sand was deep, heavy and

*A few of the destroyers' shells rained down on the cavalry in error, but the casualties were slight. (Maunsell, 217)

holding and the long trot took a good deal out of the horses. Both 13th and 14th Brigades, less their wheeled vehicles, had crossed the Nahr Falik at its mouth at about 08.00.

13th Brigade's leading Squadron, 'D' of 9th Hodson's Horse, was now fired on by dismounted Turkish cavalry. The fire was immediately returned by two machine guns, while the squadron moved round to turn the position. At the same time 'C' Squadron attacked frontally in column of troops widely extended and at increased distances. The enemy did not await the attack. 'C' Squadron pursued them. At this moment an aeroplane dropped a message saying that some 200 enemy with two guns were 400 yards ahead. 'D' Squadron's advance troop, under Risaldar Nur Ahmad, had already come under the fire of this party. The rest of the squadron rushed up and charged without waiting for covering fire. One man was killed and two were wounded. Three enemy officers, about fifty men, the two guns and twelve wagons were captured.[12]

As an example of the sort of things which were happening to both cavalry divisions on this and subsequent days, it is worth quoting extensively Hodson's Horse's regimental historian's description of the next few hours:

> '"D" Squadron was now out on the open plain over which many scattered parties of Turks and transport were retreating, and a certain amount of hostile fire was encountered. Fire was also opened on the regiment by some guns in rear, but the squadrons being opened out and the pace increased, little harm was done. Meanwhile the advance proceeded steadily and the village of Mukhalid was soon reached. Near here the right advanced patrol of "D" Sqaudron under Dafadar Mehtab Singh charged and captured or killed four men working a Lewis gun [presumably a captured one] and secured the gun. These men stuck to their gun to the last and wounded one of the patrol, besides killing a horse, before they were disposed of. As the squadron approached a troop of Turkish cavalry opened fire from some trenches on rising ground about three hundred yards north of the village, but they were promptly charged by a troop under Jemadar Nawab Ali Khan and forthwith surrendered. Some miles further on another Turkish troop was encountered covering the crossing of the Nahr Iskanderun, but they too were

charged in front and their retreat threatened by another troop on their flank, and they made haste to retreat.'*

Numbers of the Indians' horses were so knocked up by the speed at which they had been propelled forward that some of them had to be destroyed.† 'B' Squadron now took the lead from the exhausted 'D' Squadron and the advance continued unchecked across the Plain of Sharon. At 11.40 the Nahr el Hudeira was crossed and soon afterwards 13th brigade entered Liktera where a German-manned supply depot and 'a complete training establishment' surrendered without striking a blow.[13] There were also 'some lorries and a hospital sumptuously equipped'.[14] The brigade had taken well over 250 prisoners and four guns. It had ridden twenty-two miles in about four hours, mostly over heavy going. The rest of 5th Division arrived at Liktera by 15.00 hours.

*A number of other encounters took place at this time. Typical are two in which the Gloucesters took part. A patrol from 'B' Squadron with only four men was sent to a flank to discover from where some enemy guns were firing. Corporal Wiseman soon returned to report that he had found two of them in the act of limbering up. He at once charged them and the officer in command surrendered with twenty men. Whilst he was reporting this, Private Forrest 'saw a large column moving away north. He galloped straight at it, which halted on his approach and surrendered. The column consisted of thirty-seven wagons, four officers and 100 men. Private Forrest turned them about and escorted them to the rear.' The regimental historian comments: 'It was a big capture for one man!' (Fox, 262)

†Five each of the Gloucesters and 18th Lancers died while seventeen of the Gloucesters' and ten of the 18th's were evacuated and afterwards destroyed. The casualties for Hodson's Horse are not recorded but they were probably even greater. (Fox, 263; Falls, 523)

29

'I am beginning to think that we may have a very great success. The weather is perfect; not too hot and very clear; just right for my artillery and aeroplanes in pursuit. My horses are in sufficient strength to be irresistible.'

ALLENBY to his wife, 19 September, 1918

'The whole were in column of sections, moving along by moonlight in an enemy's country; no scouts; without any flanking patrols or local protection of any sort. But the enterprise was a big one; risks had to be taken, and, fortunately, there was no opposition.'

The historian of 18th Lancers
on the night of 19/20 September, 1918

'The arrival of a British cavalry brigade at [Liman's] headquarters was his earliest intimation that something really serious must have happened to his army.'

WAVELL in *Allenby*

'There is a splendid simplicity in riding straight upon the enemy's headquarters and capturing his leader which would have appealed not only to the army but to the whole world.'

The Official Historian

'Can there any good thing come out of Nazareth?'

The Gospel of St. JOHN[1]

5th Cavalry Division ordered to Nazareth – 13th Cavalry Brigade fails to capture the town or Liman von Sanders – Kelly sacked – 4th Cavalry Division crosses Musmus Pass – Howard-Vyse, 10th Cavalry Brigade, sacked

For both 4th and 5th Divisions what they feared would be the points of stiffest resistance now lay ahead: the two passes allocated to them across the Samarian hills into the Plain of Esdraelon, the ancient Armageddon. Both passes would have to be approached and crossed in the course of the night. Kelly, commanding 13th Brigade, leading 5th Division, had orders, issued on 18 September, to halt at Liktera until 18.15 so as to water, feed and rest:

> 'At 18.15 hours your Brigade will move on Nazareth, a distance of thirty miles. It is calculated that the leading regiment should reach Nazareth by 03.00 hours (September 20th). It is of the utmost importance that Nazareth is surrounded before daylight. All roads will be barricaded so as to make it impossible for a car, by rushing the posts, to enter or leave Nazareth. All individuals moving to or from Nazareth are to be made prisoners.
>
> 'The town of Nazareth will be captured, but it is not to be bombed or shelled. Hostile G.H.Q. is situated in the town and all commanders, staffs and documents are to be seized and taken care of.
>
> 'Should, for unseen reasons, the march of the brigade be so delayed that it cannot reach Nazareth before daylight, the leading regiment must be pushed on and followed as rapidly as possible by the remainder. If the fighting wheels are unable to keep up, the advance of the fighting troops is on no account to be delayed.'[2]

This raid, undertaken with the idea of riding straight at the enemy's general headquarters and capturing the Commander-in-Chief, has a prominent place in cavalry lore. It nearly succeeded. The chief reason why it did not seems to have been that it was only an afterthought. Kelly's instructions of 18 September differ significantly from those given in 5th Division's operation order of the day before. In consequence he had not been provided by General Staff (Intelligence) with native guides or with a plan, an aerial photograph or an adequate map of Nazareth. The brigade was expected under the 18 September instructions to cover the thirty miles from Liktera to Nazareth in eight and three-quarter hours. As the march involved crossing the Samarian hills and negotiating the rough hill country surrounding the town all in the dark, this does seem a very tall order. In places, indeed, the track was so bad that 'for miles the men had to march on foot leading their horses'.[3] Had Kelly been permitted to lessen his obligatory seven-and-a-quarter-hour halt at Liktera, he would have had a better chance of success. As it was, it was fortunate that the brigadier was a fluent Arabic speaker. He spent much of the lengthy halt questioning local Arabs and Jews about the track over the hills, and finally bribed two Arabs (with £5 each) to lead the way. His men were soon confronted by 'innumerable goat tracks running in every direction'. When Jarak, more than half way across the hills, was reached two squadrons of Hodson's Horse were dropped off to prevent the enemy sending down troops from Haifa to block the passage of 4th Division through the much more difficult and easily defended Musmus Pass (see p. 274). According to one reliable source 'this detachment was made under Corps orders; it is surprising, however, that [Macandrew] did not order the detachment to be made from 14th Brigade, which was closely following 13th through the pass, as every detachment from the last-named brigade would handicap its chances at Nazareth.'[4]

At about 02.15, a halt was called and the brigade ordered to close up, but beside brigade headquarters, no one was present! Guides were sent back and, after an hour, two of the three regiments had arrived, the Gloucesters having lost touch with the other two.* With these Kelly pressed on and began to climb up the lower slopes of the hills surrounding Nazareth. At around 04.15 a village was reached which the native guide declared to be Nazareth. This Kelly soon proved to be untrue. He questioned a small

*An officer of the Gloucesters wrote in his diary:

'I discovered to my horror that the Indians in front of me, whom I was following, had lost touch with the main body at the critical moment. . . . This was a terrible affair, and I pictured myself lost at the head of the regiment, and I saw Gen. Kelly with the 18th Bengal Lancers in their glory advancing on Nazareth. I was furious with the Indians in front of me, and I cursed them up hill and down dale. That being of no avail I settled down, halted, and dismounted. . . . One path led eastwards, which appeared to be the wrong way, as I knew one objective was to blow up the railway which runs to Haifa and that could only go east and west along the valley somewhere. I sent a scout up the path, but he returned saying he heard and saw nothing. Another scout reported the same of a path leading eastwards. I had meanwhile made up my mind that one of the two paths leading across the valley must be right. A scout up one of these solved the situation by picking up a packman who had fallen out from the troop ahead. We went along, soon came up with Major Mills, the 2nd in Command of the Lancers, whom I was delighted to meet – until he told me that he had also lost the Brigadier and staff, who had gone on in front. Another halt and more scouts sent out, and after half an hour we found the Brigadier near the railway, much to everyone's relief. Being late, we hurried on, leaving the R.E. officer with a specially trained demolition officer to blow up the line, which he did.' (Fox, 264)

One of the squadrons of Hodson's Horse had also gone astray in the dark.

At about 3 a.m. the demolition party blew up six separate stretches of the railway. These explosions may well have awakened the troops in Nazareth.

boy asleep in the street who told him that the place was El Mujeidel to the south-west of the town and that many Turks were in it. These sleeping soldiers had to be located and disarmed (by parts of the 18th Lancers), for otherwise they might have telephoned a warning to Nazareth.* This done, the brigade reached Yafa, one and a half miles short of the town. Here again seventy-five Turks had to be rounded up, causing more delay and forcing Kelly to detach further escorts from 18th Lancers.† By now 13th Brigade had covered fifty-four miles in twenty-four hours.

The day was beginning to dawn as the forked roads just south of Nazareth were reached. Here a squadron of 18th Lancers was sent to watch the road from Afule. This was crowded with lorries escorted by 400 Turks heading up the hill to the town. Utterly surprised, after a token fight, they surrendered.‡ The last remaining squadron of the lancers was now sent up the hill on the north-west side of the town, where a large barracks full of Turks was in a dangerously dominating position.

Left with only the Gloucesters and a few of the lancers, Kelly now trotted briskly on with these remains of his weak and weary brigade in an attempt to reach the branch roads north of the town, the only ones by which motor-cars could escape from it. At 04.25 the Gloucesters were given orders to attack. They drew swords and

*While this was being done a Turkish officer, fully dressed, emerged from one of the houses and 'very politely asked in German who we were and what we wanted. A revolver to his head and a curt order to put his hands up was the reply.' Two Turks were now impressed into service as guides. 'In close attendance and with lances occasionally pricking them forward came two men of "D" Squadron [of 18th Lancers].' (Hudson, 230)

†Prisoners were usually sent back in batches of twenty or thirty, escorted by two men.

‡The squadron discovered that some of the lorries which they had captured contained a Turkish field treasure chest stuffed with gold and notes.

> 'But how to get it away? None of the lorries was available as the drivers had drained their tanks before surrendering. There was a petrol lorry, however, to which a huge live bear was chained. A misguided sowar, meaning no doubt to be helpful, loosed him. The bear jumped on the petrol tins, sat back and defied all. Eventually he was got rid of, but only to meet a soldier's death later on. The treasure was duly handed over to the proper authority. It is said that it amounted to £20,000 in value.'

The squadron commander was given one Turkish gold sovereign. Five others became a trophy in the Mess. (Hudson, 233) galloped towards the town. On the way a machine gun and its crew of nine were galloped down and captured. The yeomen

> 'poured into the streets, which were just filled with the early morning bustle of an awakening camp. The enemy were completely taken by surprise, the soldiers, both German and Turks, mostly unarmed. Hundreds of them surrendered before they had "got the sleep out of their eyes". Lieutenant Inglis was sent with his troop to capture General Liman von Sanders, but he entered several houses without finding him, being told in all instances that the General had left the previous night! But the important documents of the enemy G.H.Q. were captured.'[5]*

Kelly's urgent application for a guide who knew the town met with no success. 'The inhabitants, recalling grisly tales of British sympathizers strung up in rows after the evacuation of Es Salt, were terrified to be seen speaking to the assailants and would give no reliable information. It seems that Liman, awakened by the shots south of the town, escaped in his sleeping-clothes. Seeing that the situation was not as bad as he at first feared, he later returned to dress and direct the fighting in the town.[6]

There now followed some hard, mostly dismounted, fighting both on the outskirts and in the streets. The Turks, strongly posted and well supplied with machine guns, generally got the better of it. Attempts were made to carry the enemy's positions but without success. Lieutenant R.U. White, with one troop of the Gloucesters, attacked an enemy post which commanded the main road into the town, killed twelve and wounded fifteen men, the remainder surrendering. At one point some German clerks 'displayed a desperate bravery which won the admiration of the British, being almost annihilated by machine-gun fire' before they gave up.[7]

The vanguard squadron ('D') of the Poona Horse of 14th Brigade (which, skirting the town, was making direct for Afule), when at 6.30 a.m. it arrived at a point some three miles south of the town came across seven lorries full of Germans and Turks escaping up the Afule road. These, led by a German officer, jumped out of their vehicles and opened machine-gun fire, killing one sowar and wounding a non-commissioned officer. The squadron

*In fact most of the papers had just been burned: a great loss to historians of the campaign.

commander returned the fire with his Hotchkiss guns, upon which the enemy lorries continued on their way only to be cut off by 13th Brigade troops. Undaunted, as the rest of 14th Brigade arrived on the scene, the enemy opened fire on them. Clarke, the brigade commander, speedily ordered up the machine-gun squadron which soon put an end to the fight. The lorries were set on fire by their occupants who then surrendered. 'D' squadron meanwhile moved on. It soon came into contact with a force which was advancing on Nazareth from Afule. The squadron immediately charged, covered by two Hotchkiss guns, killing several Turks with their lances and capturing over 200 prisoners.[8]

Soon after 10 a.m. the Gloucesters retired, covered by Hodson's Horse and 18th Lancers. At 06.50 Kelly had informed Macandrew that he had over 1,200 prisoners, but no Liman. He asked that 14th Brigade should come to his help. At 10.55 Macandrew replied that 14th Brigade's horses were too exhausted and that Kelly was to withdraw to the north of Afule. This he did.* At 1.15 p.m. Liman left Nazareth after ordering the rearguard to withdraw to Tiberias. He had with him only three staff officers and was, of course, entirely bereft of communications.

Chauvel, meanwhile, confident that Afule would soon be captured, ordered a conference there for noon, well before which time it had in fact fallen (see p. 285).

> 'He arrived early, believing,' as his biographer puts it, 'that Nazareth and Liman von Sanders would be taken by Macandrew. Perhaps he was savouring the idea of opening his conference with some such remark as: "Gentlemen, you will be glad to know that the enemy C-in-C is on his way here under escort." Impatient for news of the raid, he drove out along the Nazareth road to meet Macandrew. Chauvel's excitement can be imagined when the distant car he was watching turned out to be a large Mercedes. Liman's car! But great was his disappointment when Macandrew alighted from it to make his report.'[9]

*The Gloucesters lost in all two killed, eleven wounded, six missing and twenty-eight horses killed. Hodson's Horse suffered two killed and nine wounded. The losses of 18th Lancers are not recorded.

Later on, when Allenby had studied the full story of the failure to capture Liman, he removed Kelly from his command. From all that is known, this seems to have been a singularly harsh punishment. Kelly's mistakes seem to have been the time he wasted searching villages, dropping off squadrons other than those he had been ordered to and his failure to cut the road to Tiberias, this last probably because dawn was breaking and his force was too depleted. That it was also exhausted due to the initial speed of its advance was also counted against the unfortunate brigadier.

The 15th Brigade with the guns and wheels of the whole 5th Division did not leave the Nahr Iskanderun until dawn on 20 September. The gunners had to spend many hours making a roadway for their guns over the hills. They did not reach Afule until 11.00 p.m.[10]

* * *

While 5th Division had raced forward, getting a three and a half hour start of 4th Division, as had always been expected, since it had the easier route, Barrow's brigades had to halt at 6.30 p.m., on 19 September for three hours. The need for water, feed and rest was essential, but watering the horses took much longer than Barrow was happy about, the supply coming only from a small number of wells and an inadequate stream. 'I am doubtful,' he wrote, 'whether the time allowed of every horse getting a drink.'[11] The Division was now well behind schedule. Before it lay the easiest to defend of the passes between the plains of Sharon and Esdraelon. The Musmus Pass is fourteen miles long. Its summit is 1,200 feet above sea level. The recently repaired road through it was only wide enough for a column of half-sections of horse. 'A couple of machine guns,' as Barrow wrote, 'would have sufficed to hold us up for hours; a battalion of infantry with machine guns would have been enough to bring the whole division to a full stop.'[12] The urgent need to reach the far side of the pass before daybreak was emphasized when a captured Turkish officer disclosed that an enemy infantry force, with machine guns, was already under orders to march from Afule to block the pass at Musmus. It was fortunate that it failed to get there in time.

At 18.30 orders were issued for the division to continue its march at 22.00, Howard-Vyse's 10th Brigade leading. He was ordered to send 2nd Lancers with the two cars of 11th Light

Armoured Motor Battery and a sub-section of 17th Machine Gun Squadron to reconnoitre the pass ahead of the brigade and to hold the cross roads at Kh. Arah, a point five miles up the pass from Kerkur. 2nd Lancers' orders, received at 20.15 hours, read: 'The attainment of your objective will be reported to Brigade HQ and you will then await the arrival of the remainder of the brigade.'[13] Captain Douglas Stewart Davison, only thirty years old, was at this time in command of the Lancers because both his commanding officer and second-in-command were ill.* He was told verbally that the rest of the brigade would not move from Kerkur until he had sent back his report giving the all clear or otherwise. This meant that he might expect to remain at Arah for about four hours. He, accompanied by the armoured cars and the machine guns, moved off at 20.45 in column of half-sections. Before Arah was reached nearly 200 stragglers had been captured and sent back under escort. By 23.30 the advanced guard had taken up positions covering the crossroads.

In the meantime, at 21.30 Barrow had arrived at Kerkur, Howard-Vyse's headquarters. 'I was surprised,' he wrote, 'to find that through some misunderstanding the brigade orders had given 11.30 p.m. instead of 10 p.m. as the hour of march. Pointing out this deviation from divisional orders and emphasizing the urgency of getting on the move as quickly as possible, I hastened with Colonel Foster (GSO1) to the advanced guard.'[14] This they caught up with at 23.30. Barrow's account continues:

> 'We were considerably behind time, and as, owing to the mistake in the starting hour, the main body of the 10th Cavalry Brigade would not come up for another hour at earliest, I ordered the advanced guard forward at once, and sent the armoured cars ahead to see if the pass was clear as far as Umm el Fahm [which they soon reported it to be]. There was a certain risk in sending the advanced guard unsupported. It was one of those occasions, frequent in war, when one must be prepared to risk all "on one turn of pitch

*Another major had just arrived from England and took over command on 23 September.

Davison was a son of a former commandant of the regiment. Both father and son were fine pig-stickers, the son winning the Kadir Cup after the war. He died of a fever soon afterwards. (Barrow, 201). Captain Davison was awarded the DSO for the action of Birket el Fuleh (see p. 279).

and toss". The only thing to do was to endeavour to make good the exit to Esdraelon as quickly as possible.

'The advanced guard started and Foster and I remained to await the arrival of the main body of the brigade. Time passed without any sign or sound of it. I became anxious and told Foster that we had better go back and see what had become of it. He tried to persuade me to remain where we were, saying there was no doubt the brigade would soon come. But it was already too long overdue, and my misgiving that something had gone wrong was too strong to permit of remaining inactive. We retraced our steps towards the entrance of the pass, meeting nothing on the way until about one mile from Kerkur, where we came on a mass of horsemen dismounted in assembly position. Here we learnt that the 10th Cavalry Brigade had missed the road to the pass and was heading in a contrary direction. Although there was a bright moon, moonlight is often deceptive, and to this may be largely attributed the mistake in taking the wrong track. It was fortunate that Foster and I did not stay longer in the pass. Had we done so the whole division would have followed the 10th Cavalry Brigade. As it was, the leading regiment of the 11th Cavalry Brigade had already filed off after it. We were in the nick of time to arrest further following in the footsteps of the 10th Cavalry Brigade. Foster set off at once to find the erring column, leaving me sitting alone on a rock. It seemed ages until he returned to tell me that he had caught up the brigade three and a half miles away and had turned it back. He then got off his horse and collapsed on the ground.... For twenty minutes he lay insensible to his surroundings. Divisional H.Q. having followed the 10th Cavalry Brigade in accordance with the march orders, I was now left alone, wrapped in semi-darkness and uncertainty....

'It was essential that something should be done, and done quickly. Mounting Foster's horse, my own having vanished in the semi-darkness, I set off to try to find the 12th Cavalry Brigade. By good fortune I ran into it, massed and awaiting the order to move. I could not, however, find anyone else who was able to direct me to the brigade commander or one of his staff. Every man I questioned seemed to have seen General Wigan a minute or two earlier, information which did not bring me nearer to him. Anyone who has tried to find a particular person in a crowd at night will know that it is like looking for the proverbial needle in a haystack.

'Suddenly I ran against General Wigan himself. . . . Explaining the situation briefly, I told him that his brigade, which was to have been the rear brigade, would now be the leading brigade. He was prompt to act and within five minutes he had made his arrangements, issued his orders, and the 12th Cavalry Brigade was on the road to Musmus. It was now 1 a.m., 20th September. An officer (Captain Woelworth, MC) of the 10th Cavalry Brigade staff, reported to me somewhere about this time that the 10th Cavalry Brigade on its way back to the mouth of the pass had again gone off on a false trail.'[15]*

When El Lejjun (or Megiddo) at the exit of the pass was finally reached at 03.00,† 2nd Lancers found 100 or so Turks sitting round a fire, singing, with their arms piled.[16] These were added to the bag. They were the advance guard of the troops sent to hold the pass, the main body of which was soon to be met. At 04.30 the head of 12th Brigade arrived and 2nd Lancers were placed under Wigan's command. 10th Brigade's loss of its way accounted for over two hours' delay. If the Turks sitting round their fire had instead been manning the exit as they had been ordered, the whole division might well have been disastrously delayed.

When Chauvel drove up at dawn to see how the Musmus crossing was going, he was surprised to meet Howard-Vyse riding towards him and much distressed to learn that Barrow had just removed him from command of 10th Brigade. Though this was probably justified it was nevertheless hard on the 'Wombat'.‡ He had only recently taken over from Godwin whom Chauvel considered his 'most dashing brigadier' and who, as has been shown, had done well in the Beersheba offensive. Howard-Vyse was an

*It had been badly let down by its native guide.

†This is according to the regimental historian. The Official Historian says '4.5 a.m.'. In either case, dawn had not yet broken. (Whitworth, 137; Falls, 519)

‡The present author's father, a fellow officer of Howard-Vyse's in the Blues in 1914, was a great friend of his. He was an exemplar of the perfect gentleman.

excellent staff officer and this was his first command. He had been given it at Allenby's insistence against Chauvel's advice.[17] He was literally the only casualty incurred by 4th Division during the passage of the Musmus.

30

'My cavalry are now in rear of the Turkish Army; and their lines of retreat are cut. . . . The Cavalry H.Q. are at Armageddon at the present moment. It is called a different name on the map.'

ALLENBY to his wife, 20 September, 1918

'The whole action did not take more than five minutes and furnished a perfect little example of sound shock tactics – movement and fire at right angles to one another.'

LIEUTENANT-COLONEL THE HON. R.M.P. PRESTON
in *The Desert Mounted Corps*,
on the action near Birket el Fuleh

'Well done the Tenth! I suppose never before in the history of the world has such a number of prisoners been taken by so small a force as one regiment!'

MAJOR-GENERAL H.W. HODGSON,
commanding Ausdiv, after the fall of Jenin.[1]

Birket el Fuleh – El Afule taken – Beisan taken – Nablus taken – Jenin taken

After 2nd Lancers had watered and breakfasted, Wigan ordered Davison, with the armoured cars, machine guns and demolition troop to capture Afule and destroy the railway lines there leading to Haifa, Beisan and Jenin. Just as day dawned 2nd Lancers' four squadrons issued forth from Lejjun in the early morning mist. Davison's subsequent report stated that 'a steady fire broke out from the direction of Point 193 [see map on p.280] at 05.30 hours. . . . "C" Squadron had dismounted one troop and their Hotchkiss troop* and were engaging the enemy with fire at about 800 yards' range, while the remainder of the squadron was concentrating in rear with a view to moving round the enemy's flank. The 11th L.A.M. Battery were in action about 300 yards in front of "C" Squadron and were engaging the enemy with machine-gun fire.' Davison now saw in the diminishing mist that

*These were the only Hotchkiss rifles which came into action during the engagement.

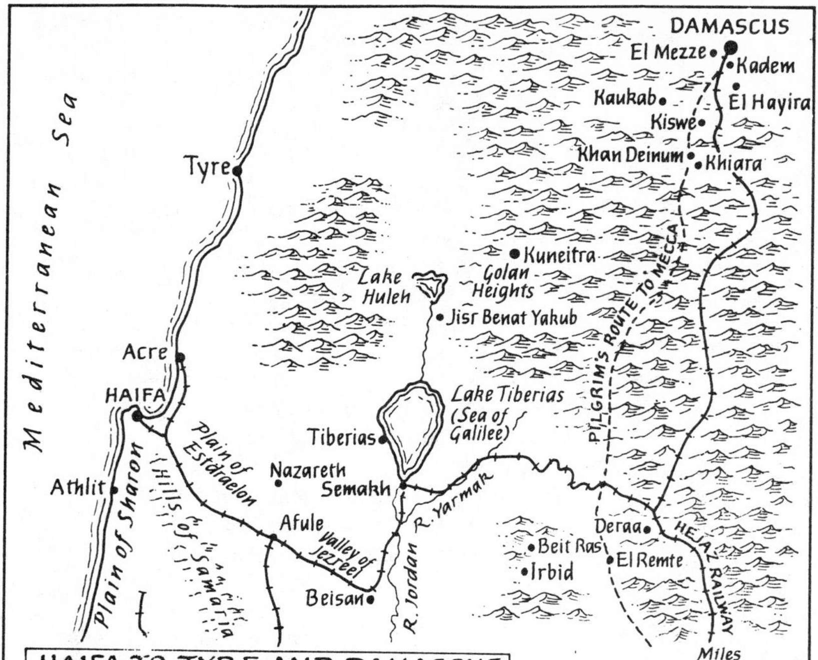

HAIFA TO TYRE AND DAMASCUS

the force opposed to him was much larger than the eighty men whom Lieutenant R.H.F. Turner, commanding "C" Squadron, had at first reported.

'The soil was black cotton in this place,' his report continued, 'and was not so bad as to stop horses galloping. The enemy's flank could be distinguished, and, as there seemed to be no obstacle to hold us up, I decided to turn his left flank and gallop the position.

'I accordingly directed the OC "D" Squadron (Captain [Edward William Drummond] Vaughan) to left shoulder, pointed out the enemy's position and ordered him to go slow for five minutes to enable me to get the machine guns into action, and then to turn the enemy's flank and charge. . . .

'By the time I had got the machine guns into action I could see that "D" Squadron was well round the enemy's [left] flank in column of troops, and almost immediately they wheeled head left, formed squadron and charged in open order, rolling up the Turkish front line, capturing or killing every man. R.M. Mukand Singh, on seeing "D" Squadron, mounted his troop of "C" Squadron and charged from the front with good results.'[2]

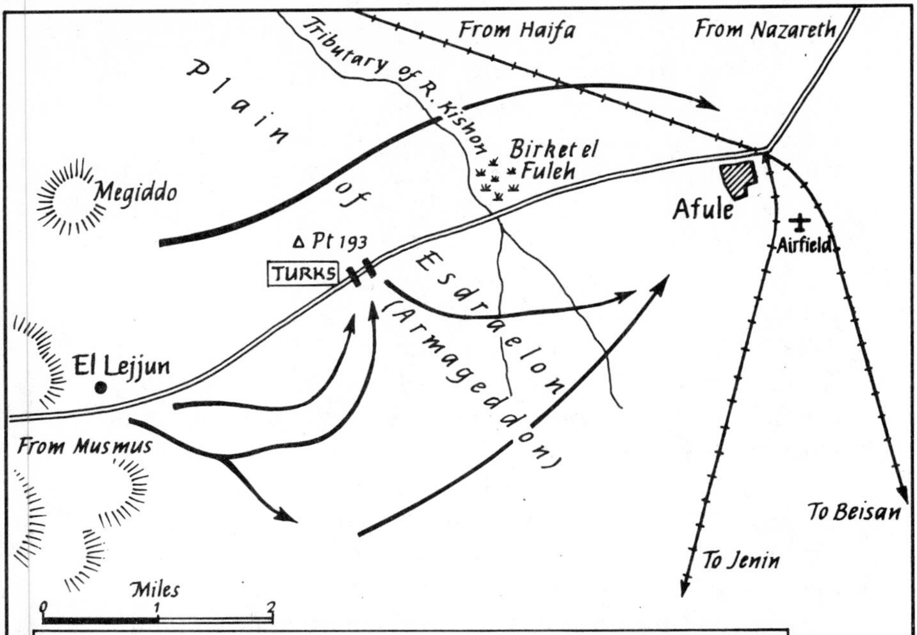

ACTION AT BIRKET EL FULEH, 20 SEPTEMBER, 1918

Captain D.E. Whitworth, commanding 'B' Squadron, which was moving, as originally ordered, to guard the right flank, when he found that 'bullets were coming unpleasantly close', increased his pace

> 'to a hand gallop and edged towards the right with the intention of locating the enemy's left flank. Almost immediately,' he wrote in his subsequent report, 'I saw "D" Squadron moving out in my direction, and I guessed that the C.O. intended making an enveloping movement from the right. There was no time to receive orders, so I decided to co-operate with "D". Just then the squadron ran into a wire fence hidden in the jowar [millet] which covered that part of the plain. I went on ahead and left it to my second in command (Ressaidar Jang Bahadur Singh)* to reform the squadron, which he did with great celerity.
>
> 'Bullets were coming thick and fast now, and I imagined that the squadron had had pretty heavy casualties; added to

*He was awarded the Indian Distinguished Service Medal for this action.

this I was in a blue funk of striking an uncrossable nullah, for the map showed a tributary of the Kishon between me and the enemy. From our experience of the inaccuracy of the Palestine maps, I might have equally expected a hill as a nullah, but in the flurry of the moment I became obsessed with the fear of sharing the fate of Sisera's host on the same field, and the words "that mighty river Kishon" kept ringing through my head.*

'Up to now I had not caught a glimpse of the enemy. Looking back I saw that B was crossing D's line, and I galloped back to lead the squadron off more to the right. There seemed to be plenty of the men left, and the formation was still tolerably good. We were moving at a good fifteen miles an hour by now. It was just then that I caught sight of the Turks. They were formed in two lines in column with a distance of two hundred yards [between them]. The men saw the enemy at the same moment, and I was just able to direct the leading troops on the rear line of the enemy before they all saw red and broke into a hell for leather gallop. It now became a question of who could get in first, but I was still in mortal terror that the Turks' determined stand might be fortified by the knowledge that a deep nullah lay between themselves and us. We had overtaken our ground scouts long ago. However, I need not have worried, for we were in for it now. I could hear yells from D on my left rear.

'Before I realized it we were right on top of the enemy, and it was only when I saw a young Turk deliberately aiming at me that I realized that I was still holding my map in my right hand, and had forgotten to draw my sword. The little brute missed me and ran under my horse's neck, and tried to jab his rifle in my stomach. I had just time to draw and thrust over my left knee. The point got him somewhere in the neck and he went down like a house of cards. My next opponent was a moustachioed warrior, probably an NCO, who was hopping about with a fixed bayonet, and a blood-thirsty expression. I suppose he had never met an opponent going at fifteen miles

*'They fought from heaven; the stars in their courses fought against Sisera.
'The river of Kishon swept them away, that ancient river, the river Kishon.'
(Judges, V, vv. 20, 21)

an hour; at any rate he had not begun to point when I poked him in the ribs.

'All the Turks on the left flank had now had enough of it, but a few ran out of the ranks on the right and fired down the line at us. It did not take long to polish them off. I made for one who had murderous intentions, but changed his mind at the last moment. My point caught him plumb between the shoulders, and the shock nearly dislocated my arm; perhaps he was wearing a pair of steel braces. During this time I was naturally preoccupied with saving my own skin and saw little of what was going on around me. I only remember my orderly Lal Chand, on my second horse, Advocate, who had the regimental flag furled round his lance, dragging along the ground a Turk who had stuck on the point, and I distinctly saw Jemadar Gobind Singh, VC, deliver a magnificent cut at an opponent, but as his sword was not made for this sort of thing, the Turk was not damaged.

'The men stopped killing directly all resistance ceased, but rallying them was an awful business. I was just beginning to do so when I saw that D had just reached the front line, which must have been considerably shaken by B's success on their supports.

'While we were rounding up the prisoners I became aware of a fat officer doubling about in circles. He finally fetched up in front of me and fell on his knees, shouting "Spare my life. I am a Syrian," to which I answered, "All right, old cock. Go and join your pals over there." This evidently conveyed nothing, so I tried less colloquial English, and then found that he understood none, but had prudently acquired his one sentence against a rainy day.

'All was over now bar the shouting, which, rather to my surprise, was provided by the prisoners, who burst into song as they were marched to our rear.

'After rallying the squadron, I sent a patrol back along the route of the charge to give first aid to our wounded, and we then found to our immense astonishment that, although about half-a-dozen horses had been hit, not a single man was wounded. One horse which had been hit by a bullet which passed between the back tendons of the off hind did not even go lame. All the hits were in the legs or belly, and I have noticed the same report in the case of other cavalry actions. It therefore seems that a machine gunner who has the misfortune to be charged by cavalry should, if he is cool enough to remember it, aim high.

'My batman, Nagington, by profession a stable boy from Auteuil, was not with the squadron, but saw the charge from the mouth of the Pass. I asked him that evening what he thought of the show, and he answered "I thought that Advocate went fine." To him it was obviously a species of race.'*

Because Whitworth, on Davison's orders, had earlier detached one of his troops as right flank guard, he had only thirty lances with him, but his Hotchkiss troop, dropping its weapons and packs, joined in the charge 'so as to make the numbers appear greater than they were'.[3]

The enemy force was, of course, that which had been ordered by Liman personally at 12.30 p.m. on 19 September to occupy the Musmus Pass, the advance party of which had been discovered singing around the fire earlier on. It consisted of over 500 Turks of 13 Depot Regiment and was entirely fresh. Why it took so long to march the fifteen miles which it had to cover is a mystery. Its three machine guns kept firing until the horsemen were fully upon the gunners, but they were clearly unnerved by the pace of the charge, for their fire was inaccurate to a surprising degree. The Lancers had only one man wounded (from 'C' Squadron) and twelve horses killed. This astonishingly small casualty list was probably due to the steady machine-gun fire of the armoured cars from 500 yards of the enemy first line, and of the machine-gun section further back,† as much as to the speed of the charge. The demoralizing

*This account is extensively quoted here not only because of its clarity but also because it is so redolent of the outlook of a typical cavalry officer of the Indian Army of the time.

Captain Whitworth had gained the Military Cross at Cambrai. He served with 2nd Lancers from 1911 to 1935 and in both World Wars, ending up as brigade commander in India in 1942. In 1943 he was awarded a CBE and he died, aged eighty-four, in 1974. Captain Vaughan was awarded the Military Cross for this action near Birket el Fuleh. He commanded 3rd Indian Motor Brigade, Mechili, North Africa, in 1941, when he was taken prisoner. In 1944 he escaped from Italy and commanded the Delhi Area and District from 1944 to 1947. He died, aged fifty-nine, in 1953.

†Both of these ceased firing just in time to avoid hitting the charging lancers.

effect of attacks from both front and flank is never slight. Though small in scale, this action near Birket el Fuleh was an almost perfect example of a speedily delivered shock attack, supported by automatic fire against an enemy force given no time to select ground from which to meet it. Forty-six Turks were speared or otherwise killed and 470 taken prisoner.

* * *

Within forty minutes of this action 2nd Lancers was advancing at the gallop on Afule where a certain amount of initial opposition was encountered. The Deccan Horse of 5th Cavalry Division had also at about the same time galloped into the village. By 8 a.m. the station had been captured, as well as the aerodrome with three aeroplanes intact. A fourth aeroplane got away in the confusion, but a fifth, unaware of the changed ownership, landed. Its crew was captured. Ten locomotives, fifty wagons and a complete hospital were taken in Afule.* So was a large petrol store and much else besides. There was also found a considerable stock of champagne and hock. An officer of 2nd Lancers wrote home: 'On the morning after the charge outside Afule, having settled down to a Boche cigar and a bottle of ditto hock, I would not have changed places with President Wilson himself!'[4] Left behind at Megiddo to guard the pass until Ausdiv arrived was the Middlesex Yeomanry.

> 'It was natural to conclude,' wrote Barrow, 'that in its movement on Beisan the division would have swept the open

*Trooper Reeves of the Staffordshire Yeomanry, it is said, at this time had a temperature of 104° from a touch of the sun and was sent off on a horse to the hospital a few yards away. Being almost insensible, he rode in error towards an enemy body. This he at once charged with his sword drawn, whereupon it surrendered. When he arrived back at Afule with eight German officers and seventy-two Turkish prisoners his temperature had risen to 105°! (Kemp, 72; Hudson, 241).

At about this time, a Jat recruit of 6th Cavalry on flank guard came across an enemy party with a machine gun. Charging it, he wounded one man with his lance, whereupon all surrendered. He brought in one German officer, ten German other ranks and sixty-nine Turks. (Hudson, 241).

These single-handed feats well illustrate the state of the enemy's morale at this time.

valley [of Jezreel] clear of all hostile bodies. But it is never safe to conclude anything in war. The officer commanding the Middlesex [Lieutenant-Colonel Lawson], going some distance ahead of his regiment in order to examine the ground over which his men would have to ride, found himself suddenly confronted by a Turkish detachment of several hundred Turks, which, retiring from Jenin, had come in rear of the division by a track which enters the valley east of Jenin. He was made a prisoner but the period of his captivity was short. He explained the situation to the Turkish officers and showed them on his map how hopelessly they were placed with British troops on all sides of them. This strange conference, held in indifferent French, ended in the position of captured and captors being reversed; and the officer led back to his regiment 800 armed Turkish officers and men as prisoners of war. This must be nearly a record capture by a man single-handed.'[5]

By noon the whole of the 4th Division was concentrated at Afule. At 1 p.m. it marched for Beisan, led by 10th Brigade, under its new commander, Lieutenant-Colonel Wilfrith Gerald Key Green, Jacob's Horse. Left behind to hold Afule until evening were the 19th Lancers. One of its squadron commanders wrote:

'We hoped to be able to get a bit of sleep in the afternoon, as we had had none for thirty-six hours and had another night march, but not a bit of it; as soon as the division had moved off fugitives started crowding down the road from Jenin [ten miles south of Afule], hoping to find the line of retreat through Afule open, so we had to reinforce the picquet on the road to disarm them, and send them in by batches; eventually practically the whole regiment was occupied over these prisoners. By the evening we had about 2,000 of them collected and were heartily sick of them. The aerodrome contained a German canteen full of good "Fizz" and some wonderful Hock [see above]. We pinched a camel for the squadron mess and loaded it up with two dozen each.'

The regiment, with 18th Machine Gun Company and a party of 4th Field Squadron, Royal Engineers, left Afule at 7.30 p.m. They marched, leading their horses most of the way, about twenty miles over extremely rough country. This included 'a colossal nullah about 200 feet deep' which it took two hours to cross. This 'did for our camel, who took a toss which only three bottles of wine survived, a serious loss as they proved to be the only drink stronger than water that we had for six weeks.' The detachment arrived at Jisr el Mejami on the Jordan, nine miles north of Beisan at 5 a.m. on 21 September. There the numerous Turks guarding the bridges 'simply melted away'. By 8.30 both rail and road bridges had been prepared for demolition in case, which was already increasingly unlikely, the enemy should want to use them. A few rails were removed and the detachment settled down to guard the line which connects the Hejaz with Palestine. This night march across unknown territory was a remarkable performance by men who had had no sleep for the three previous nights and who had only recently been subjected to the deadly climate of the Jordan Valley in summer. The detachment remained in position until 23 September, when the Central India Horse relieved it. They had covered well over ninety miles since the start of their march.[6]

By 6 p.m. the rest of 4th Division had reached Beisan, having trotted the greater part of the seventeen miles from Afule. About 800 prisoners were captured on the way and a further 700 on arrival. Amongst the numerous other seizures made at Beisan were two 5.9 in. howitzers, facing *eastwards*! By now 4th Division had covered at least seventy miles in thirty-four hours without the men unsaddling.* It had lost only twenty-six horses, though the rest were nearing the end of their tether. Most

*By comparison, John Hunt Morgan of the American Confederate cavalry marched with 900 men and four guns ninety miles in thirty-five hours in July, 1862. His more famous colleague, 'Jeb' Stuart, in June, 1862, covered 110 miles in about forty-eight hours when raiding behind George Brinton McClellan's army at the beginning of the 'Seven Days'. Later, in October, he marched about 126 miles in sixty hours. On both occasions his force did not exceed 1,800 men. He was without accompanying wheeled vehicles. In chasing Tantia Topi during December, 1858, 210 men of the 17th Lancers marched 148 miles in 120 hours. (See Vol. 2, 217).

Barrow's men had with them their twelve 13-pounder guns, thirty-six machine guns and 108 Hotchkiss guns. Further the first twenty miles were roadless and sandy. It may well be that 4th Cavalry Division's pace and distance constitute a world record for cavalry marching.

Preston, 211, says that 4 Division marched eighty-five miles in the thirty-four hours. The actual mileage is likely to have been between seventy and eighty-five.

of the transport was more than fifty miles behind and took many hours to catch up.

Having sent out picquets along the road to Afule, 4th Division awaited the influx of the defeated Turkish armies from the south. It was not until after dark that the first formed enemy body, about 200 strong, emerged. The direction of its approach had earlier been reported from the air. Consequently, the outpost in its path, consisting of most of the Dorset Yeomanry, was reinforced by a squadron of Central India Horse. At 10 p.m. the head of the column appeared and boldly attacked the outpost. At one point, indeed, the picket line was pierced, 'when Risaldar Dyal Singh with a troop of "D" Squadron [CIH] made a dashing charge in the moonlight', preceded by a few rounds from the Hants Battery. Into the risaldar's hands fell 158 prisoners.[7] There was virtually no more fighting that night. Until well after daylight on 22 September the Turks whom the infantry had routed streamed and stumbled blindly towards Beisan. Stupefied by fatigue, hunger and thirst, they surrendered at the first challenge. Among the thousands captured was the whole of 1st Turkish Cavalry Regiment with three of its officers' wives in tow. The musicians of another regiment's band came in carrying their instruments.*

* * *

5th Australian Light Horse Brigade, when last referred to, had just entered Tul Karm. (See p. 263). Its next objective was Ajje in the hills from where it was to cut the railway in the Dothan Pass. To get there extremely rough and trackless country had to be traversed. Some units were delayed as much as three hours by steep declivities. Consequently the troops had become much scattered and strung out over more than twelve miles. Eventually, at 7 a.m. on 20 September, two squadrons of 14th Regiment succeeded in reaching and breaking up the railway line. The rest of the brigade was not reassembled at Tul Karm till 7 p.m. and soon afterwards

*During the night the 18th Lancers to the north of Nazareth were attacked by a body some 700 strong which had marched from Haifa making for Tiberias. After a confused, brisk action with the bayonet on foot, the Turks streamed away in confusion, but not before forty-three had been killed and 180 captured, together with eight machine guns. (An excellent and amusing account of this action in which 18th Lancers sustained not a single casualty is to be found in Hudson, 234–7.)

orders were received to move on Nablus early the following morning. These orders superseded the earlier ones which had stipulated an advance on Jenin. Accompanied by 2nd Light Armoured Motor Battery, the brigade moved speedily along the Tul Karm to Nablus road, beat down the last resistance outside Nablus and took the town. The 14th Regiment made touch with XX Corps troops at Balata. The brigade was then ordered, having carried out most of its tasks as an independent cavalry body, to rejoin Ausdiv at Jenin.

* * *

Hodgson's Ausdiv, meanwhile, less 5th Brigade, had followed in the wake of Barrow's 4th Division in the morning of 19 September. By the time it arrived at El Lejjun its strength was reduced to only the three regiments of 3rd Brigade. One regiment of 4th Brigade was acting as escort to Corps headquarters, while 11th and 12th Regiments were escorting various transport columns. At 10 a.m. on 20 September 3rd Brigade had arrived at El Lejjun, where Descorps' advance headquarters, consisting of a fleet of fifty or so motor cars, soon joined them. Since the swiftness of the advance precluded telegraphic communications and to supplement the not entirely reliable wireless, Chauvel had organized a special force of light horse and yeomanry officers to act as gallopers between the divisions and his advanced headquarters. 'But,' as the official Australian historian put it, 'so swift was the progress that, before the first day had closed, these officers had ridden their horses to a standstill, and their service was limited.'[8]

At 2.45 p.m. Chauvel received a report from an aeroplane that large numbers of retreating Turks were marching northward from Jenin. At 3.35 Hodgson ordered Wilson's 3rd Brigade to leave 8th Regiment at El Lejjun and with 9th and 10th Regiments, Notts Battery and the four cars of 11th Light Armoured Motor Battery to push on with 'all speed and greatest possible boldness' to Jenin and 'capture the hostile forces now retreating north and north-east of that place'.[9] The importance of the town was great. Situated on the only railway and metalled road behind the enemy's centre, the main Turkish columns were bound to try to escape through it.[10] 10th Regiment led the way with six machine guns of 3rd Machine Gun Squadron. Starting at a steady trot, Lieutenant-Colonel

A.C.N. Olden, commanding 10th Regiment deemed it, in his own words, 'necessary to increase the pace in order to get astride the roads leading east and north-east from the town by dusk.' Hodgson and Wilson gave their permission for this and Olden was told that he need not bother about keeping in touch with 9th Regiment which was following him. 'This naturally increased the gap [between the regiments] . . . to about two and a half miles, which ordinarily would have been considered too great, but, under the circumstances, justifiable.'[11] Thus, moving at a hand gallop, the regiment and Notts Battery covered eleven miles in seventy minutes.

Three miles short of Jenin 2nd Lieutenant P.W.K. Doig, commanding the right flank troop, discovered a large body of the enemy resting in an olive grove. The troop, 'acting with great boldness and gallantry, immediately deployed, drew swords and charged right into the Turks, wounding several and taking the whole force prisoner. Lieutenant Doig methodically formed them up and marched them to the rear.'[12] They numbered over 1,800 men, including a number of Germans, with 400 animals. This was the first occasion on which Ausdiv employed its swords. Wilson, 3rd Brigade commander, was delighted with its performance.

> 'If we had had no swords,' he wrote in his report, 'the procedure would have been a careful approach, then probably a fire fight and we could not have got into Jenin that night. Probably the 6,000 extra prisoners [captured] that night would have evaded us or had time to organize. Later on the same evening our men, galloping up the streets of Jenin, demoralized the enemy much more quickly than a dismounted approach with fire would have done. The quickness of it meant practically no casualties to us.'[13]

Other parts of the regiment blocked all the exits from Jenin and twenty-seven motor lorries were captured intact with their personnel. When the town was surrendered there were some 3,000 prisoners to be dealt with. The looting by local Arabs therefore went largely unchecked, but a bullion wagon was saved and so were 120 cases of German champagne. With vast dumps of stores set alight by the retreating troops, 'the moon rose upon a scene of frenzy like one of the sacks of German towns in the Thirty Years'

War – save that here the peaceful inhabitants played the part of despoilers instead of that of victims.'[14]*

At 9 p.m. Lieutenant R.R.W. Patterson of the machine-gun squadron, whilst watching the road to Nablus, saw a long column approaching from the south. With him he had two machine guns and only twenty-three men. He was wondering what to do when his sub-section lance-corporal, T.B. George, suggested that it might be safer to bluff it out than retire.[15] Patterson agreed. He opened fire over the heads of the unsuspecting Turks when they were in the narrowest part of the defile and summoned them to surrender.

> 'By a lucky chance,' wrote the official Australian historian, 'a German nurse who spoke ready English was marching with the officers at the head of the column, and Patterson, advancing, told her he was supported by a large force immediately behind [which he was not]. She interpreted Patterson's bluff to the officers. And, after a brief discussion, the column of 2,800 troops and four guns surrendered to the twenty-three Australians.'[16]

When, as sometimes happened, the enemy turned out to consist entirely of Germans things were very different. Soon after Jenin had been entered, a small party of two machine guns 'with about six rifles supporting them' saw some 100 Germans approaching. When called upon to surrender these replied with a fusillade of bullets. They were at once engaged by the machine gunners and again called upon to give up their arms, but their answer was the same. 'A third time,' reported Olden, 'this was repeated, but on being again called upon, they surrendered, not however until many of their number had been killed or wounded.'[17]

Lieutenant-Colonel Preston gives a graphic account of 3rd Brigade's activities during the night of 20 September:

*German officers in Jenin refused to believe that the cavalry had ridden so far in so short a time. They 'declared that we must have been landed at Haifa, covered by our "Wonderful Navy", as they called it.' (Preston, 213)

'On the morning after the capture of Jenin, a British armoured car, probing cautiously up the road leading towards Nablus, was surprised to meet a little car of the Australian YMCA, the driver of which had pushed through across Samaria in advance of Onslow's [5th] brigade and the infantry, bringing cigarettes and other luxuries for the troops upon Esdraelon.' (Gullett, 729)

'As was expected, after dark the enemy began to retire from his positions at Nablus and Samaria, and all night long his battalions marched down the road, through Jenin, and out on to the plain. These were not fugitives, but formed bodies of troops, retiring to the Nazarene hills, where they had a partially prepared defence line extending from the sea to Lake Tiberias. It was rather an eerie experience to watch these troops, trudging wearily along the road in the bright moonlight, all unconscious of the keen eyes of their enemies on every side of them. As each detachment got well out into the plain, at a given signal the waiting squadrons sprang from their hiding places and charged down upon it. One can imagine the terror of the Turks, nodding with half-closed eyes as they trudged along, when their senses were suddenly assailed by the thunder of hoofs all round them, and by the sight of wild horsemen, exaggerated in size by the moonlight, charging down upon them from every side. Small wonder that there was little resistance. Many flung themselves on the ground, shutting eyes and ears to the horrid nightmare, and calling on Allah to deliver them. Others threw down their rifles and held up their hands. Each lot was quickly hustled out of sight, and the squadrons returned to their lairs, to await the coming of the next.'[18]

By the morning of 21 September the 10th Regiment, with some assistance from 9th Regiment, found itself with 8,107 prisoners, including several divisional commanders with their staffs, five guns, many machine guns, millions of ammunition rounds, numerous lorries and motor cars and one unburnt aeroplane out of a total of twenty-four. The 11th and 12th Regiments of 4th Brigade and 14th Brigade from 5th Cavalry Division soon arrived to relieve 3rd Brigade from the embarrassment of looking after so vast a haul. As these troops arrived an extraordinary sight met their eyes.

'The whole plain seemed to be covered with prisoners, motor cars, lorries, wagons, animals, and stores, in an inextricable confusion. In and out of this mass,' wrote Preston, 'the sorely tried Australian troopers pushed their way, sweating and swearing, every now and then riding savagely at the hordes of

> natives hovering on the outskirts of the crowd like a flock of vultures, and looting the stores that strewed the ground; anon pressing into the throng again, to round up a group of straying prisoners. Over all presided the stocky figure of the brigadier [Wilson], like the leader of a gigantic school picnic, unhurried and efficient.'[19]

It took ten hours for 8th Regiment to escort about 7,000 prisoners to a hastily made compound at El Lejjun. The state of many of the 14,000 who were now assembled there was appalling. Their feeding was undertaken by lorry-borne supplies which came from Tul Karm through the Musmus Pass. Every ounce of food in this convoy was handed over to the prisoners, the cavalry relying on their three days' rations. Though they did not actually go hungry, it was clear that until further supplies arrived no further advance could be contemplated. Next day, 22 September, 8,000 prisoners were escorted to Kerkur by the Hyderabad Lancers of 15th Brigade. 'The Germans, marching at the head of the column out of the dust and evidently far better nourished, kept their discipline and goose-stepped when a general's car passed them.'[20]

During these dramatic days the British aeroplanes dealt mercilessly with their helpless quarry. With both bombs and machine guns they swooped down on the doomed residue of VII and VIII Armies as they tried to escape by the only routes still partially open to them – those towards the Jordan crossings. Every available aeroplane was pressed into service. During one four-hour period two machines arrived over the objective every three minutes. A full moon made night bombing possible. On Chetwode's XX Corps front, in one five-mile-long gorge, part of Wadi Farrah, the airmen had caused the destruction or abandonment of eighty-seven guns, fifty-five lorries, four motor cars and 932 other vehicles. 'The work of the airmen, following on the blow of the infantry, had taken all the fight out of the Turks before they reached the cavalry cordon beyond.'[21]

By the evening of 20 September, 'within,' to quote Allenby's despatch, 'thirty-six hours of the commencement of the battle, all the main outlets of escape remaining to the Turkish VIIth and VIIIth Armies had been closed. . . . The first phase of the operations was over.'[22]

Chauvel found time that evening to drop his wife a line. In it he wrote:

'I have had a glorious time. We have done a regular Jeb Stuart ride. I wrote to you last two days ago from just north of Jaffa, where I had my whole Corps, except Chaytor, hidden in the orange groves, and I am writing now from a hill close to Lejjun (Megiddo) overlooking the plain of Armageddon (now Esdraelon) which is still strewn with Turkish dead "harpooned" by my Indian Cavalry early this morning, and from my tent door I can see Nazareth, Mt Tabor, El Afule, Zerim (Jezreel), Mt Gilboa and Jenin. . . .

'We have been fighting what I sincerely hope will be the last "Battle of Armageddon" all day. . . .

'All this is miles behind the enemy front line. . . . It is the first time in this war that the G in GAP Scheme* has really come off and I am feeling very pleased with myself.'[23]

* * *

With the details of the actions of the two infantry corps after XXI Corps had punched the gap for the cavalry to push through this work is not concerned. It is enough to say that, as soon as the success of the initial attack was assured and XXI Corps was seen to be swinging right-handed on its great wheel towards the hills, Allenby, on the morning of 19 September, ordered Chetwode's XX Corps to advance that night on Nablus. Against troops who at first 'showed no signs of demoralization' and across terrain which was 'very rugged and difficult',[24] to quote Allenby's despatch, both Corps pushed forward. They met considerable resistance which at times involved losing ground to counter-attacks. 'While the cavalry were occupying the line of butts,' as Wavell puts it, 'the infantry were steadily driving their prey towards them.'[25] By the evening of 21 September virtually all organized enemy resistance had ceased. XXI Corps had reached a line in the Samarian Hills of which Attara and Bela were the centre, while XX Corps had fought its way to the north-east of Nablus. The last serious infantry engagements of the campaign had been fought. XXI Corps suffered 3,378 casualties. Of these only 446 of all ranks had been killed. The two divisions of XX Corps lost 225 killed out of a total casualty list of 1,505.

*Contemporary army slang for getting cavalry through a gap in the enemy's lines.

42. General Chauvel, escorted by a squadron of the 2nd Light Horse Regiment, as bodyguard, riding through Damascus, 2 October, 1918, and followed by units representing his three cavalry divisions including men from the United Kingdom, Australia, New Zealand, India and France

43. Allenby, 'saluting some civilians on a balcony… Mysore Lancers lining the street', probably in Damascus

44. The Es Salt raid, 1918 (see p.312)

45. Charge of the 2nd Lancers at Birket el Fuleh, 20 September, 1918 (see p.279)

31

'On the morning of September 23rd Chauvel held the plain from Haifa to Beisan. The Seventh and Eighth Turkish Armies were destroyed.'

The Australian Official Historian[1]

The 'Haifa Annexation Expedition' – Acre taken – Haifa taken

On 22 September Chauvel received a visit from Allenby at El Lejjun. By then Chauvel had counted some 15,000 prisoners. 'No bloody good to me!' said Allenby, laughing. 'I want 30,000 from you before you've done.' Allenby then asked, 'What about Damascus?' to which Chauvel replied, 'Rather!' First, though, essential to Allenby's further progress was the capture of Haifa as a base for the landing of supplies by sea. Chauvel therefore ordered 5th Division to take the town and also Acre, ten miles to the north, on 23 September. But air reports declared that Haifa was being evacuated. Consequently Chauvel ordered his chief artillery officer to jump into a Rolls Royce, fly Allenby's Union Jack on the bonnet, enter the town, read out a proclamation to the peaceful inhabitants and proclaim himself Military Governor. He was to take as escort 12th Light Armoured Motor Battery and 7th Light Car Patrol. Eight miles to the south of Haifa a detachment of sixty-nine of the enemy surrendered without firing a shot. Their presence proved that the air report was wrong. Haifa was still held. Brigadier-General King decided nevertheless to go on, 'trusting,' as the Official Historian puts it, 'that the demoralization of the enemy would have spread to the garrison, though it had not yet been engaged'.[2]

Three miles from the town the motorcade was fired on by artillery and machine guns.

> 'The batteries,' writes Preston, 'had evidently registered carefully, for almost the first salvo hit the general's car, knocking it into the ditch and smashing the flag. The general himself, with his staff, had to take cover in the same ditch, and quickly, too, and there they lay, getting the proclamation covered with mud, till the armoured cars succeeded in retrieving them. . . . The "Haifa Annexation Expedition", as

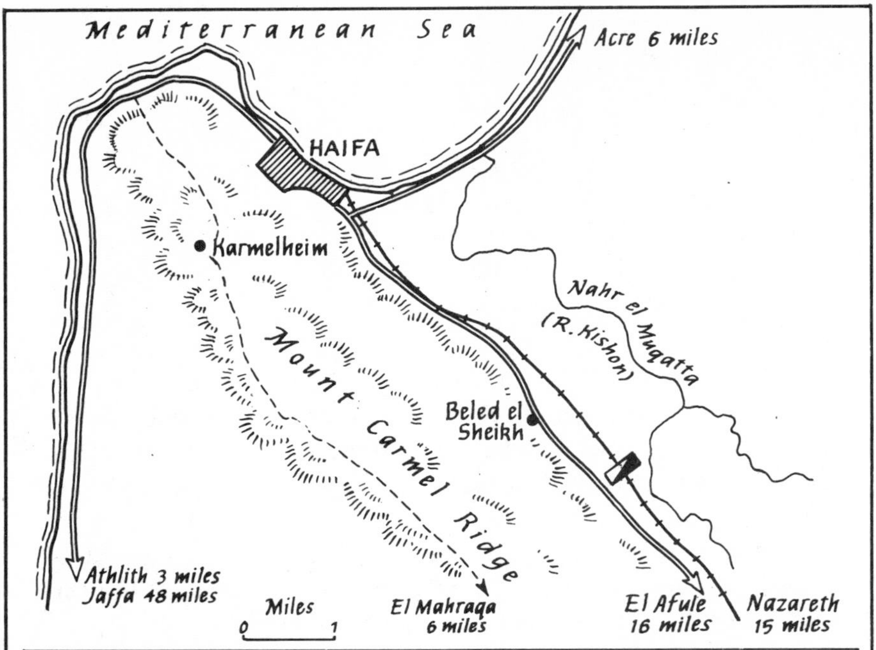

'HAIFA ANNEXATION EXPEDITION', 23 SEPTEMBER, 1918

it was irreverently called, returned to Afule in somewhat chastened mood, but fortunately without any serious casualties.'[3]

In view of this unexpected rebuff, it was necessary to carry out an attack in force on the 23rd. At 5 a.m. Macandrew ordered 13th Brigade, less 'B' Battery HAC and accompanied by 11th Light Armoured Motor Battery, to take Acre, which it did without difficulty, capturing two guns and 260 prisoners. 14th and 15th Brigades, the latter less Hyderabad Lancers which was still escorting prisoners, left Afule about the same time. Ten miles south of Haifa 14th Brigade remained in reserve while 15th Brigade, accompanied by 'B' Battery, proceeded along the main road to Belled el Sheikh, where the vanguard squadron of Mysore Lancers at about 10 o'clock came under fire from machine guns and rifles hidden among the vineyards on the lower slopes of the Mount Carmel ridge. The map shows how this ridge dominates and protects the town. It rises to 200 feet on its western side and to the east its foothills approach within 2,000 yards of the Bay of Acre. The terrain between is much broken by the swift and swampy Nahr at Muqatta or River Kishon. Towards the northern end of the Carmel ridge sits a prominent religious 'pension' known as Karmelheim. Four guns placed here also fired on the vanguard, as did six others near a small mound about 800 yards from the town's eastern extremity. An aeroplane at this point dropped a message which confirmed that Haifa was strongly held.

Brigadier-General Harbord now sent off one squadron of the Mysores to a point four and a half miles east of the town, with orders to advance on it from the north-east. At the same time, another squadron was despatched, with two machine guns, to climb up the ridge from a point south of Belled el Sheikh, march along it and deal with the Karmelheim guns. Meanwhile 'B' Squadron came into action north-east of Belled. The two remaining Mysore squadrons with two machine guns waited on the railway, while the Jodhpore Lancers prepared to launch the charge upon which Harbord had decided. At 11.50 the Sherwood Rangers from 14th Brigade came on the scene. One of its squadrons was ordered to follow the Mysore squadron on to the Carmel ridge.

To give the two flanking detachments time to reach their objectives, the brigadier, anxious that the attacks on the heights and in the plain should be more or less synchronized, postponed the start of the Jodhpores' mounted charge till 2 p.m. During the interval several patrols went out to reconnoitre the ground. They were prevented, however, from examining the river bank west of the disused railway line to Acre by fire from eight machine guns located just to the north of the river and by two artillery pieces in the foothills. One Mysore patrol did in fact reach the river bank, but its horses became bogged down and had to be abandoned to a death by drowning. Much of the ground over which the Jodhpores were to charge remained therefore largely *terra incognito*.

Punctually at 2 p.m. the Jodhpores advanced at the trot in column of squadrons in line of troop column. In support were 'B' Battery, four machine guns and the two remaining Mysore squadrons. These last took up position along the Acre railway embankment about 500 yards from its junction with the main line to Haifa and dismounted ready to give covering fire. The Jodhpores soon met intense machine-gun and rifle fire. Two of the reserve machine guns were galloped forward to augment the fire support, and the Jodhpores increased their pace, suffering little loss. As the river bank was neared, its precipitousness became alarmingly clear, but this obstacle was as nothing compared to the next one. The two leading ground scouts, forcing their horses down to the water's edge, were immediately swallowed up in a quicksand, proving that the river was everywhere impassable.* The regiment, therefore, less one squadron which tried to find a way over to the right, was ordered by Lieutenant-Colonel H.N. Holden, the senior Special Service Officer, to swing sharply to the left, cross the narrow wadi beside the railway and, deploying as it advanced, charge the machine guns and riflemen on Carmel's lower slopes. This the leading squadron achieved at the gallop, spearing all the machine gunners and scattering the riflemen. It had been a most critical manoeuvre, for the enemy's fire from front and flank at the moment of the check was concentrated and fierce, and accounted for a number of the squadron's horses. Its success opened up the defile formed by the road and railway, clearing the way to Haifa for the rest of the regiment.

Whilst the leading squadron rallied west of the road, the following squadron galloped up it, wheeled half-right, charged and captured the two machine guns sited on the mound east of the road, as well as two artillery pieces in the palm groves nearer the shore. Almost simultaneously, Holden led his last remaining squadron straight into the town at the gallop.† The two Mysore squadrons which had given dismounted fire to the charge, now, as soon as the Jodhpores had begun to mask their fire, mounted and also galloped into the town. Here occasional shots were fired from behind walls and from houses and a few Turks were ridden down in the streets. At the far end of the town the Jodhpore squadron which had tried and failed to find a way over the river, now rejoined the rest. It had wheeled about, followed along the lower slopes of Carmel, skirted the southern edge of the town and reached the Jewish colony, where it captured two guns which had

*This stretch of the river was marked 'unfordable' in the secret pre-war military report on the area. This information seems not to have been included on the maps available to the Lancers.

See also *Judges*, iv, and note on p. 282 for the fate of Sisera's hosts in the River Kishon. That knowledge of this was not to hand is less surprising!

†The 15th Field Troop, Royal Engineers, happened to be alongside the Jodhpores when the charge was ordered. On the invitation of the commanding officer, they armed themselves with lances and swords from casualties and rode in the charge. None of them had ever handled the *arme blanche* before, but it was claimed that at least one Turk was killed by a sapper. The likelihood that this is the sole occasion in which sappers took part in a cavalry charge is considerable. (Wavell, 281)

remained in action some ten minutes after the rest had ceased firing.

The Mysore squadron, which had originally set off well to the right flank, had been held up and forced to dismount by two guns and some machine guns two and a half miles north-east of Haifa, but as soon as its commander saw the Jodhpores begin their charge he decided to co-operate and ordered his men to mount and gallop into the enemy in their front. The Turks here were holding the sandhills along the seashore near the eastern mouth of the river. They greeted the Mysore charge with a terrific burst of fire, but 'the moral effect of the lancers' galloping straight at them was such,' according to one reliable account, 'that their fire was wild and the position was quickly taken. In this charge two guns, two machine guns and 110 prisoners were captured.'[4]

The Mysore squadron on the left, meanwhile, had made its way up a very difficult gorge losing a number of horses from enemy fire, exhaustion or lameness. By the time it was ready to tackle the Karmelheim position there were left only fifteen men for the actual charge, its two machine guns and its Hotchkiss section being detached to the left flank to give covering fire. At almost the same moment that the Jodhpores were charging in the valley, these fifteen lancers galloped over excessively stony ground and captured one 150-mm. naval gun, two mountain guns and seventy-eight prisoners. Part of the Sherwood Rangers' squadron arrived in the nick of time to support this brave attack and was able to take up the pursuit and capture a further fifty Turks.

The brigade's total bag at the end of the fight amounted to two German and twenty-three Turkish officers and 664 other ranks. How many of the enemy were killed is not known. In guns, two naval, four 4.2 inch, six 77 mm, four 10-pounder camel-, were taken, as well as ten machine guns. A mass of ammunition and numerous good, much welcomed horses, were also secured. The brigade's casualties numbered one Indian officer (Lieutenant-Colonel Thakur Dalpat Singh, commander of the Jodhpore Lancers), two other ranks and sixty-four horses killed and six Indian officers, twenty-eight other ranks and eighty-three horses wounded. An astonishing fact was how seldom machine-gun bullets succeeded in stopping galloping horses, though numbers of them succumbed to their wounds some time later.

By 3 p.m. the battle was over and victory complete. A vital new supply base had fallen into British hands. Four days later the landing of supplies started.* Without doubt this was the most successful mounted action of its scale fought in the whole course of the campaign. It was won by a weak brigade of only two regiments and a single 12-pounder battery pitted against about 1,000 well-armed troops who had so far seen no action. These, skilfully deployed, occupied a naturally formidable defensive position with an impassable river on one side of a narrow defile and a steep hill on the other. That they had already received news of the general rout is certain and this may well have affected their behaviour, but there is little evidence to show that they put up less than a respectable resistance. The speed and daring, dash and boldness of the two Indian Imperial Service regiments, in conjunction with the skilful flanking movements devised by Holden, were what made the action such a success. The speed and good order demonstrated by the leading squadron of the Jodhpores when it was forced to change direction under heavy fire, were other vital ingredients in what was almost certainly the only occasion in history when a fortified town was captured by cavalry at the gallop.[5]

*Although Haifa was a long way short of 100 miles from the large supply depots which had been established at Ludd, the terrain was so bad for wheeled traffic that it was easier and quicker to send stores back to the Suez Canal, 200 miles by rail, and then to put them on ships for the voyage to Haifa, rather than to forward them by road. (Massey: *Triumph*, 14)

32

'The destruction, moral and material, worked by the airmen, had been far greater than Allenby had expected.'

The Australian Official Historian

'[On 22 September Liman] received a delightful telegram from the Military Mission, which is worth mentioning because it shows how little the situation was understood in Constantinople, even though Turkish headquarters had been constantly informed of what was happening. He was asked to offer a prize for the sack-race at the military sports shortly to be held. Liman remarks that he took no interest at that moment in sports and least of all in a sack-race, which was a painful reminder of his own position.'

Quoted by the Official Historian

'The operations may stand as a pattern to cavalry upon the flank of a beaten and demoralized foe.'

The Official Historian on the actions at Abu Naj and Masudi on 23 and 24 September, 1918

'The most fiercely fought action of the whole pursuit. . . . A grim little fight of which the Australians are justly proud.'

WAVELL of the action at Semakh on 25 September, 1918

'Once the VIIth and VIIIth Armies ceased to exist, all eyes turned eastwards over Jordan towards Turkish IVth Army.'

LIEUTENANT-COLONEL REX OSBORNE in the *Cavalry Journal*, 1923

'The vulture appearance of the Arabs, who were willing that we should do the fighting and they the looting, will not be readily forgotten.'

LIEUTENANT-COLONEL K.D. RICHARDSON commanding 7th ALH Regiment, 28 September, 1918

Actions south of Beisan, 22–23 September, 1918

'The Bull is going on well. . . . They say he is nearing Bashan, so he is nearly home!' [Psalm XXII, v. 12: 'Many bulls here compassioned me: strong bulls of Bashan here beset me round']

BRIGADIER-GENERAL ARCHIBALD HOME,
BGS Cavalry Corps, BEF, France
in his diary on 22 September, 1918[1]

4th Cavalry Division: Makhadet Abu Naj – Masudi – Ausdiv: Semakh – Tiberias occupied – Chaytor's Force: Jisr el Damiye, Es Salt and Amman captured

While troops of 5th Division were engaged in securing Haifa and Acre and dealing with the thousands of prisoners in the western sphere of operations, 4th Division was ordered, in its commander's words, 'to occupy the fords over the Jordan to the south of Beisan and to stretch a hand towards the Anzac and other troops which, under Chaytor's command, were following the enemy up the Jordan valley.'[2] The urgency for this was only fully realized in the afternoon of 22 September when an air report (which proved more reliable than that which had indicated that Haifa was being evacuated) revealed that there was a serious hole in the net which had been thrown around the enemy west of Jordan. Chaytor had taken the Jisr el Damiye crossing at 1 a.m. on that day [see p. 312], but between him and 4th Division at Beisan there were some twenty miles along which the more northerly crossings were unguarded. Indeed the air report disclosed that formed enemy bodies were escaping by the Makhadet Abu Naj ford some six miles south of Beisan.*

Gregory's 11th Brigade marched at 6 a.m. on 23 September, with Jacob's Horse on the east bank and the Middlesex Yeomanry and 29th Lancers on the west bank of the river. The latest air report was confirmed when at 0830 hours the leading squadron of 29th Lancers was fired on as it approached the ford. Barrow's report tells what happened next:

*The Official Historian suggests that the failure to close this gap 'by a movement southward along the Jordan on the part of the 4th Cavalry Division simultaneous with the approach of Chaytor's Force to Jisr el Damiye was the sole blot on an operation the main lines of which had hitherto been perfection itself. It must be remembered, however, that the division had had to picquet the road from Beisan half-way to El Afule, that it had been encumbered with prisoners and that rations had not been delivered at Beisan till [late] on the 21st.' (Falls, 538).

'Reconnaissance disclosed the fact that a strong Advanced Guard of about 1,000 infantry and some thirty machine guns, with a few mounted men, were holding an advanced position covering the ford. The position ran through dense scrub, with its centre occupying a mound and a few houses. The mound was garrisoned by some 300 infantry with fifteen machine guns. Captain M. H. Jackson, with two squadrons 29th Lancers was detailed to clear up the situation.'[3]

Jackson at once moved forward. He placed his Hotchkiss section on a small mound 800 yards west of the position, ordered one squadron to attack the enemy's left flank and the other to make a wide enveloping movement so as to launch an attack from the rear. The flank squadron was checked by heavy fire, but the second squadron, employing speed and surprise to the fullest extent, was spectacularly successful. It galloped right through the enemy's position and shattered its defenders, taking 800 prisoners, including one divisional commander, and eighteen machine guns. Thus the rearguard of Colonel von Oppen's Asia Corps, comprising some of the best troops available to Liman, was decisively crushed.

At much the same time as 29th Lancers were fighting on the west bank, the leading squadron of Jacob's Horse,* a little bit short of the ford, encountered another large enemy force with numerous machine guns marching northwards.

'The squadron fell back some 400 yards and,' according to the regimental historian, 'took up a position in a nullah roughly about the edge of the bush [through which until then it had been advancing]. The supporting squadron came up on its left and with its assistance the enemy's advance was held up. Owing to the scrub it was very hard to get much idea either of what was in front or of the ground, and the heavy hostile machine-gun fire made movement extremely difficult.

'It was possible to see the dust made by the 29th across the river, but no more, and one could helio to Brigade Headquarters. It was evident, however, that without artillery to help, or some strong enfilading fire from across the river, it was hopeless to try and advance.'[4]

*Two troops had been dropped off to watch the crossing near Beisan and two troops were escorting prisoners back to Afule. Consequently there were only about 180 men available.

The regiment therefore heliographed headquarters for artillery support. Gregory passed the message on to the Hants Battery at Beisan, which had saddled up in expectation of being sent for, and now trotted the whole six miles and arrived at 11.00, men and horses alike smothered in dust and sweat. It at once came into action from the west bank. Barrow's report tells of the immediate result:

> 'It was not until the battery had fired its first round that the enemy disclosed his own batteries, of which he had apparently two,* posted on the east bank of the river and about 1,500 yards south-east of the ford. [These] quickly got on to Hants Battery, which, owing to the nature of the ground, was compelled to come into action in the open. Every gun of the battery was hit [by splinters], but no damage was done to personnel; but so hot and accurate was the hostile fire that the gunners had temporarily to leave their guns.'[5]†

In the meantime 'C' Squadron of the Middlesex Yeomanry had forded the river at Makhadet Fath Allah, about a mile further south, driving before it the body of Turks which Gregory had ordered it to attack. The squadron then 'drew swords and charged the enemy's guns, capturing four, without sustaining any casualties.'[6] The gunners fled before the mounted assault and the yeomen removed the breech-blocks from the guns.

Meanwhile Jacob's Horse had crept forward partly on foot and partly mounted. By 14.20 a squadron of the 29th had crossed by a ford 'which must have been five feet deep' and joined up with Jacob's Horse at about 16.00, both regiments all the time subjecting the Turks to heavy rifle and machine-gun fire, forcing them back 'as fast as the ground admitted, for a succession of formidable nullahs, in which the Turks had taken up their first position, had to be crossed. It was as well that no attempt had been made to charge, for disaster would have inevitably resulted on ground like this.'[7]

The Turkish rout which followed was complete. The banks of the Jordan were bestrewn with the dead, nearly 4,000 prisoners

*In fact only one 5.9mm battery.

†Preston, 231, says that 'Gregory ordered a troop of cavalry out into the open to try to draw the fire of the Turkish guns and so enable the battery to withdraw and take up a concealed position.'

were taken and 'an enormous quantity of warlike stores' was abandoned. 'It had been a very hot day, the going bad, and,' as Barrow put it, 'the horses had been without water since leaving Beisan; the brigade therefore went into bivouac [beside the Abu Naj ford] at 1700 hours.'[8] The Middlesex Yeomanry held outposts during the night and Jacob's Horse remained on the east bank.

In the course of the day an officer's patrol of about one troop of 29th Lancers had been sent out into the hills to the west of Jordan with the object of gaining touch with the Worcesters (XX Corps Cavalry Regiment). This regiment, which had been acting as the connecting link between the right of 10th Division and the left of 53rd Division, had been, since the 20th, experiencing an exciting and exhausting time. On the 21st, for instance, two of its squadrons had made a spirited charge at Askar, taking a large body of Turks by surprise under fire from numerous machine guns. Whilst they were collecting some 250 prisoners (the enemy machine gunners now firing on friend and foe alike), 'six of our own aeroplanes, flying very low, emptied their drums and bomb racks into the village. . . . It was not the fault of the airmen,' wrote one who was there, 'for the squadrons had omitted to light the coloured flares provided for just these contingencies, but it is difficult to remember such details during a cavalry charge. The flares were at last discovered and touched off and the airmen flew away, leaving captors and captives in comparative peace.'[9] By the morning of 24 September the Worcesters had taken 4,300 prisoners.* The patrol of 29th Lancers, led by Lieutenant W.K. King, made contact with the Worcesters seven miles east of Tubas at 6 a.m. on 24 September. It had then returned to 11th Brigade, riding clean through the retreating enemy, covering altogether some fifty miles of most difficult unmapped country. It brought the valuable but not completely accurate news that, 'but for a few terrified refugees skulking here and there in rocky gullies, not a Turk now remained west of Jordan between the Dead Sea and the Lake of Tiberias.'[10] 'This long-distance patrol,' suggests Osborne, 'was probably the most difficult of the very limited number carried out in Palestine.'[11] That it was feasible at all well demonstrates the state of Turkish demoralization at this time.

*The Worcesters lost one man killed and three wounded, as well as three horses killed and seven wounded in this action.

Barrow's report for 24 September reads in part:

> '11th Cavalry Brigade [having waited for its rations to come up] moved down both sides of the Jordan [at 11.30] with the object of mopping up the retreating Turkish Army which was retiring on Beisan, apparently ignorant of the fact that our cavalry had been in possession of the place since the 20th. Almost immediately on leaving bivouac the Middlesex Yeomanry came in contact with a hostile Advance Guard about 1,200 strong, supported by numerous machine guns.'[12]

This was making for the ford at Masudi. 'A' Squadron was ordered to gallop for the ford. As they did so

> 'machine gun after machine gun opened fire on them. Just as the squadron was about to charge, deep, impassable ravines suddenly appeared, which probably saved them from annihilation. A party of 300 Turks arrived at the ford first and also lined a wadi running parallel to the Wadi Sherar. Under protection from this covering party they also started to cross the Jordan two miles south of Masudi. "B" Squadron', according to the regimental history, 'was sent along the line of hills on the west of the Jordan to turn this force. It was held up by difficult ground and the squadron leader, Capt. Bullivant, MC, went on alone to make a personal reconnaissance. His riderless horse galloped back to the squadron and he was found dead – shot through the head. "A" Squadron crossed the river and pursued the enemy column to the foothills. Only 300 of the enemy had crossed at Masudi and they were now in barren and waterless country.'[13] (See map on p. 247)

'B' Squadron was now ordered by Lieutenant-Colonel Lawson, commanding the Regiment, to head off the main body, numbering, perhaps, 5,000 men. Dismounted, it opened a devastating fire upon the head of the column, marching six abreast, forcing it to turn away south-eastward. Gregory, learning from Lawson of the situation, at once sent the 29th Lancers and a section of the Hants Battery to his aid. Lieutenant-Colonel Patrick Barclay Sangster, commanding the 29th, directed two guns to fire on the main enemy body, joined a few moments later by the second section of the battery. All four guns played upon it from under 3,000 yards. The Turks fled southward in confusion. Some 3,000 prisoners, 'all poor miserable creatures only too glad to have done with fighting,'[14] were speedily rounded up. The Turks who had already crossed the river were now pursued by 'A' Squadron of the Middlesex and Jacob's Horse, which, due to an error, had been detained at Beisan and had thus only just arrived on the scene. A comparatively small body escaped to the hills, while the approaches to either side of the river were again a veritable shambles. A further 1,000 prisoners were captured by 10th Brigade on the west bank further north. In all over 5,000 prisoners were taken this day by 4th Division.* What exactly 11th Brigade's casualties amounted to for the period 22 to 25 September is difficult to assess. Certainly those of Jacob's Horse were the highest of the regiments engaged. Those of 29th Lancers are known: one British and three Indian officers and fourteen men.

* * *

By 25th September all Palestine west of the Jordan had been cleared of the enemy except for the railway village of Semakh at the southern end of Lake Tiberias. This was being held partly for the purpose of covering the removal of the large dump of supplies which had been accumulated there. When Liman had left Nazareth he drove at once to Semakh. He later wrote: 'Only one course remained open to me. The Tiberias sector from Lake Hule to Semakh must be held with all the means at our disposal to prevent the pursuit overtaking us.'[15] It was clear that Liman desperately hoped to establish a rearguard line from Deraa along the Yarmuk River through Semakh and Nazareth to Acre. Without it he would certainly not have time to organize the defence of Damascus. He therefore added extra German machine gunners to the defences and, giving command to a German officer, ordered him to hold the place to the last man.

On 24 September a squadron of the Central India Horse had

*One of the problems encountered in dealing with mixed German and Turkish prisoners was the marked tendency of the latter to attack the former. See Cobham, 189, for a good exposition of this problem as experienced by the Worcesters.

reconnoitred Semakh and found it strongly held.* Hodgson therefore ordered 4th Brigade, under Grant of Beersheba fame (who on the 21st had hastened from hospital to resume command), to capture the village at dawn next morning, the 25th. 4th Brigade, however, consisted at this time of only 11th Regiment and part of 12th Regiment, 4th Regiment still protecting Descorps' headquarters and five troops of 12th Regiment being away escorting prisoners. Grant was told that 15th Regiment from 5th Brigade was on its way from Jenin but could not arrive before daylight. Hodgson gave him the choice of awaiting this reinforcement or going on without it, but if he found the village too strongly held he was to await it. The CIH had been fired on by one field gun and a local resident of the nearby Jewish village reported that 'about 120 Turks and Germans – the latter not in uniform'† occupied the new stone railway buildings with 'not more than four machine guns'. Grant decided, therefore, as he wrote in his report,

> 'that he had sufficient force to capture the place and that there would be less casualties if the place were rushed just before dawn than if the attack were delivered in daylight. The question of time,' he continued, 'was of particular importance as the principal *rôle* assigned to the brigade was the protection of the Yarmuk valley bridges and tunnels.‡
>
> 'Orders were issued that the 11th A.L.H. Regiment were to attack, mounted, from the south-east just before dawn, under the covering fire of the machine guns, and the 12th would be held as a reserve.
>
> 'Jisr Mejamie is six miles south of Semakh and there were two bridges to cross at the Jordan and Yarmuk rivers respectively; so two hours were allowed for the march. Dawn being expected at 04.50, the brigade marched from Jisr

*This squadron also attempted to blow up the railway to the east of the village. 'The attempt was opposed and was unsuccessful, for just as preparations were on the point of completion the mule carrying the gun-cotton broke loose from its driver and disappeared into the blue. Lieutenant Coster, who had been attached to the squadron for this operation, and one Indian soldier were killed.' (Watson, 397)

†All of which information was incorrect.

‡Which in the event it was not called upon to carry out. Hodgson had understood that 4th Division was to do the job. Due to this muddle, one of the bridges was blown up by the enemy. It took many weeks to repair.

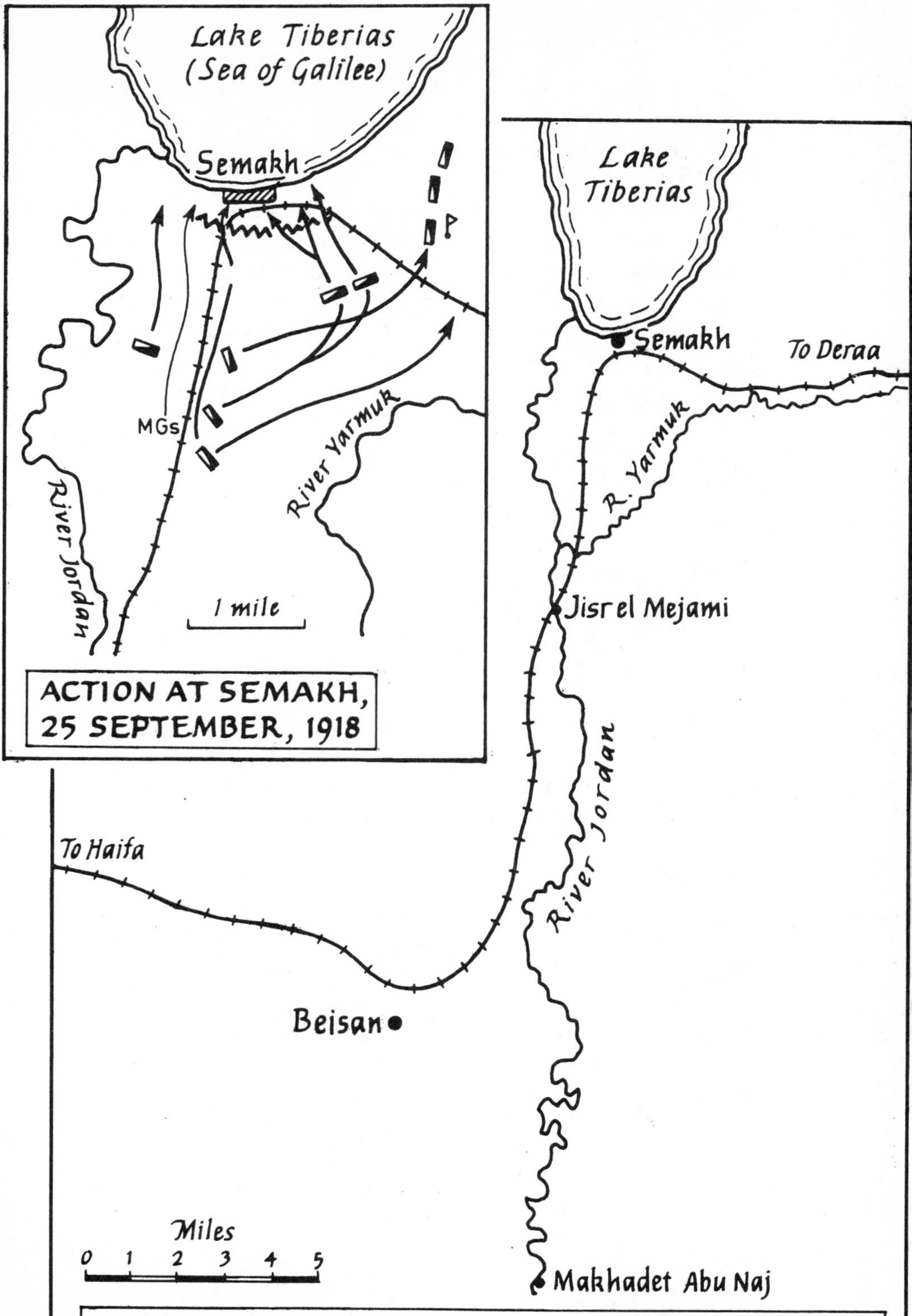
Lake Tiberias
(Sea of Galilee)
Semakh
MGs
River Jordan
River Yarmuk
1 mile
ACTION AT SEMAKH,
25 SEPTEMBER, 1918
Lake Tiberias
Semakh
To Deraa
R. Yarmuk
Jisr el Mejami
River Jordan
To Haifa
Beisan
Miles
0 1 2 3 4 5
Makhadet Abu Naj
ADVANCE TO LAKE TIBERIAS, SEPTEMBER, 1918

Mejamie at 02.30. A British officer and three Indians were provided by the C.I.H. as guides. A squadron was also sent by that regiment along the road on the west bank of the Jordan to act as a flank guard.

'After crossing the bridge over the Yarmuk river, the brigade deployed into column of squadrons, each in line of troop columns, and moved parallel to the road east of the railway line. The 11th A.L.H. Regiment with one section of machine guns attached, were leading with one squadron as advanced guard, followed by brigade headquarters, signal troop, machine-gun squadron and 12th A.L.H. Regiment.'[16]

At 04.25, while it was still quite dark, intense machine-gun and rifle fire burst out on a front of half a mile on either side of the railway. Major E. Costello, commanding 'A' Squadron, who was riding with the regiment's commanding officer, Lieutenant-Colonel J. W. Parsons, shouted to him 'What orders, Colonel?' To which Parsons replied without hesitation 'Form line and charge the guns'.[17] He then swung 'A' and 'B' Squadrons sharply to the right and all twelve guns of the machine-gun squadron, six either side of the railway, came into action firing at the flashes in front. 'A' and 'B' Squadrons (Costello and Major J. Loynes)* trotted to within a short distance of the railway to Deraa and, in Grant's words, 'charged mounted with drawn swords in two lines of half squadrons, with about 200 yards distance between lines. As they charged they yelled, which enabled the machine gunners to know their whereabouts and when to cease fire as it was still quite dark.'[18] Though some of the horses fell into pits dug to cover the enemy's position, these two squadrons, galloping for nearly a mile across a perfectly flat plain, broke clean through the left flank of the defences. They then rode on to the eastern end of the station buildings, except for two troops of 'A' Squadron which diverged to the left and entered the village from the south-west. Their horses showed up clearly in the moonlight and the German machine gunners shot down numbers of both them and their riders. Both station and village were found to be strongly held. 'So the two squadrons dismounted, left their horses in a wadi near the pump house and attacked on foot' with the bayonet.[19]

*'A headstrong veteran of the South African campaign, approaching sixty years of age'. (Gullett, 732)

Meanwhile one of the troops of 'C' Squadron, which had been the leading squadron, was detached to escort the machine guns, while the others moved to a hill on the right to watch the railway and road there.* As soon as the charge had been driven home the machine guns, having put out of action most of those of the enemy opposite them, were galloped forward to the west end of the village.

'When dawn appeared,' wrote Grant, 'the enemy had taken up a stand in the village and station buildings. He was fighting in a most determined manner with automatic rifles and bombs, beside rifles, and here most of our casualties occurred.' From engines and tenders, from behind stone walls and from house windows, the enemy machine gunners and riflemen fought with exceptional stubbornness, employing an ample store of hand-grenades to very good effect. With the details of the street fighting – a rarity in the campaign – we are not concerned. It ceased, as Grant put it, 'at 05.30, when all the garrison of the railway buildings had been killed.'[20]

The Australian casualties numbered seventy-eight and nearly 100 horses.† The field gun and ten machine guns were taken. Of the enemy, ninety-eight were killed and twenty-three officers and 341 other ranks captured. Of these 150 were Germans. Most of the Turks had taken little part in the battle.‡

Since it was obviously Allenby's policy to run considerable risks so as to obtain quick decisions, the heavy casualties incurred at Semakh, greater than in any other action of the pursuit, can

*At 05.10 they joined in the other two squadrons' fight. 'C' Squadron of 12th Regiment was at the same time sent in to the west of the village and also took part in the dismounted fighting there.

†11th Regiment alone lost two captains, one lieutenant and eleven other ranks killed. Four officers and twenty-five other ranks were wounded. The accuracy of the moonlight shooting against the opening gallop is shown in the loss of sixty-one horses killed and twenty-seven wounded.

‡Grant says that treacherous use of the white flag caused several of the casualties.

During the fight one motor boat escaped, while another was hit by a Hotchkiss rifle, burst into flames and sank.

The bearer division of the Cavalry Field Ambulance with camel cacolets had followed the brigade from Jisr el Mejami.

'A very liberal supply of *arak* (native spirit) accounted for the stiff fight put up by the enemy.' (Osborne, III, 148)

perhaps be justified. A daylight attack against such determined German resistance might well have resulted in at least as great casualties. From the cavalry point of view, it is worth noting that the mounted attack, made over unseen country in the dark, is the only one of its sort that was undertaken in the whole course of the war.

Grant's brigade, in conjunction with 3rd Brigade, now occupied Tiberias against slight opposition, taking nearly 100 prisoners and thirteen machine guns. Liman's last hope of making a stand had thus been denied him.

* * *

During the first three days of this momentous week of 19 to 25 September Chaytor's Force [see p. 248], of which Anzac Div formed the mounted element, made no forward move against IV Army to the east of Jordan. But, by active demonstrations, Djemal, its commander, was kept uncertain of Allenby's intentions. The breakdown of communications meant that news of the débâcle to the west of Jordan reached him slowly. Liman says that he warned him on 20 September to make good his retreat at once and had next day actually ordered him to do so.[21] Whether this is true or not, Djemal hesitated. He was probably chiefly motivated by a desire not to abandon II Corps, between 4,500 and 6,000 strong, which was withdrawing northwards from Maan and other posts on the Hejaz railway. He did not start to withdraw till 22 September, thus, in the event, ensuring the maximum danger for his own army without being able to save II Corps. To add to the awful nature of his predicament, Faisal's Arabs were in strength between him and Deraa, the Hejaz railway was cut and at long last the mass of Bedouin were beginning to rise against the Turks.

Since Liman's expectation had been that Allenby would make his main attack on the right of the line, considerable strengthening of IVth Army's defences had taken place during the late summer. When, therefore, Chaytor's Force advanced, the New Zealanders leading and the West Indian infantry playing an useful part, a number of sharp fights followed. Since, with only a few minor exceptions, all the action took place dismounted in mountainous country, it is sufficient to say here that the Jisr el Damiye crossing of the Jordan was taken on 22 September and that Es Salt was occupied for the third and last time by the New Zealanders on the 23rd and that after a truly pertinacious defence by the Turkish rearguards, Amman fell the following evening. Those numerous Turks who escaped did so over many a waterless mile, harassed and cruelly murdered by the revengeful Arabs. Very few of them survived. Even fewer reached Damascus over 100 miles away.

Chaytor's next task was to intercept II Corps under Ali Bey Wakaby at Amman. He was also ordered on the 25th to occupy the Wadi el Hammam, the sole water supply point for the corps should it manage to slip past Amman. This he effected on the 26th. On the 28th the enemy was located by aeroplane at Ziza near Kastal (see map p. 216). A message dropped to the corps commander read: 'Surrender your force. We hold all water you can reach. You cannot march northward now. If you surrender put a large white flag on the station buildings. If you do not surrender you will be bombed by our aeroplanes [in the morning of 29th].'[22] Wavell explains what happened next:

> 'The Turks were willing enough to surrender, but unwilling to lay down their arms until the British force was large enough to protect them from the hordes of savage Bedouin who surrounded them. [Only 5th Regiment was up at this time.] The other two regiments of 2nd A.L.H. Brigade came up just before dark, but, as the attitude of the Arabs was still threatening and their numbers [were] large, they joined forces with the Turks, who were allowed to keep their arms. British and Turks together held a line of defence round the station and kept off the Arabs during the night.'[23]

Throughout that night of tension the Turks engaged in bursts of machine-gun and rifle fire to hold off the Arabs. The light horsemen, outnumbered by eight to one, had, of course, no fear of the Turks and 'could be heard cheering on their activities. "Go on, Jacko," they would shout, "give it to the blighters." What was grim tragedy to the Turks was farce to the Australians.'[24] At one moment a Hotchkiss gun had to be turned on the Bedouin, none of whom was part of Faisal's army. They were all of that Beni Sakhr tribe which had let Allenby down at the time of the second raid east of Jordan (see p. 225). They eventually dispersed and in the morning of the 30th well over 4,000 grateful but exhausted Turks were marched prisoner into Amman.[25]

In the nine days since Chaytor's Force had moved forward, it had lost only twenty-seven killed, 105 wounded and seven missing. It had captured 10,322 prisoners, fifty-seven guns, 132 machine guns, eleven railway engines, 106 trucks and a vast amount of material. IVth Army was shattered. Anzac Mounted Division's work was finished. The ravages of malaria now descended upon it at the moment of victory.

* * *

The battles of Megiddo* were over. The Commander-in-Chief cabled to the Australian Government: 'The completeness of our victory is due to the action of the Desert Mounted Corps under General Chauvel.' He concluded a message to his army with these words: 'Such a complete victory has seldom been known in all the history of war.'[26] In some ways it looked forward to the German *blitzkrieg* of 1940. It was certainly the exact opposite of the horrendous static warfare in France where it had no parallel even in the final partial breakthrough.†

*These included officially the battles of Sharon and Nablus.

†Though such comparisons are open to obvious criticism, it can be claimed that the measured rate of opposed advance achieved between 19 and 21 September was fractionally faster than that of the Israeli tanks in 1967 and twice as fast as the breakout from Normandy in 1944 and the 1941 Barbarossa offensive. (Dupuy, T. N. *Numbers, Predictions and War*, 1979, quoted in Badsey, 352).

33

'Seventh and Eighth Turkish Armies have been destroyed. Fourth Army is retreating on Damascus via Deraa. Desert Mounted Corps will move on Damascus. . . . Divisions should move on as broad a front as ground admits. On arrival at Damascus a defensive position will be taken up on high ground commanding the town.'

Telegraphic order by Desert Mounted Corps, 26 September, 1918 (7 p.m.)

'In that desolate, treeless country, with its ruined cities and with the stench of death about them, Barrow and his men longed for the waters of Damascus almost as much as did their wretched enemy.'

A. H. HILL, Chauvel's biographer

'It is an open secret that General Allenby had been urged by the amateur strategists of Downing Street to make a cavalry raid on the city [of Damascus], supported by the forces of the Emir [Faisal], but he had steadily refused to commit his cavalry to this hazardous enterprise until he had dealt with the Turkish army. Now, however, the way was clear and he determined to push on with all speed.'

* * *

'[10th Cavalry Brigade] tried to dispose in too contemptuous a fashion of a strong flank-guard from the Fourth Army.'

WAVELL in *Allenby* on the action at Irbid.

'The capture of Damascus was the climax of the campaign, [but it] brought no feeling that strenuous exertions were nearly at an end or that demands for sacrifice were soon to cease.'

CYRIL FALLS in *The Official History*

'The last phase of the Palestine campaign was fought and won in the incubation period of malignant malaria which is ten to fourteen days.'

MAJOR C. HERCUS, DADMS of Anzac Division

Arrangements for taking Damascus

> 'Many brave men who, as if with charmed lives, had fought through from the early days in Gallipoli, died miserably in the hour of victory.'
>
> GULLETT in the *Australian Official History*[1]

Irbid – Kuneitra – Sasa – Kaukab – Kadem – Kiswe – El Mezze – Barada Gorge – El Hayira – Damascus entered – Khan Ayash

When on 26 September Allenby conferred with his three corps commanders at Jenin, the question was not what to do, but how to do it. Everyone knew that the ancient Syrian capital of Damascus was the next objective and that few very formidable bodies of the enemy stood in the way. At most there were believed to be 40,000 Turks and Germans either in the city or retreating on it. Of these the numerous remnants of IV Army, withdrawing through Deraa, formed the largest part. Disorganized though its units were, it was possible that given time they could form a force capable of much delaying any further advance, particularly as most of them had not yet been heavily engaged. The tempo of the pursuit must therefore be fully maintained.*

Having been warned by Allenby on 22 September, Chauvel was able to issue his detailed orders without delay. Barrow's 4th Division was to march on Deraa via Irbid, try to intercept the retreating IV Army and, should it fail in this, to pursue along the ancient Pilgrim's Way and the Hejaz railway to Damascus in full cooperation with Faisal's Arabs, who were already operating against Deraa from the east. Hodgson with Ausdiv from Tiberias and, following him, Macandrew with 5th Division from Nazareth were to take the direct road to the city, through Kuneitra. Barrow would have to march 140 miles, Hodgson and Macandrew not less than ninety-five. Barrow left, therefore, a day earlier than the others. Speed and good timing would result, it was hoped, in all three divisions arriving before Damascus more or less simultaneously. In the event they were well up to the appointed time.

Practically all transport was left behind except ammunition

*Meanwhile Bulfin's and Chetwode's XXI and XX Corps were taking over from the cavalry the country so far occupied and XXI Corps was ordered to send one or two divisions along the coast to Beirut.

wagons and ambulances.* Only two days' rations for the men and one for the horses were allowed. When these were exhausted, living off the country was the order of the day. At the Corps Commanders' conference Allenby had turned to his senior administrative officer and asked: 'And what of the supply situation?' 'Extremely rocky, sir,' was the dampening reply. 'Well you must do your best,' was the Commander-in-Chief's dismissive response. As Wavell has put it, he then

> 'proceeded to impress on his commanders the need for boldness and rapid action. He was meticulous in his administrative arrangements before an operation, but in a pursuit like this was prepared to drive his troops forward on the shortest of rations. He realized better than anyone that his most dangerous enemy at the time was disease rather than the Turk and that his best weapon was speed. He knew that malaria, which his protective measures had so hardly held at bay in the Jordan valley and elsewhere, was bound to take toll of his troops once they had passed into the lines of the Turks, who had taken no measures of protection. He told a senior Medical Officer that the knowledge that in fourteen days from the opening of the campaign malaria would begin to sap the strength of his force acted as a spur to his determination to press forward as rapidly as possible.'[2]

* * *

Early on 26 September Barrow's 4th Division began its march towards Deraa. He ordered Green's 10th Brigade to move from Jisr el Mejami to Irbid and to try to get in touch with Lawrence's Arabs.[3] The distance to be covered was only sixteen miles, but the going was bad, the maps were inaccurate and a partially broken bridge delayed the brigade to such a degree that the leading troop of 2nd Lancers which had left at 08.00 did not reach the bank of a deep gorge three miles short of Irbid until 4 p.m. It was there fired upon from the village of El Bariha which it then galloped,

*4th Division 'carried its supplies with it on wheels'. (Osborne, III, 281). It was 'obliged to requisition meat and forage on its march, and even then the troops had finished their iron rations when they reached Damascus.' (Falls, 562).

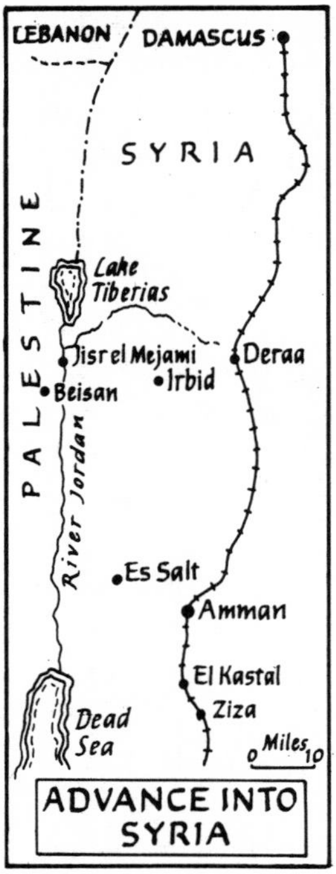

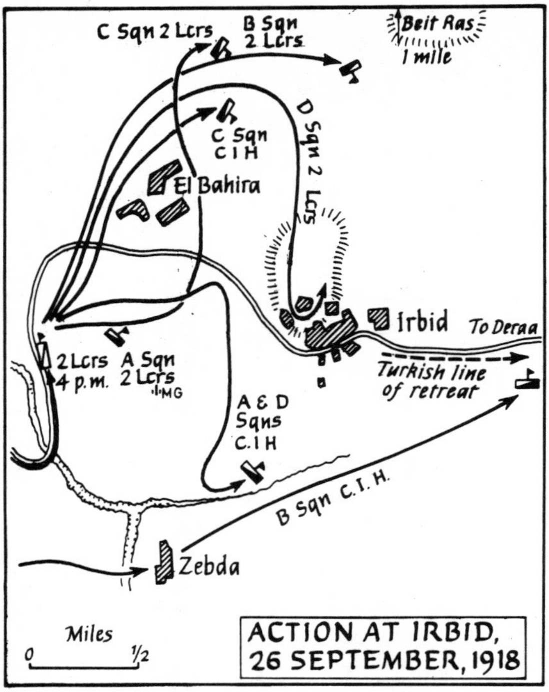

dislodging the few Turks within it. Major G. Gould, commanding the regiment, had been told by a friendly Arab sheikh that the 2,000 Turks in Irbid would probably be willing to surrender.[4] Consequently, and anxious to gain Irbid's water supply before nightfall, he decided to attack at once, without a reconnaissance and without waiting either for the Central India Horse or the Berks Battery to come up. The enemy's position was formidable. It extended for over a mile northwards from the village along a high ridge surrounded on its western side by an open glacis slope.

The lines of attack taken by 'B', 'C' and 'D' Squadrons are shown on the map above. As they moved off intense machine-gun fire was poured on them from the ridge. Owing to this, 'B' Squadron failed to get round by the north and instead took up a position 1,200 yards from the village and brought its Hotchkiss rifles into action in support of 'D' Squadron's attack. In fact, the historian of the 2nd Lancers, Captain Whitworth, who commanded 'B' Squadron in the action, says that he had 'only a Hotchkiss Rifle troop in hand when the attack was ordered'.[5] He tells what happened next:

' "D" Squadron having brought their Hotchkiss rifles into action at Bariha prior to moving N.E., came under heavy fire as they moved southward to the attack and Captain Vaughan (see p. 280) led them in a most gallant charge on the machine guns. The ground was almost impassably stony. When within 200 yards of the enemy, Captain Vaughan realised that he was not as far round the flank of the position as he ought to have been. There was very rough ground between him and the northern end of the ridge the enemy was holding, which itself appeared too steep to gallop in the face of the intense fire which was now going on. He accordingly swung the squadron half-right into the village, which protruded further to the west than the line of the ridge. His advanced troop was too far to recall and he thought he might obtain a footing in the village until reinforced.'[6]

This advanced troop galloped straight up the hill. Its commander, Ressaidar Raj Singh, actually reached the enemy position before being shot down, as were his sowars, almost to a man.

Whitworth's account continues:

'When giving the signal to head for the village, Captain Vaughan was wounded in the knee and his horse was mortally wounded but carried his rider as far as the village square. Here the squadron commander found he had with him Jem. Basti Singh, Ress. Kheri Singh (both wounded), some half dozen men and four wounded horses. There were roads running S., E., and N., by the latter of which they had entered. On reconnoitring those on S. and E. they came under heavy machine-gun fire. Having too few rifles to hold on and seeing no signs of other squadrons, Capt. Vaughan ordered a retirement, adding that anyone who reached the regiment should report what had happened. Clambering over or through the houses they escaped from the village westward being under continuous fire. Capt. Vaughan had his left arm shattered by machine-gun bullets when 150 yards from the village, and 50 yards further on, his left thigh broken. He was carried, with magnificent devotion, by his Indian orderly to a place of safety within the advanced line of the C.I.H., just before the battery opened fire [see below]. (For this act of gallantry the orderly, A.L.D. Nand Singh, was awarded the I.D.S.M.)'[7]

While Vaughan's squadron was thus engaged, 'C' Squadron under Captain D. S. Davison* had mistaken Bariha for Irbid. He was therefore too late to help in the attack. When he saw Vaughan's failure he placed his squadron under cover 1,200 yards from Whitworth's 'B' Squadron, the Hotchkiss rifles of which 'now came in for serious attention from the enemy. They were lying in a stony ploughed field without cover and one man and two horses were killed.' Whitworth therefore moved them out of action and eventually collected his squadron under cover of a ridge between Biet Ras and Irbid. His patrols from here were met by heavy machine-gun fire.[8]

The two subsections of 17th Machine Gun Squadron which were accompanying 2nd Lancers on the march had been meant to support the three squadrons' attack. From a knoll to the south of Bariha the four guns were heavily fired upon by numerous enemy machine guns from the south of the ridge. This prevented them from getting closer than 2,700 yards, which, as Whitworth put it, 'militated against effective fire on the actual point of attack'.[9]

The brigadier came on the scene at 16.15 just after 2nd Lancers' attack had begun. When he saw large numbers of the enemy hurrying to take up positions on the ridge to oppose it he realized the urgent need to bring the Berks Battery into action. 'But,' as Whitworth put it, 'the defile behind was all against rapid debouchment; and it was 17.10 before the battery, which was in rear of the M.G. Squadron and C.I.H., could open fire, too late to be of assistance to the 2nd Lancers' attack. Shooting was difficult owing to the failing light and the uncertainty of the location of our troops.'[10]

The brigadier had meanwhile been hurrying on the Central India Horse, but due to the bad going over the boulder-strewn ground, the misunderstanding of a verbal order and the fact that one squadron, like 'C' Squadron of 2nd Lancers, mistook Bariha for Irbid, the regiment could neither help the lancers nor, as Green desired, cut the Deraa road beyond Irbid before darkness descended.

*Captain Davison (see p.275) had been attached to the regiment from 4th Dragoon Guards in 1914, as adjutant. He had been in command of it for a short period, until Major Gould took over only three days before the Irbid action.

'The night,' wrote Whitworth, 'passed quietly, the 2nd Lancers remaining in position and the remainder of the brigade concentrating S.W. of the village. Owing to the uncertainty of the situation, saddles could only be removed for a short time by portions of the force and the weary animals got no rest and no water for the night after the fatigue of the action added to the most exhausting march that the brigade had yet performed.'[11]

During the night the enemy, overestimated by the locals as being about 3,000 in strength, withdrew unmolested toward Deraa. They left behind them nine tons of barley and a small herd of cattle which 4th Division took along on the march. These were godsends, for both men and horses were by now approaching starvation.

In this, the first check met with by 4th Division since it rode through the Turkish trenches a week earlier, 10th Brigade suffered forty-six casualties. Of these twelve other tanks of 2nd Lancers were killed and Captain Vaughan and twenty-nine wounded.* How many horses were killed is not recorded. That there were many is not in doubt.

This small-scale action was, as Barrow put it, 'a complete failure'.[12] The moral of it is that even when pursuing a dispirited enemy it was courting disaster for a small cavalry force to take on an unestablished number of men in a strong position, without previous reconnaissance and without adequate fire support.[13]

* * *

Next morning, the 27th, the Dorset Yeomanry took the lead. Its leading troop was met outside El Remte by heavy machine-gun fire. It withdrew to cover and dismounted, whereupon 300 Turks moved towards it and deployed. About a third of these boldly attacked, supported by the fire of four machine guns. The rest of the Dorsets quickly came up and the attack dissolved.

'The Turks, in fact,' as an eye witness related, 'retired at the double, losing fifty men, and one squadron pursued them into the village, where the white flag was then hoisted. In spite of this the leaders were fired on as they entered. One officer was shot dead, one was wounded and several men were hit. This

*'For this light loss [Captain Vaughan's squadron] undoubtedly had above all to thank the very bad marksmanship of the Turks.' (Falls, 580).

treacherous behaviour entailed the systematic clearance of every house in the village.'

The leading squadron of the Central India Horse was now sent forward to help the yeomen, while the three other squadrons pursued the enemy as they fled towards Deraa, five miles away. Two of these squadrons charged a body on the march, speared numbers of them, captured the four machine guns and nearly 100 prisoners. The third squadron rode on to gallop down and rout another party of Turks occupying a prepared position overlooking Deraa. A further four machine guns and about 100 prisoners were taken here.[14] The day's fighting was over by noon.

Because the horses were exceptionally fatigued and being uncertain whether Deraa was occupied by Turks or Arabs, Barrow called a halt for the rest of the day. Next morning Brigadier-General Green was met outside Deraa by Lawrence, from whom he learnt that the Sherifial irregulars had entered the town the previous afternoon. The unspeakable atrocities inflicted on the wounded and helpless Turks which met the division on its entry sickened and shocked the toughest of the cavalrymen.*

The division began next day the direct pursuit towards Damascus 'with,' as Wavell put it, 'the Arabs hanging on to the right flank of such Turkish forces as were still in being between Deraa and Damascus'.[15] The remnants of IVth Army outpaced Barrow's troops,† but by midnight on 30 September 11th Brigade had

*Barrow describes the revolting savagery of the scene at the railway station: 'A long ambulance train full of sick and wounded Turks was drawn up in the station. . . . The Arab soldiers were going through the train tearing off the clothing of the groaning and stricken Turks, regardless of gaping wounds and broken limbs and cutting their victims' throats. . . . I asked Lawrence to remove the Arabs. He said he couldn't "as it was their idea of war". I replied, "It is not our idea of war and if you can't remove them, I will." He said, "If you attempt to do that I shall take no responsibility" and at once gave orders for our men to clear the station. . . . All the Arabs were turned off the train and it was picquetted by our sentries.' (Barrow, 211).

†Gullett says that Barrow 'repeatedly asked Chauvel to detach a force which, striking north-east from the main road followed by Hodgson [Ausdiv], might cut across the front of the Turks before they gained Damascus. But the delay at Jisr Benat Yakub [see below] made this for the time impracticable.' (Gullett, 744). Neither Barrow nor Falls nor any other authority mentions this, but it seems likely to be true.

reached Khan Deinun, some eleven miles from the Syrian capital, and the rest of the Division was closing up. Its guns were heard by Ausdiv to the west.

* * *

While 4th Division was chasing the fugitive IVth Army, Ausdiv and 5th Division were following up the even more demoralized remnants of VIIth and VIIIth Armies, by the direct road to Damascus* with Ausdiv in the lead.

Leaving the Tiberias area early in the morning of 27 September, a strong rearguard, consisting chiefly of lorried German machine gunners, was found defending the half-destroyed bridge across the Jordan at Jisr Benat Yakub. Searching for alternative fords and being forced to take part in some hard dismounted fighting, the 3rd Brigade, supported by 4th and 5th Brigades, eventually crossed the river – but not till next morning, the enemy having withdrawn during the night in motor-lorries.

By 1 p.m. on the 28th Kuneitra had been reached. From the town's roofs a mass of white flags made the place look like 'a Chinese laundry on a Monday'.[16] From there a great plain, broken only by wadis, stretched away to Damascus, thirty-seven miles distant. During the following night (the 19th) further opposition was encountered at Sasa. After some dismounted fighting and by some skilful outflanking by the 10th Regiment, all in the dark, the enemy was nevertheless again able to escape in lorries. Early in the morning of the 30th 4th Regiment of 4th Brigade charged into a large column, cutting off a party of 180 Turks. 'Spreading out across the plain, now covered with fugitives, the regiment,' according to the Official Historian, 'continued to harry them, groups of three or four horsemen riding at bodies of Turks ten times their own strength and calling upon them to surrender.'[17] A field gun, several machine guns and 350 prisoners were gathered in.

About eleven miles short of Damascus, near Kaukab, a further apparently strong position was encountered. To attack it the whole

*The first part of which consists of the highlands known today as the Golan Heights, whole areas of which are lava country through which mounted movement is scarcely possible.

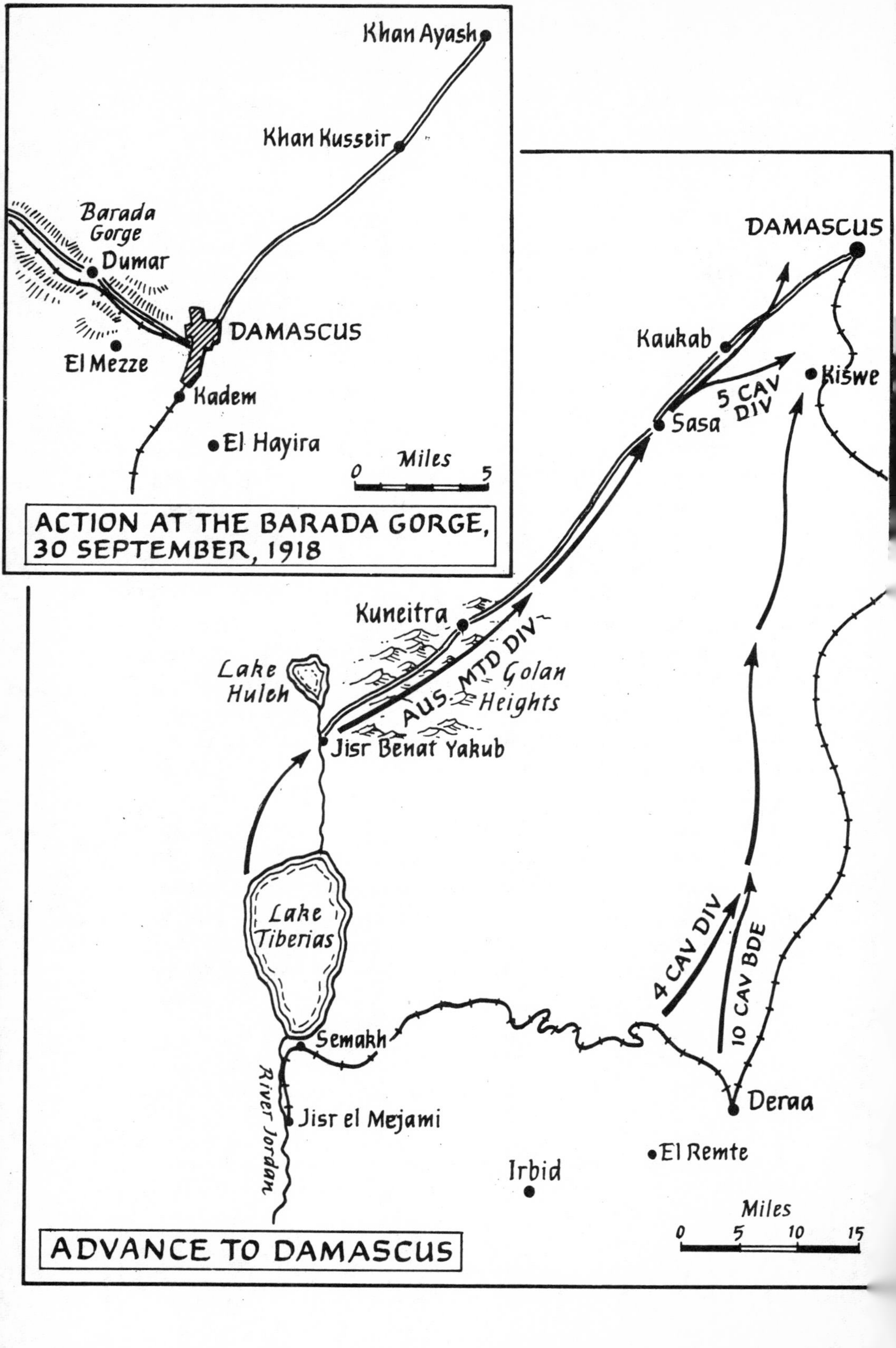

ACTION AT THE BARADA GORGE, 30 SEPTEMBER, 1918

ADVANCE TO DAMASCUS

of Ausdiv (less the Gloucesters*) deployed in 'artillery' formation and trotted down into a two-mile-wide valley in two wings, each regiment in column of squadrons in lines of troop columns. The two horse artillery batteries gave support, while the divisional staff rode between the two wings, Hodgson riding at the head of his division controlling them by means of gallopers. The spectacle was magnificent: one which had seldom been seen in modern war and one which was never to be seen again. The expected blast of machine-gun fire as the brigades charged up the opposing slope did not come. The Frenchmen of the *Régiment Mixte* of 5th Brigade had 'struck in at once past the right flank of the enemy and so menaced his rear, while the batteries [Notts and 'A', HAC] freely shelled his machine-gun positions.'[18] Thus assailed, with no artillery to support them, and at the sight of the galloping horsemen, the enemy broke without firing a shot, leaving behind a dozen machine guns and twenty-two prisoners.† The Australians and French lost not a man.

* * *

While this dramatic if bloodless engagement was taking place, 14th Brigade of MacAndrew's 5th Division north of Kiswe intercepted and took prisoner most of the leading half of a column of some 2,000 Turks including the remnants of 3rd Turkish Cavalry Division with its commander and staff. The rear half of this column broke back, but 13th and 15th Brigades dealt with them, taking numerous prisoners and heading the rest off towards 4th Division, whose guns were seen, just before dusk, firing from Khiara northwards. Thus were the two widely separated portions of Descorps once again on the point of being joined together in the outskirts of Damascus.

*From 13th Brigade two troops of the Gloucesters under Captain Lord Apsley were sent by Barrow that afternoon to seize the high-powered wireless station at Kadem, through which communication with Constantinople and Berlin was maintained and from which much German propaganda in the Middle East emanated. As they approached, the Germans blew up the great aerial masts and were in the process of destroying the buildings when Apsley's men charged them hoping to save some of the equipment. They put three Germans and seven Turks to the sword before the remainder surrendered, but a strong German force then appeared and forced the two troops to retire at speed.

†Osborne, III, 280, says seventy-two.

* * *

Hodgson, meanwhile, had pressed on from Kaukab to get astride the enemy's line of retreat along the Beirut and Homs roads. At El Mezze 5th Brigade, in the lead, was faced by sixteen machine guns firing from long-prepared positions. Dismounted, the brigade, led by the *Régiment Mixte*, could make no head against these until at about 1 p.m. Notts Battery came into action. 'The effect was almost instantaneous; the machine guns ceased firing and the crews abandoned their guns and fled; such,' wrote Lieutenant-Colonel Osborne, 'is the effect of even weak artillery upon demoralised troops.'[19]

The brigade now scrambled over the hills and just before dark reached the southern heights overlooking the precipitous, hundred-yard-wide Barada Gorge through which the road and railway to Beirut and Baalbek run beside the torrential Barada River. There the French regiment with six machine guns of 2nd New Zealand Machine-Gun Squadron saw from the cliff top in the failing light columns of thousands of troops and other fugitives with masses of transport including railway trains. The machine gunners opened up. Sergeant M. Kirkpatrick of the squadron describes what he called

> 'one of the most frightful tragedies of the campaign. . . .
>
> 'Part of our squadron, reaching the brink of the precipice, quickly took up positions almost invisible to the dense mass of enemy below. The head of the column was felled, and, as the unfortunates behind kept pressing forward, they were mown down as by some invisible scythe. Horses and men went down together in hundreds and died in one tangled bleeding mass. Many fell into the river and were drowned. The Germans fought desperately from the tops of lorries and from a train with their machine guns, but, seeing not where to fire, their shots were wild, and they too went down in the slaughter. The water in the M. G. jackets hissed, and bubbled, and steamed. The barrel in one of the guns was so hot that it bent like a crooked stick. Australian Hotchkiss guns and rifles joined in the work of destruction. Above the rattle of the machine guns and the roar of the river, the cries of anguish and despair swept up from this valley of death.'[20]

46. Looking westward from Damascus can be seen the narrow Barada Gorge through which ran the Abana River, a metalled road, a single-line railway and a telegraph line. Through this gorge six railway trains and a column of 14,000 Turks tried to escape. The head of the column was caught by machine-gun fire from the hill on the left of the photograph. The gorge was turned into a shambles (see p.331)

47. ‘A’ Squadron, 10th Australian Light Horse Regiment, Tripoli, 1919

There being no way forwards, the milling chaotic masses tried to flee back to the city. 4,000 of these were scooped up by 14th Regiment. Two miles to the west 9th Regiment also reached the edge of the gorge. There they found another packed column into which the Australian machine gunners and Hotchkiss riflemen 'fired and fired till the road was littered with bodies of men and animals and the wreckage of transport wagons. Four hundred dead were later found on the road and it took several days to burn the vehicles in order to clear the pass.'[21] Both the French and the Australians when they first came on the scene were fired on by German machine gunners who held them up for a considerable time, and when 9th Regiment tried to enter Dumar on the northern side of the gorge (see map, p. 324) by a narrow rocky track it was prevented from doing so 'by a number of machine guns placed in and around the village'.[22] Orders were now received to hold the position through the night.*

* * *

Chauvel now gave orders for the encirclement of Damascus next morning, 1st October. Soon after the Poona Horse received its orders its advance troop near El Hayira, supported by its Hotchkiss rifle section, charged a Turkish column of about 300 men, killing numbers of them before the rest of its squadron could come into action. More than 200 prisoners were taken. Simultaneously with this charge 'a small body of Hedjazian cavalry rounded a corner of a hill and charged the rear. This was the first time that the regiment had come in close contact with our Hedjaz allies.'[23]

The scouts of Wilson's 3rd Brigade, meanwhile, failed to discover any mountain tracks which would take it around to the north so as to seal the Homs road. He therefore asked and received permission to disregard Allenby's strict order not to enter the city (given for political reasons). At first light, aware that thousands of the enemy must be concentrated there, but counting on the moral effect which would be produced by hundreds of galloping horsemen,

*In the morning Dumar was found to have been evacuated.

Astonishingly enough in spite of almost continuous machine-gun fire a train from Beirut managed to pass into the gorge in the dark unscathed! It was captured next day by 10th Regiment when the advance was continued. (Darley, 157. See also Falls, 573, fn. Olden, 276)

he gave the order to advance. The decision was, under the uncertain circumstances, a truly bold one. A troop of 3rd Brigade scouts under young Lieutenant Foulkes-Taylor who had led the gallop into Es Salt six months earlier (see p. 226), struggled through the chaos in the gorge and, followed by 10th Regiment, sped on towards the city centre. On the way the horsemen passed the main Turkish barracks where, in the words of Major Olden, the 10th's Second-in-Command, who led the regiment,

> 'an immense body of Turks were observed on the open parade ground [they numbered about 12,000 and were later taken prisoner by 4th Brigade when it entered the city]. They seemed to be in a state of great confusion and indecision. . . . Suddenly a burst of rifle fire was directed against our men. . . . The pace of the scouts was momentarily checked, until the leading troops of "C" Squadron appeared close up, and then with drawn swords the column charged down the road at the gallop.
>
> 'No further opposition came from the Turks in the barracks; but as the column neared the centre of the city, dense masses of people were seen to fill the streets and squares. . . . A very large proportion carried firearms of some kind and these they now proceeded to discharge in the air at the same time uttering frenzied cries and shouts.'

Forcing a passage through the milling crowds, Olden drew up at the Serai or Town Hall, 'the steps of which were lined with officials and notables'. He called for an interpreter and a Greek at once volunteered. '"Where is the Governor?" was the first enquiry. "He waits for you in the Hall above," came the answer.' Olden and two other officers carrying their pistols in their hands then entered the hall and found there the Governor of Damascus* who rose

*Emir Said Abd el Kader was a grandson of Abd el Kader, the Sultan of Algiers, who had kept the French at bay for fifteen years before being overthrown, captured and exiled to Damascus. Djemal, on his departure from the city had installed him the previous day.

'and, coming forward with outstretched hands, said in Arabic:-

"In the name of the City of Damascus, I welcome the first of the British Army."

"Does the city surrender?" he was asked.

"Yes; there will be no further opposition in the city."

"What, then, is all the firing in the streets?"

"It is the civil population welcoming you."

"Who are all the uniformed men with arms?"

"They are the *gendarmerie*. What are they to do?"

"They may retain their arms for the present, prevent looting by the Arabs and otherwise maintain order. As for the shooting in the streets, issue orders that it must cease immediately, as it may be misunderstood. You will be held responsible for this."

"You need not fear," replied the Emir. "I will answer for it that the city will be quiet."'

Telling the Governor to keep for the Commander-in-Chief the lengthy speech upon which he now embarked, Olden politely refused the proffered refreshment and obtained guides to lead him to the Homs road. By 6.45 a.m. the march was resumed.[24]*

* * *

With the troubles which confronted Chauvel when he took over temporary command in the city, especially those connected with Lawrence, this work is not concerned. It is enough to say that on 2 October he had to organize a show of force by marching a large number of horsemen, guns and armoured cars through the streets.† Only then was order restored to the city. 'The bazaars were opened,' wrote Chauvel later, 'and the city went about its normal

*Damascus had a day or two previously thrown off Turkish rule and German control, Liman having gone to Baalbeck on 25 September to try to reorganize his shrinking forces. There had been sporadic fighting between Germans and Turks and between the Turks and the Christian parts of the population.

†Chauvel, Barrow, Macandrew and Hodgson headed, with their staffs, as well as the guns and armoured cars, one squadron from each regiment, one battery from each division, the New Zealand machine gunners and a squadron of 2nd Brigade. The French were also represented.

business; but I had to find a whole regiment of police, the gendarmeries not proving to be of much value.'[25]*

* * *

Wilson's 3rd Brigade when it reached the Homs road chased the retreating enemy as far as Khan Ayash, seventeen miles north of Damascus. It found that the German machine gunners still had much determination left in them. Each time they were rushed they abandoned their weapons, but so copious was the supply that those who escaped capture came into action again a few miles on. There was some street fighting and the use of bayonets in Khan Kusseir as well as effective enemy fire near Khan Ayash. There the very last action fought by the Australian light horsemen in the campaign saw two squadrons of 9th Regiment charging a body of enemy cavalry. It was also the very first time that any of the light horsemen since they had been armed with the sword 'had the chance to try conclusions with the Turkish cavalry who were armed with sword and lance, and it was expected,' as one who was there reported, 'that they would put up a fight, [but] they surrendered without the slightest show of resistance.' The bag totalled ninety-one officers, 318 cavalrymen, 1,064 infantrymen, eight German machine gunners, twenty-six machine guns, three artillery pieces, twelve automatic rifles, 264 rifles and 285 animals. These were all taken within an hour of the regiment leaving its bivouac on 2 October – a fine ending to the Australians' final engagement.[26] The 9th also captured the standard of the 46th Turkish Regiment. This was the sole enemy flag taken in action by the Australians during the war. Since 3rd Brigade had now outstripped its artillery, was running out of machine-gun ammunition and was short of food, Wilson

*Chauvel also had 'to find supplies and forage for his corps with its 25,000 horses; there were 300,000 Damascenes who must be fed and whose government must be put in order and about 20,000 Turkish prisoners to be cared for.' He had been provided with no political adviser by Allenby who had given him only 'sketchy instructions' as to what to do, while 'Lawrence's preoccupation with Arab politics when he should have been providing an efficient link between Chauvel and the Arab government' was distinctly unhelpful. (*Chauvel*, 182)

Upon the complexities of the political situation, involving Syrian Arab, French and British interests, all in conflict, many books have been written, none more thoroughly unreliable than that great work of literary art, Lawrence's *Seven Pillars of Wisdom*. Robert Graves's *Lawrence and the Arabs* is almost equally untrustworthy.

decided that he must fall back and bivouac, and this he did, much to the relief of the exhausted men and horses.

* * *

From 26 September to 2 October Descorps had captured 662 officers and 19,205 other ranks. This brought the total since the commencement of operations on 19 September to over 47,000. The actual battle casualties of the corps from that date numbered only 533.* Of the 360 guns taken by the whole EEF about 140 fell to Descorps.[27] From 15 September to 5 October only 1,021 horses out of a total strength of 25,618 were killed in action, died or were destroyed. This is less than 4%. A further 259 were reported as missing. In the same period only 3,245 were admitted to veterinary hospitals or mobile veterinary sections.† Of these 904 were re-issued as cured during the period, while the majority of the rest were later made fit again for active service. Colic, diarrhoea, galls, general debility and 'fever' were the chief causes. This wonderfully small amount of wastage shows how admirable horsemastership had become in Descorps, particularly as the average weight carried was probably never less than twenty stone. Barrow's 4th Division had the lowest wastage in the corps. The prize goes to Captain Gerald Michael Fitzgerald whose squadron of Sikhs in 19th Lancers reached Damascus with every one of its horses intact.

Of the three enemy armies, virtually the only totally disciplined formations still in being on 2 October were about 700 men of Colonel von Oppen's Asia Corps troops which had managed to pass through the Barada gorge before it was closed and the 146th Regiment under Lieutenant-Colonel von Hammerstein-Gesmold, which had escaped along the Homs road. The total remnants of the three armies numbered perhaps 17,000 of whom, according to Allenby, only 4,000 were at this time effective rifles. The rest were 'a mass of individuals, without organization, without transport and without any of the accessories required to enable it to act even on the defensive.'[28]

*Officers: sixteen killed, forty-eight wounded, one missing. Other ranks: 109 killed, 317 wounded, thirty-two missing. Of the missing a few were later found in hospitals, together with some who had gone missing in earlier actions.

†The returns were made weekly. This figure therefore certainly includes a considerable number of horses discarded between 15 and 19 September as being unlikely to stand the strain of the long marches ahead.

34

'The obstacles to the occupation of the rest of Syria were now distance, supply and disease.'

* * *

'This advance northward seemed rather an anti-climax – like walking up a few rabbits after the drives in which the birds had come high and fast.'

WAVELL in *Allenby*

'[Allenby] realized the soldier's dream of what a victory should be like. . . . The picture is almost unmatched as a brilliant conception brilliantly executed.'

B.H. LIDDELL HART in *Through the Fog of War*[1]

Advance to Aleppo – malaria – influenza – Haritan – end of war with Turkey

Urged on from London* and himself not loath to finish off the remnants of the Turkish armies before Liman could manage to reorganize them, Allenby lost as little time as possible in propelling his triumphant if weary cavalrymen forward to Aleppo, some 200 miles away. Another reason for acting with speed was the onset of the expected outbreak of malignant malaria. From the moment the troops emerged into unprotected enemy terrain, hordes of mosquitoes laden with the deadly parasites descended upon them from the Turkish positions. The incubation period of ten to fourteen days meant that those men who had not already been infected in the Jordan valley (and they were the majority, due to the stringent precautions taken) would start to suffer about 29 September – and so it proved. On the approach to Damascus men began to be attacked. Unlike the benign tertian form of the disease, the victims are suddenly prostrated in high fever, 105° and 106° Fahrenheit being frequently reported, and they often become delirious and

*Immediately the success of the breakthrough was known in London, Wilson, the CIGS, had pressed Allenby to make a cavalry raid on Aleppo. Without a large-scale landing at Alexandretta, Allenby rightly refused to consider the idea.

sometimes maniacal. Numerous cases of men falling from their saddles in such a state occurred before Damascus was reached. Immediate treatment with quinine was the only way of preventing complete prostration or in many cases death and this could not often be applied on the march.[2] Anzacdiv suffered particularly badly, especially the New Zealand Brigade, as a result of its advance into swampy ground near Jisr el Damiye.* The plight of 4th Division was no better, for it had been forced to spend several days near Beisan, one of the unhealthiest places in all Palestine. Ausdiv suffered rather less, though its one night beside the Jordan at Jisr Benat Yakub had not helped. 5th Division on the other hand had been fortunate in generally avoiding the mosquito-ridden districts. Admissions to hospital from Descorps were 1,246 for the week ending 5 October. The following week they were 3,109. Nearly four times as many men of the corps died in the overcrowded hospitals of Damascus as had been killed between 19 September and the capture of the city. The number of malarial cases dropped abruptly from 9 October onwards.

On about 6 October, to compound the awful vexations and gross overwork of the force's medical officers, numbers of whom themselves succumbed, the wave of pneumonic (or 'Spanish') influenza which was sweeping over large areas of the world struck with great vigour. At first it was difficult to distinguish it clinically from malaria. The Indians stood up to the malarial and influenzal onslaughts hardly better than the men of European blood. The reductions in manpower available for future operations were very serious. A side effect was that the problem of looking after the horses became acute. It was a common sight to see one man leading as many as eight horses.

The most worrying aspect of the situation was the appalling state of the tens of thousands of Turkish prisoners, most of whom were suffering not only from malaria and influenza but also from typhus, enteric and pellagra, a severe skin infection due chiefly to poor diet. To prevent these spreading to the British troops was a principal object of the medical authorities. On top of their own patients Descorps field ambulances had to take in 2,000 seriously

*The evacuations from malaria between 21 September and 10 October in 1st Brigade were 239 and in New Zealand Brigade 316, but those of 2nd Brigade, which had moved directly from the protected area into the hills, numbered only fifty-seven over the same period. (Report by Major Hercus, Anzacdiv's DADMS, in Powles, 262)

sick and wounded Turkish soldiers. Equally grave was the deplorable state of the many thousands of prisoners in the compounds set up at Kaukab and El Mezze. For some time seventy were dying every day, but soon after Lieutenant-Colonel T. J. Todd of the 10th Regiment was put in charge, the death rate fell to fifteen a day. He introduced strict sanitary precautions, peremptorily requisitioned supplies, recruited Syrian doctors and billeted the weakest men on local houses.

* * *

On 6 October 5th Division followed by the 4th occupied Rayak and Zahle without opposition. Next day the armoured cars found the port of Beirut unoccupied. This was a considerable relief to Allenby as supplies could now be speedily landed there, cutting his communication lines by hundreds of miles. On 8 October the Worcesters (XXI Corps cavalry regiment), spearheading 7th Infantry Division, entered the town to be received with acclamation by its inhabitants. The next stage of the advance was to the line Homs–Tripoli. Tripoli was occupied by the infantry on 13 October, providing yet another supply port, and Homs by 5th Division on the 16th. Both towns were free of the enemy. By now 4th Division had become so immobilized by disease in its ranks that it had to fall out of the advance.

Even 5th Division, the healthiest of Descorps' divisions, was much reduced by sickness, its regiments numbering a total of only about 1,500 sabres. Further, in 15th Brigade, since the Hyderabad Lancers were still detached on the lines of communication, only its two other regiments were present. With the Division marched two horse artillery batteries numbering between them only four officers and eighty men,[3] as well as three armoured car batteries and three light car patrols. This gave Macandrew the strongest column of motorized units yet employed in the theatre. A squadron of the RAF was also at his disposal.* Nevertheless the decision to push

*As had been done during the advance to Damascus, RAF motor cars accompanied advanced headquarters, in this case 5th Division's. These carried men who selected and marked landing grounds at each halting place, providing forward air bases which proved highly useful both for reconnaissance and attack. These were the first instances of successful co-operation between air and land forces in mobile warfare.

The Royal Naval Air Service and the RFC had been merged in the RAF on 1 April.

on from Homs to Aleppo, still 120 miles to the north, was a very bold one. The best of Liman's surviving units had had time to reorganize and it was believed that some 20,000 men, perhaps 8,000 of whom were capable of resistance, were assembling there. Indeed Allenby actually cancelled the advance on 20 October, but Macandrew (described by Cyril Falls as 'that vigorous and headstrong commander'[4]) telegraphed to Chauvel that he believed that there was 'no opposition worth thinking of at Aleppo' and that he proposed 'advancing with armoured cars' on the town. Chauvel passed on this request to Allenby who replied on 21 October: 'Yes. It is the wish of the Commander-in-Chief that Aleppo be taken as soon as possible.'[5] He was encouraged to make the decision because air reports had indicated that troops were leaving the city by train and road in some numbers.

Hama, halfway between Damascus and Aleppo, was reached by the armoured cars on 21 October. Next day they chased an enemy rearguard carried in lorries, which they encountered halfway between Hama and Aleppo, for fifteen miles, killing twenty-four men and capturing an armoured car and two lorries. On 23 October Macandrew sent into the city a summons to surrender. This the commander, Mustapha Kemal, politely refused, saying that he did not 'find it necessary to reply to your note'.[6]

During the two following days the armoured and patrol cars were occupied in reconnoitring the enemy positions south of the city while waiting for 15th Brigade, which they had greatly outdistanced, to come up. When it arrived in the afternoon of 25 October, news was received that the Arab force numbering some 1,500 which had been marching parallel with 5th Division along the railway line had that same day managed to enter Aleppo, inflict casualties on the Turkish garrison and force not only its withdrawal but also that of the two strong divisions which were holding the road to the south.* Kemal's retirement was also influenced by reports that a strong column of motor vehicles was advancing on the city. These were in fact lorries carrying not infantry but supplies.

When at 7 a.m. on 26 October Harbord moved the two regiments of his Brigade forward to clear the ridge west of Aleppo, on which the previous evening some Turks had been seen, he was

*These, under Major-General Ali Fuad Pasha, constituted a new Corps which Mustapha Kemal's organizing power had brought into being.

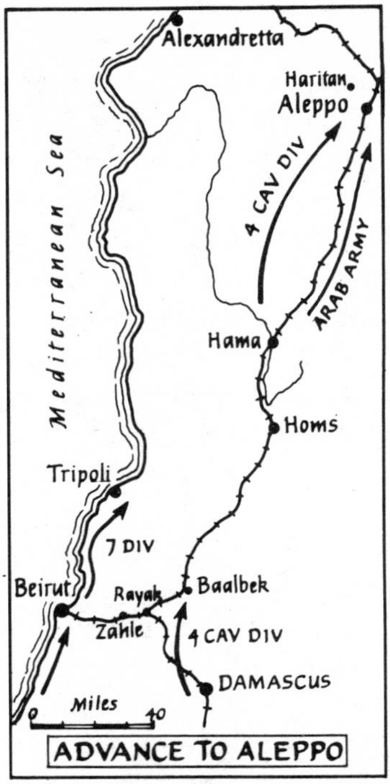

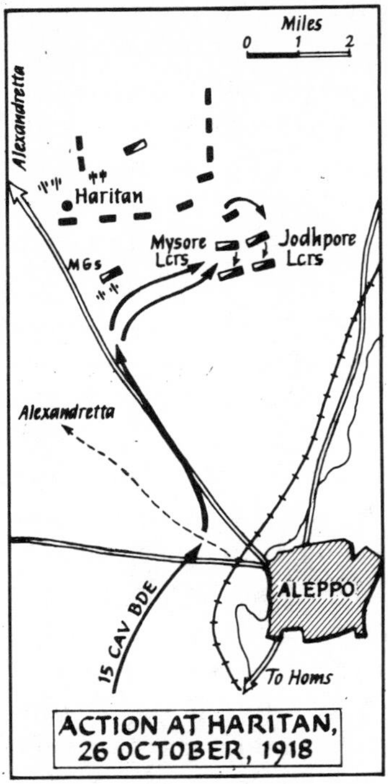

not yet sure that the city had in fact been evacuated. He therefore marched around its west side and reached the Alexandretta road at 9.45. Air reports told him that there was a body of 300 Turks on that road eight miles to the north-west and that 'about 1,000 scallywags of all descriptions with two small guns'[7] had left Aleppo at 7.30. Pushing forward at a trot the advance guard of two squadrons of the Jodhpores and a subsection of 15th Machine-Gun Squadron were suddenly fired upon by numerous riflemen and shelled by artillery near Haritan. They therefore fell back to a position about one and a half miles south-east of that village, dismounted and placed the machine guns on the right of the road.

Harbord now coming on the scene decided to attack without delay. He ordered the Mysores to move round east of the ridge on which the enemy was posted with its guns north-east of the village and to charge from the east. The other two squadrons of the Jodhpores were to move in support 'as a mopping-up party'.[8] The two regiments were by this time in the campaign weak in numbers. Together they could find no more than 500 men. The remaining four machine guns now arrived and were ordered to remain with the advance guard and support the mounted attack. While these positions were being taken up the 12th Light Armoured Motor Battery, having driven through Aleppo, appeared at 11.30. It was ordered to drive along the road in co-operation with the mounted attack. Unfortunately the battery leader's car broke down, which forced it to return and through some misunderstanding the other three cars also retired.

Meanwhile the senior Special Service Officer with the Mysores, Major W. J. Lambert, galloped on and, making a personal reconnaissance, found that the enemy position extended a good deal further to the east than he had supposed. He therefore moved his lancers in that direction so as to gain the enemy's left flank. Major Lambert tells what occurred next:

> 'The enemy had spotted this movement and turned the fire of several guns on the advancing cavalry; but as the cavalry kept moving at an appreciable pace, no casualties were suffered. . . . The formation of the advance was first of all line of squadron columns and afterwards line of troop columns with squadrons at open column distance.
>
> 'When a favourable opportunity offered, the Mysore Lancers were ordered to charge, pivoting on the Hotchkiss guns which were ordered to support the charge with fire action. Unfortunately, a shell fell into the middle of the Hotchkiss guns, and somewhat disorganized them, blowing three ammunition mules to pieces.
>
> 'At 12 noon the charge got well home, the horsemen galloping through lines of infantry, mostly Germans, lying down. Fifty of the enemy were killed and over twenty taken prisoners, and a great many more put up their hands, but, when they found how weak we were, started firing into the rear of the horsemen.
>
> 'The Jodhpore Lancers were, unfortunately, too far in the right rear to act as an effective support, and as they came up ran into a line of German infantry. Colonel Hyla Holden was shot dead at the head of the regiment, and other casualties were suffered. He had no scouts out in front of him, as far as I remember. A leader runs grave risks without local protection.
>
> 'The enemy were in far greater strength than previous information indicated, and as the further movement east had taken the Mysore Lancers beyond supporting distance of the machine guns, and the want of weight prevented them going

ALEPPO, ENTRANCE TO THE CITADEL.

26th October, 1918.

on and penetrating far into the enemy's position, the order was given to rally to the rear behind the Hotchkiss guns. They then took up a dismounted position and were joined by the Jodhpore Lancers [at about 1 p.m.].

'This action forced the enemy to reveal his whole strength, which was estimated at about 3,000 infantry, 400 cavalry, with 8 to 12 guns of various calibre, and from 30 to 40 machine guns and automatic rifles.

'With this force they advanced as if to attack the position we were holding, but when within 800 yards halted and began to dig themselves in. It was then realised that they were taking up a defensive position, fearing a further attack, and it was consequently decided to remain in observation on the line then held.

'Desultory fire was kept up till about 9 p.m. The whole night the enemy could be heard moving about. We remained with a thin line of outposts in touch with them.

'By daylight the enemy had been withdrawn.

'The 14th Cavalry Brigade arrived about midnight and took over the observation line at 6 a.m. on the 27th.

'The enemy's casualties were estimated at 100. Our casualties were 4 British officers, 1 Indian officer, 16 Indian other ranks killed; 6 British officers, 6 Indian officers, 44 other Indian ranks wounded, and 3 Indian other ranks missing. The total casualties were, therefore, 21 killed, 56 wounded, and 3 missing. On the 27th the Brigade was relieved by the 14th Brigade, and marched back to a bivouac area north of Aleppo, where it came into Divisional Reserve.'[9]

It is clear that Harbord made an error in relying too much on air reports and too little on ground reconnaissance. It is also clear that he was unlucky in the very last engagement of the campaign to have come up against virtually the only rested, fed, re-equipped and properly disciplined enemy force at Liman's disposal, and that it happened to be commanded by Mustapha Kemal in person.[10] Considering that 15th Brigade had no artillery with it and that the machine guns were mostly out of range of the left flank of the enemy, the casualty list of eighty is smaller than might have been expected.

During the following two days the enemy force fell back some eleven miles and took up a strong position with perhaps as many as 7,000 or more rifles manning it. These probably outnumbered Macandrew's division by six to one. To dislodge them, Ausdiv was therefore ordered up from Damascus on 27 October, but before it even reached Homs the last shot of the campaign had been fired. An armistice with Turkey was signed on 31 October.

* * *

A measure of Descorps' achievement since the breakthrough on 19 September is the fact that three of its divisions had captured more than three times their own number in prisoners. The total is in doubt, but including Anzacdiv's tally, it was not less than 80,000.* The three divisions, other than Anzacdiv in Chaytor's Force, had advanced in thirty-eight days more than 300 miles, many of their

*For September and October the number that passed through the corps' cages was 78,462. The individual counts made by the divisions add up to 83,700. Chauvel put the total nearer 100,000. He took into account those captured in Damascus and elsewhere who subsequently died or who were employed in various jobs in the depleted units of the EEF.

units marching over 500. Most of 5th Division, indeed, had covered 550 miles, losing, incidentally, only 21% of its horses.* Only 649 out of the EEF's 5,666 casualties had been sustained by Descorps.

> 'The greatest exploit in history of horsed cavalry, and possibly [certainly!] their last success on a large scale, had ended,' as Wavell put it, 'within a short distance of the battlefield of Issus (333 B.C.), where Alexander the Great first showed how battles could be won by bold and well-handled horsemen.
>
> 'It had taken just four years to conclude the war with Turkey; it took nearly five more to conclude peace. Which proves the staying-power of the pen over the sword.'[11]

Thirteen years later Allenby wrote to Murray, 'I think that the success obtained in the East justified the risk in the main theatre.'[12] This seems to have been the sole occasion on which Allenby committed himself to a firm judgement upon the historical importance of the Palestine campaign.

*In the course of the great advances upon Damascus and Aleppo, it was impossible to send back sick horses to base. Consequently those that failed to keep up had to be destroyed on the spot. Another reason for wastage in horse flesh was that the artillery was empowered to draw upon the mounted troops to make good casualties in the gun teams.

The high standard of horsemastership in virtually all the regiments is well illustrated by the wastage in the Deccan Horse. On 17 September the regiment's strength was 461 sabres and 464 riding horses. By 26 October the wastage numbered fifty-eight men and 141 horses (chiefly from exhaustion or laminitis). In this regiment as in others the surplus men who found themselves horseless usually managed to keep up by capturing ponies, donkeys and even camels from the villages. (Tennant, 86–7).

EPILOGUE

In looking back on this remarkable campaign, there are some points of general interest especially worth making. First, Allenby was very lucky to have under his hand large numbers of well-trained officers and men in all the EEF's branches. Unlike those on the Western Front who, trained to the pre-war standard, had been expended after the first year, those in Egypt and Palestine had a much lower percentage of casualties. Thus, due to the long non-operational periods when training could be carried out, which were unknown in France and Flanders, there were sufficient pre-1914 officers and NCOs left alive to give instruction. Against this there were at times influxes of very raw Indian recruits for whom there was less training time. Yet for most of the campaign the third of the mounted arm which consisted of Indians could be fully trained by the time they went into battle. In the infantry, where nearly two-thirds were from the sub-continent, things were on occasion less easy.

Of the other factors which contributed to the value of the horsemen and which vastly conduced to their success were the utterly extraordinary qualities which the horses exhibited. Their most ardent, enthusiastic advocates were astonished to discover their quite unexpected powers of resistance to the extremes of climate in which they had to operate. Even the sweltering heat of the Jordan Valley in summer hardly affected them at all. Yet, marvellous and unforeseen though this was, even more so was the amount of activity which they were able to undertake without frequent watering. What the three mounted divisions managed in the pursuit after Third Gaza is truly amazing: continuous hard work with few halts over three complete days and nights – a good seventy-two hours – without a drop to drink! Some went for even longer. The Lincolnshire Yeomanry's mounts were still alive and almost kicking after eighty-four hours of total thirst, while the Dorsets covered no less than sixty miles with minimal halts, without water. On top of this, only very small quantities of fodder were available.

In some ways nearly as remarkable were the horses' powers of endurance when better watered and fed. As has been seen, at Magdhaba a whole division marched overnight for twenty miles, surprised and overpowered a staunchly posted detachment and then covered more than a further twenty miles before regaining its base – all in under thirty hours. In two raids east of Jordan and on the first advance to Jerusalem neither horribly rugged hill country nor most trying weather conditions managed to immobilize the horses. At the beginning of the Megiddo battles one division rode seventy miles in thirty-four hours. In the course of the great pursuit to Damascus and Aleppo, against admittedly not very formidable opposition, the 5th Cavalry Division marched almost 600 miles in thirty-eight consecutive days. Each of these overwhelming exploits equals the best of any in the whole of the history of warfare – and all of them were accomplished by non-regular mounted regiments.

THE HORSES STAY BEHIND

In days to come we'll wander west and cross the range again;
We'll hear the bush birds singing in the green trees after rain;
We'll canter through the Mitchell grass and breast the bracing wind;
But we'll have other horses. Our chargers stay behind.

Around the fire at night we'll yarn about old Sinai;
We'll fight our battles o'er again; and as the days go by
There'll be old mates to greet us. The bush girls will be kind.
Still our thoughts will often wander to the horses left behind.

I don't think I could stand the thought of my old fancy hack
Just crawling round old Cairo with a 'Gyppo on his back.
Perhaps some English tourist out in Palestine may find
My broken-hearted waler with a wooden plough behind.

No; I think I'd better shoot him and tell a little lie:—
"He floundered in a wombat hole and then lay down to die."
May be I'll get court-martialled; but I'm damned if I'm inclined
To go back to Australia and leave my horse behind.

Trooper Bluegum.

When hostilities came to an end, because it would have been uneconomic, especially in view of a shortage of shipping, it was decreed that some 22,000 army horses in the Middle East should be sold off for use or as meat. At least 4,000 fell into the hands of the fellaheen and the traders in Cairo's streets. They suffered the usual cruelty and hardship imposed by the ignorance and heartlessness of their new masters. In the early 1930s Mrs Dorothy Brooke, wife of Major-General Geoffrey Brooke, who had been given command of the cavalry division in Egypt, was so outraged at seeing overworked horses bearing British army brands covered with sores and with their ribs showing through, that she started the Brooke Hospital for Horses in Cairo and set about buying as many as she could afford. Still today the resources of this fund are used to treat, without charge, the descendants of those animals slaving away towards inevitably early deaths in the city.

APPENDIX 1

DESERT MOUNTED CORPS OPERATION ORDER

12th September 1918

Secret.

1. *Information.* (a) On Z day at Zero hour the Army is taking the offensive, pivoting on its positions in the Jordan Valley (Chaytor's Force), and attacking on the front between the high ground east of El Mughaiyr and the sea, with the object of inflicting a decisive defeat on the enemy and driving him from the line Nablus–Samaria–Tul Karm–Caesarea. . . .

3. *Orders to Troops.* On Z day:-

(a) (i) *The 5th Cavalry Division* (plus 12th L.A.M. Battery and No. 7 Light Car Patrol, less 2 gun cars) to be in a Position of Readiness in rear of the 60th Division by 5 a.m.

(ii) *The 4th Cavalry Division* (plus 11th L.A.M. Battery and No. 1 Light Car Patrol) to be in rear of 7th (Indian) Division by 6 a.m.

(iii) *Australian Mounted Division* to be in the Sarona area vacated by 4th Cavalry Division by 7 a.m.

(b) When the XXI Corps has opened the way for crossing the Nahr el Faliq and Kh. ez Zerqiye marsh:-

(i) *5th Cavalry Division* will advance on:-

1st Objective, the line Tell edh Dhrur (exclusive) – Hadera–the Sea – securing in the first place the crossings over the Nahr Iskanderune.

Thence moving with utmost speed to a position north of El 'Affule by the Hadera – Ez Zerganiya – Kh. es Shrah – Abu Shudhe road.

Touch must be maintained with the 4th Cavalry Division on its right.

Should any portion of the enemy's force retreat in the direction of Haifa, only sufficient troops to keep touch with them and protect the line of communication should be detached.

The passage of the remainder of the Corps through the Musmus Defile must be protected from the north by a detachment left in the neighbourhood of J'ara for this purpose. This detachment will not be recalled without reference to Corps.

In advancing on the El 'Affule road the Haifa railway should be cut and dispositions should ensure a detachment visiting Nazareth, with a view to capturing influential prisoners and important documents.

The division will be prepared to operate towards Jenin and Beisan according to circumstances. . . .

(ii) *4th Cavalry Division* will advance on:-

1st Objective, the line Qaqun–Jelame–Tell edh Dhrur (inclusive), keeping touch with the 5th Cavalry Division on its left.

The advance will be continued with the utmost speed *via* the Jelame–Kh. es Sumra–El Lajjun road to El 'Affule. On no account will the division be diverted from its objective by the presence of hostile troops in the Tul Karm–Qulunsawe–Et Tire area; these will be dealt with by the XXI Corps.

The railways from El 'Affule to Jenin, Beisan, and Haifa, will be cut as soon as possible.

On arrival at El 'Affule a detachment will be sent to seize the roads and railway bridges over the Jordan at Jisr el Majami, moving *via* Nein, Endor, Sirin. The railway bridge at Jisr el Majami' will be prepared for demolition but will not be destroyed so long as it can be held.

From El 'Affule the advance will continue as early as possible on Beisan with a view to closing the roads converging on that place from the Jordan Valley and Nablus.

(iii) *Australian Mounted Division* (less 5th A.L.H. Brigade) will move across the 'Auja into the position of readiness vacated by the 4th Cavalry Division directly that division is clear.

Thence it will move to a position on or about the Nahr Iskanderune on receipt of orders from Corps headquarters; thence follow the remainder of the Corps.

It will be prepared to send a detachment of one brigade or less from El Lajjun to block the roads and railway passing through Jenin and to gain touch with the 5th A.L.H. Brigade, which is to rejoin the division in this area.

(iv) *R.H.A.* – Directly the initial barrage under which the infantry advance and which commences at an hour known as "XXI Corps Zero Hour" is complete, the R.H.A. batteries will rejoin their divisions under divisional arrangements, after reference to G.O.C.R.A. XXI Corps.

4. *Dividing Line.* – The dividing line between 4th and 5th Cavalry Divisions will be Tabsor–El Mughaiyi–Tell edh Dhrur–Qannir–Kefrein–Buseile.

5. *Demolitions.* – Demolitions on railways should be limited to such as can be easily repaired.

6. *Naval Co-operation.* – Torpedo-boat Destroyers are co-operating in the attack and later in the advance of this Corps along the coast. Care must be taken to display the Red and Yellow Troop Flags as prominently as possible for the information of the Navy. . . .

8. *Traffic.* – If the 4th Cavalry Division follows the 5th Cavalry Division along the Coast road, it will follow immediately behind the fighting troops, the transport of 5th Cavalry Division giving way and keeping clear of the roads. . . .

11... *Time.* – Watches will be synchronized daily, commencing from receipt of this order, by the time received by Wireless from G.H.Q. every day at 8.15 a.m. and 6 p.m.

12. The date of Z day and Zero hour will be communicated verbally to those concerned. . . .

C.A.C. GODWIN,
Br.-General, General Staff, Desert Mounted Corps

(Falls, 720–23)

SPECIAL INSTRUCTIONS FOR DESCORPS IN AMPLIFICATION OF OPERATION ORDER OF 12 SEPTEMBER 1918

1. . . . In view of the long marches which you have to make and the necessity of conserving your full strength to carry out the important rôle assigned to you in the enemy's rear, you must on no account allow your troops to be drawn into the infantry fight south of the N. Falik, nor, after the passage of this stream, to be diverted from your objective by the presence of hostile troops in the Tulkeram–Et Tireh area, which will be dealt with by XXIst Corps.

The advance of your Corps from its positions of readiness will be regulated by the progress of XXIst Corps, and you will be responsible for arranging that the line Zerkiyeh Marsh–mouth of N. Falik is crossed at earliest possible moment. In your approach march, care must be taken that there is no interference with the movement of units and formations of XXIst Corps engaged in the attack; it is particularly important that the fire of the Corps Artillery should not be masked.

2. Should any portion of the enemy's forces retreat in direction of Haifa, you will detach only sufficient troops to keep touch with it and protect your L.-of-C., as it is vital that as large a proportion of your force as possible should be available to carry out the rôle assigned to you, which is to place your troops about Afuleh and Beisan where the enemy's railway communications can be cut at their most vital point, and whence you will be in a position to strike his columns if they endeavour to escape in a northerly or north-easterly direction.

The action of your troops must be characterised by the greatest vigour and rapidity, as it is essential that they should reach Afuleh and Beisan before the enemy can withdraw his rolling-stock and material or assemble troops for the defence of the railway.

3. On arrival at El Lejjun you will detach a Brigade to block the roads and railway passing through Jenin and to gain touch with the Cavalry Brigade attached to XXIst Corps (i.e. 5th A.L.H. Brigade of Ausdiv.), which will be directed on Jenin from the south if the situation permits.

4. From Afuleh a detachment will be sent to seize the road and railway bridges over the Jordan at Jisr Mejamie. The railway bridge should be prepared for demolition, but not destroyed so long as we are able to hold it.

5. Demolitions on the railway should be limited to such as can easily be repaired.

(Osborne, III, 355–7)

APPENDIX 2

NOTE ON THE BEERSHEBA CHARGE PHOTOGRAPH
(See p. 154)

This photograph purports to show the charge of the 4th Brigade at Beersheba. Grant who led it stated in 1920 that he could not 'take exception to the claim . . . that it is a photograph of the actual charge'. Members of his staff agreed with him. They all said that the terrain and the formation shown were correct. Ian Jones, many years later, spoke to thirteen men who rode in the charge, all of whom confirmed that the formation was as shown.

In 1967, however, the Director of the Australian War Memorial summarized the official view thus:

> '1. The formation in which the charge was made was different from that shown [contradicting Grant]. . . .
> 2. The ground was quite bare and the galloping horsemen quickly became enveloped in large clouds of dust. In the photograph the ground appears to be fairly well covered with herbage and there is practically no dust.
> 3. [The charge was made] in a westerly direction in the late afternoon so that the shadows would fall behind the horses.
> 4. The final stretch . . . was downhill whereas the horsemen in the picture are galloping uphill.'

Further, the AWM's view is that the photograph was taken during a later re-enactment staged for an official film cameraman at Belah on the coast of Palestine. Grant, however, stated that the terrain there did not match that at Beersheba.

With respect to paragraph 2 above, Jones states that 'reports of dust raised . . . came from men looking into the sun – a situation which exaggerated its density. Men watching from the side, with the sun over their shoulders, could see the action quite clearly. It also seems likely that the combination of reddish brown dust and reddish sunlight minimized the effect of dust in the photograph.

'Regarding the "herbage", a veteran of the action, Vic Smith, commented in 1981 that part of the charge lay over the stubble of a recently harvested Bedouin crop. Untrimmed copies of the photograph show traces of furrows in the foreground.'

On the question of the direction of the sun Jones states that the charge was made to the north-west and not to the west. He adds that, according to the Curator of Astronomy at the Melbourne Science Museum, 'on that day at Beersheba the sun set sixteen degrees south of west. This placed it more than sixty degrees to the left of the charge, as shown in the photograph.'

On the fourth paragraph of the official statement, Jones says that 'the "uphill" effect in the photograph was created by a slightly tilted camera. With true verticals restored, the formation is heading down a very slight slope.' He adds that when he walked over part of the charge course 'and photographed the hills south-east of Beersheba' he found that 'in character and detail, they matched those seen in the photograph.'

The man who purports to have taken it came forward in 1967 with a formal statement. It reads as follows and has the stamp of authenticity:

> 'I, Eric George Elliott, state that on 31st Oct. 1917 I was a range finder with the 4th A.L.H. Brigade which was resting in reserve in a depression between two ridges about four miles east of Beersheba. At approx. 1430 hrs. I was called to H.Q. and instructed to accompany a party of officers and take some ranges and prepare range charts. The party moved out to the front on the Brigade position and halted behind a knoll approx 3,300yds from the town of Beersheba.
>
> 'I was given my points to range on by the M.G. [machine-gun] and Artillery officers in the party. This I did and reported the ranges verbally to each officer and noted them on a rough plan from which to prepare my range charts.
>
> 'The officers then moved back to Brigade area and left me to prepare my range charts and deliver them to the respective officers with a copy to Brigade H.Q. I was about half way through this task when I happened to look in the direction of the Brigade area and saw dust rising indicating that there was some movement taking place. I hastened to complete my job under the impression that this was what my charts were required for, at the same time keeping my eye on the Brigade area. To my surprise it seemed to be within minutes, I saw horsemen in extended order coming over the crest of the ridge. I packed my gear, and then came another line of troops in the same order. I then moved around to the other side of

the knoll, and by this time the third line appeared. Bewildered by what was happening I just lay there and gazed in astonishment. As the front line drew nearer I saw that their bayonets were drawn and that they were approaching at a hard gallop. Having a camera in my knapsack I got it out and took a shot, got on my horse and went as fast as I could further out to a flank and then back to H.Q. There I learned that my charts were not now needed as the whole plan had been changed and that an attempt was being made to capture Beersheba with a Cavalry Charge by the 4th and 12th Regts.'

However, Jeff Cutting, the Administration Officer, Art Section of the Australian War Memorial, writing to the present author in 1993, states that he was 'told of a conversation some years ago with a Mr Pat Gallagher whose late father had ridden in the charge at Beersheba and in the re-enactment charge on the coastal plain near Dier el Belah.' On seeing the photograph he asserted that 'it was of the Belah romp [*sic*] and indicated where he could see himself in it.' Mr Cutting adds that the official file 'dealing with an inquiry into this affair' has eluded his search and that he feels that it is 'out of the normal records management system and not available'. He points out that Trooper Idriess, who observed the charge, writing in *The Desert Column*, claimed to have seen much dust and that the light was pretty poor. (Idriess, 251-2). Readers of the present work will remember that Idriess was writing a long time after the event and they may perhaps treat his evidence with caution.

Mr Cutting goes on to say:

'You will observe in the alleged "Charge" photograph that dust is not a feature, furthermore, the sunlight, as evident from the shadows, is coming from the wrong direction if this formation was moving towards Beersheba in the north-west (the light is coming from the left rear of the horsemen i.e. South/South-East) and is too harsh and high in the sky for a sunset. [See diagram on p. 352].

'Also in this image there are no signs of casualties or of the bombardment by shrapnel shells and nothing can be seen of the road down which the charge was driven.

'Near or at sunset on a dusty day would produce a quality of light that would be semi-diffused and of such reduced

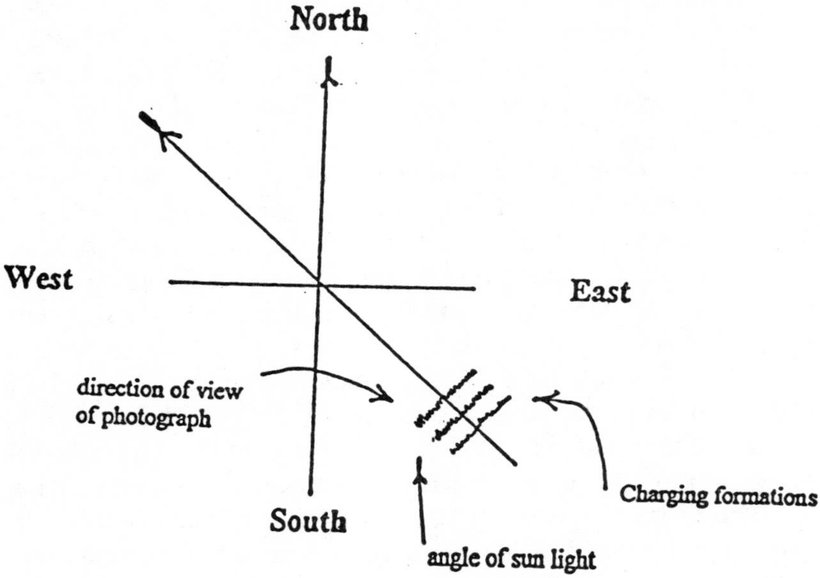

intensity that the use of fast camera shutter speeds to record action scenes (as the "Charge" photograph does) would be near impossible. Even today with medium and fast film speeds and fast lenses (wide aperture) it would be a difficult task. The "Charge" photograph if it is purported to have been taken near dusk at the Battle for Beersheba with its rendering of fast galloping horses is a technical improbability.'

With so much conflicting evidence it is, alas, impossible to say with absolute certainty that this photograph is the only one ever taken of a mounted charge actually in progress against an enemy. Nevertheless, unless Elliott has invented his statement, the present author inclines towards the view that this is exactly what it is.

ABBREVIATIONS USED IN THE FOOTNOTES AND SOURCE NOTES

Some important sources which have been consulted but not quoted from are also indicated here. They may be useful for further reading.

Adderly	Adderly, Hon. H.A. *The Warwickshire Yeomanry in the Great War*, 1922
Advance	[Official] *The Advance of the Egyptian Expeditionary Force* (incl. Allenby's three despatches), 1919
Alan-Williams	Alan-Williams, Lieut A.C. 'Charge of the Warwickshire and Worcestershire Yeomanry at Huj, November 8th, 1917', MS notes, 27 November, 1917, and TS, n.d., IWM
AWM	Australian War Memorial
Badcock	Badcock, Brevet Lt-Col G.E. *A History of the Transport Services of the Egyptian Expeditionary Force,* 1916–1918, 1925
Baker	Baker, Lt-Col Sir Randolf, Bt, M5 letters, 1915–17, IWM
Barrow	Barrow, Gen. Sir George de S. *The Fire of Life*, 1942
Bean	Bean, C.E.W. *The Story of Anzac from the outbreak of War to the end of the first phase of the Gallipoli Campaign,* 4 May 1915, [1921], 1933
Berrie	Berrie, Lt Geo. L. *Under Furred Hats (6th A.L.H. Regiment*), 1919
Blackwell	Blackwell, F. and Douglas, D.R. *The Story of the 3rd Australian Light Horse Regiment* [1952]
Blagg	Blagg, 2nd Lieut S., letters, IWM (PP/MCR/220)
Blaksley: IWM	Blaksley, 2nd Lieut, Dorset Yeomanry, letter home 3 Mar., 1916 and accompanying typed note 'The Dorset Yeomanry Charge at Agagia, 26th February, 1916', IWM

Blenkinsop	Blenkinsop, Maj.–Gen. Sir L.J. & Rainey, Lt-Col J.W. *History of the Great War based on Official Documents: Veterinary Services*, 1925
BM	British Museum
Bond	Bond, Brian 'Doctrine and Training in the British Cavalry, 1870–1914', (ed.) Howard, Michael *The Theory and Practice of War: Essays Presented to Captain B.H. Liddell Hart on his 70th Birthday*, 1967
Bostock	Bostock, Henry P. *The Great Ride: The Diary of a Light Horse Brigade Scout: World War I*, 1982
Bourne	Bourne, Lt-Col G.H. *The History of the 2nd Light Horse Regiment, Australian Imperial Force: August 1914–April 1919*, n.d.
Brittain	Brittain, Pte H.C., letters 1/1 West Somerset Yeomanry, 85/43/1, IWM
Browne	Browne, Brig. J.G. and Bridges, Lt-Col E.J. ((ed.) Miller, Maj. J.A.T.) *Historical Record of the 14th (King's) Hussars, 1900–1922*, 1932
Browne: Ops	Browne, Lt-Col J.G. 'Operations of the Mounted Troops of the E.E.F.', *Cav. Jnl*, XI, 1921
Buchan	Buchan, John *A History of the Great War*, 4 vols, 1921–2
Cardew	Cardew, Maj. F.G. *Hodson's Horse, 1857–1922*, 1928
Cav. Jnl	*The Cavalry Journal*
Cav. Trng	[Official] *Cavalry Training* (various dates)
Chauvel	Hill, A.J. *Chauvel of the Light Horse: A Biography of Gen. Sir Harry Chauvel*, 1978
Chetwode	Chetwode Papers, (including 'Notes on the Palestine Operations', 21 June, 1917), IWM
Churchill: *GW*	Churchill, W.S. *The Great War*, 3 vols, 1933
Cobham	'C' [Lord Cobham] *The Yeomanry Cavalry of Worcestershire, 1914–1922*, 1926
Connell	Connell, John *Wavell: Scholar and Soldier*, 1964

Darley	Darley, Maj. T.H. *With the Ninth Light Horse in the Great War*, 1924
Dening	Dening, Maj.–Gen. R., 18 (KGO) Lancers, MS diary, IWM
Desert	'Scotty's Brother' *The Desert Trail: with the Light Horse through Sinai to Palestine*, 1919
DNB	*Dictionary of National Biography*
Falls	(ed.) Falls, Capt. Cyril *Military Operations, Egypt & Palestine (the Official History)*, II, *From June 1917 to the End of the War*, Part 1, 1930; Part 2, 1930
Field Notes	[Official] *Field Notes: Mesopotamia* (Gen. Staff, India Feb., 1917)
Fletcher	Fletcher, A.W. MS 'Egypt, Palestine, 1916–18', IWM
Foster	Foster, Lt-Col W.J., in collaboration with Browne, Lt-Col J.G., G.S.O.1, Anzac Mounted Division, 'Operations of the Mounted Troops of the E.E.F.' *Cav. Jnl*, XI, 1921
Fox	Fox, Frank *The History of the Royal Gloucestershire Hussars Yeomanry, 1898–1922; the Great Cavalry Campaign in Palestine*, 1923
Gardner	Gardner, Brian *Allenby*, 1965
Godrich	Godrich, Sgt E.V. 'Q.O.W.H. Diary of a Yeoman 1914–1919' (Typescript. Property of his son, Dr John Godrich)
Gullett	Gullett, H.S., *The Australian Imperial Force in Sinai and Palestine (Official History of Australia in the War of 1914–18)*. 1923
Hansard: (C)	*Hansard's Parliamentary Debates*, House of Commons
Hatton	Hatton, S.F. *The Yarn of a Yeoman*, n.d.
Hogue	Hogue, Oliver *The Cameliers*, 1919
Home	Home, Brig.-Gen. Sir Archibald, *The Diary of a World War I Cavalry Officer* (ed. Briscoe, Diana), 1985
HRAVC	Smith, Maj.-Gen. Sir Frederick *A History of the Royal Army Veterinary Corps, 1796–1919*, 1927

Hudson	Hudson, Gen. Sir H. *History of the 19th King George's Own Lancers, formerly 18th King George's Own Lancers and 19th Lancers (Fane's Horse), amalgamated in 1921*, 1937
Husni	Hussein Husni Amir Bey *Yilderim*, trans. Capt. G.O. de R. Channer, n.d. (n.p.), Pt 3, AWM
Idriess	Idriess, Ion L. *The Desert Column: Leaves from the Diary of an Australian Trooper in Gallipoli, Sinai and Palestine*, 1932
Inchbald	Inchbald, Geoffrey *Imperial Camel Corps*, 1970
IOL	India Office Library
IWM	Imperial War Museum
JAHR	*Journal of the Society for Army Historical Research*
James	James, Lawrence, *Imperial Warrior: The Life and Times of Field-Marshal Viscount Allenby, 1861–1936*, 1993
Jones	Jones, Ian, 'Beersheba: the Light Horse Charge and the Making of Myths', *Jnl of Australian War Memorial*, III, Oct., 1983
Kemp	Kemp, Lt-Cmdr P.K. (ed.) *The Staffordshire Yeomanry (Q.O.R.R.) in the First & Second World Wars. . .*, 1950
Kress	Kress von Kressenstein, Col Baron 'The Campaign in Palestine from the Enemy's Side' in *Between Caucasus and Sinai* (Bund der Asienkämpfer, Berlin, n.d.) in *RUSI*, LXVII, 1922
Lambert	Lambert, Lt-Col W.J., Royal Deccan Horse, 'A Cavalry Fight near Aleppo, October 26, 1918', *Cav. Jnl*, July, 1924, 315–20
LHC	Liddell Hart Centre
Liddell Hart: *Fog*	Liddell Hart, Sir Basil Henry *Through the Fog of War*, 1938

MacMunn	(ed.) MacMunn Lt-Gen. Sir George & Falls, Capt. Cyril *Military Operations, Egypt & Palestine (the Official History)*, I, *From the Outbreak of War with Germany to June 1917*, 1928
Massey: *Jerusalem*	Massey, W.T. *How Jerusalem was won, being the record of Allenby's campaign in Palestine*, 1919
Massey: *Triumph*	Massey, W.T. *Allenby's Final Triumph*, 1920
Maunsell	Maunsell, Col E.G. *Prince of Wales's Own, The Scinde Horse, 1839–1922*, 1926
McGrigor	McGrigor, A.M. typescript of diary, IWM
Meinertzhagen	Meinertzhagen, Col Richard *Army Diary, 1899–1926*, 1960
Moore: *Vet.*	Moore, Maj.-Gen. Sir J., *Army Veterinary Service in War*, 1921
Murray	Murray, Rev. R.H. *The History of the VIII King's Royal Irish Hussars, 1693–1927*, 2 vols, 1928
Murray: *Des.*	*Sir Archibald Murray's Despatches, June 1916 to June 1917*, 1920
Nalder	Nalder, Maj.-Gen. R.F.H. *The Royal Corps of Signals: A History of its Antecedents and Development*, 1958
Newell	Newell, J.Q.C., 'Learning the Hard Way: Allenby in Egypt and Palestine, 1917–19' *Journal of Strategic Studies*, Vol. XIV, No. 3, Sep., 1991
Olden	Olden, Lt-Col A.C.N. *Westralian Cavalry in the War – The Story of the 10th Light Horse Regiment, A.I.F., 1914–1918*, 1921
Ops: EEF	[Anon] 'Operations of the Mounted Troops of the Egyptian Expeditionary Force', *Cav. Jnl*, X, 1920
Ops: V. Cav. Div.	'Operations of Vth Cavalry Division in Palestine during the advance to Aleppo – September and October, 1918', [typescript: Commander of 'B' Squadron, 18th King George's Own Lancers], NAM 6709/33

Osborne: I	Osborne, Lt-Col Rex 'Operations of the Mounted Troops of the E.E.F., *Cav. Jnl*, XI, 1921
Osborne: II	Osborne, Lt-Col Rex 'Operations of the Mounted Troops of the E.E.F.', *Cav. Jnl*, XI, 1922
Osborne: III	Osborne, Lt-Col Rex, 'Operations of the Mounted Troops of the E.E.F.', *Cav. Jnl*, XIII, 1923
Parfit	Parfit, J.T. *Servia to Kut: An Account of the War in the Bible Lands*, 1917
Perkins	Perkins, Capt. C.H., MC 'A troop leader in the Royal Bucks Hussars', MS kindly lent by Capt. Perkins
Pirie-Gordon	Pirie-Gordon, H. *A Brief Record of the Advance of the Egyptian Expeditionary Force*, 1919
Poona	[Anon] '34th Poona Horse: Narrative account of Cavalry Engagements in 1918 in Palestine', *Cav. Jnl*, X, 1920
Powles	Powles, Lt-Col C. Guy *The New Zealanders in Sinai and Palestine*, 1922
Preston	Preston, Hon. R.M.P. *The Desert Mounted Corps: an Account of the Cavalry Operations in Palestine and Syria, 1917–1918*, 1921
Rawlinson	Rawlinson, A. *Adventures in the Near East*, 1923
Reid	Reid, Frank *The Fighting Cameliers*, 1934
Robertson: *S&S*	Robertson, FM Sir William *Soldiers and Statesmen 1914–1918*, 2 vols, 1926
Smith	Smith, Maj.-Gen. Sir Frederick *A History of the Royal Army Veterinary Corps, 1796–1919*, 1927
Smith: *Hist.*	Smith, George Adam *Historical Geography of the Holy Land*, 1894
Smith: *Sub.*	Smith, Rev. R. Skilbeck *A Subaltern in Macedonia and Judaea, 1916–17*, 1930
Stonham	Stonham, C. & Freeman, B., (ed.) Judd, J.S. *Historical Records of the Middlesex Yeomanry, 1797–1927*, 1930

Tallents	Tallents, Maj. H. *The Sherwood Rangers Yeomanry in the Great War 1914–1918*, 1926
Teichman	Teichman, Capt. O. *The Diary of a Yeomanry M.O., Egypt, Gallipoli, Palestine and Italy*, 1921
Terraine: *S and F*	Terraine, J. *The Smoke and the Fire: Myths and Anti-Myths of War, 1861–1945*, 1980
Twentieth MG Sqn	[Anon.] *Through Palestine with the Twentieth Machine-Gun Squadron*, 1920
Tylden	Tylden, Maj. G. *Horses and Saddlery. . .*, 1965
Vernon	(ed.) Vernon, P.V. *The Royal New South Wales Lancers 1885–1960, Incorporating a narrative of the 1st Light Horse Regiment, A.I.F., 1914–1919*, 1961
Watson	Watson, Maj.–Gen. W.A. *King George's Own Central India Horse: The Story of a Local Corps*, 1930
Wavell	Wavell, Gen. Sir A. *Allenby: A Study in Greatness*, 1940
Wavell: *Pal.*	Wavell, F-M Earl *The Palestine Campaigns*, 3rd ed., 1931
Weir	Weir, Brig.-Gen. G.A. 'Some Critics of Cavalry and the Palestine Campaign', *Cav. Jnl*, Oct., 1920
Whitworth	Whitworth, Capt. D.E. *A History of the 2nd Lancers (Gardner's Horse) from 1809 to 1922*, 1924
Wilson	Wilson, Brig.-Gen. L.C. and Wetherell, Capt. H. *History of the Fifth Light Horse Regiment (Australian Imperial Force) from 1914 to June, 1919*, 1926
Wilson: *Pal.*	Millgate, Helen D. (ed.) *Palestine, 1917, Robert Henry Wilson*, 1987
WO	War Office
Wylly	Wylly, Col H.C. *The Poona Horse, 1817–1931*, Vol. II, 1933

SOURCE NOTES

PREFACE

1 3 May, 1917, Baker, Lt-Col Sir R. L., Bt, letters; IWM
2 Martin, Lt-Col A.G. (formerly 6 Dragoons, German Army) 'Cavalry in the Great War – A Brief Retrospect', *Cav. Jnl*, XXIII, 1933, XXIV, 1934
3 9 Dec., 1915, Boyes–Stone

CHAPTER 1 (pp. 1–20)

1 Berrie, 15; Bean, 18; quoted in Darley, 193; *Chauvel*, 128; Hatton, 187; Essame, 119
2 MacMunn, 1
3 Nevakivi, J. *Britain, France and the Arab Middle East, 1914–1920*, 1969, 7
4 Bean, 20–36
5 Bean, 40–41; Blackwell, 16
6 Wilson, 13
7 Bourne, 7
8 Bean, 44
9 Gullett, 30
10 Bean, 43
11 Bean, 59; Gullett, 662
12 Gullett, 31
13 Olden, 13
14 Gullett, 33
15 Bourne, 8; see also Darley, 2
16 Hamilton to Senator Millen, quoted in *Chauvel*, 43
17 7 Feb. 1916, Murray-Robertson Corres., BM Add. Ms 52463
18 Gullett, 34; Gullett, 111; Badcock, 307; *Chauvel*, 75; Idriess, 202; Information from Dr Prys Morgan, whose father, an officer of the Pembroke Yeomanry, related the story to him; Bostock, 3[illegible]: Barrow, 191
19 Maunsell, 196
20 Gullett, 648–50
21 23 May, 1915, Blagg; Idriess, 96–7
22 Blackwell, 15
23 23 May, 1915, Blagg

24 Wilson, 125–6
25 Idriess, 162
26 Bean, 48
27 Gullett, 55
28 Gullett, 533
29 Wilson, 68, 90
30 Darley, 82
31 Vernon, 87
32 Bean, 60–3
33 Reid, 27
34 Inchbald, 12
35 Reid, 5
36 Inchbald, xviii; Blenkinsop, 154; Badcock, 81
37 Gullett, 213
38 Blenkinsop, 157; Gullett, 290

CHAPTER 2 (pp. 21–26)
1 Wavell: *Pal.*, 21–2; Buchan, I, 501; Maunsell, 217; *Field Notes*, 83
2 Wavell: *Pal.*, 21
3 MacMunn, 4
4 Wavell: *Pal.*, 21
5 Quoted in Bean, 142
6 Gullett, 260
7 *Field Notes*, 88; Wavell: *Pal.*, 21
8 *Field Notes*, 88; Teichman, 144; Near Rafa, 27 Feb., 1917, Wilson, 92; see also Vernon, 127
9 23 Oct., 1917 MacMunn, 277; Falls, 37
10 Preston, 21–22; Falls, 38–9
11 Olden, 126
12 Kress, 504
13 Bean, 140–165
14 Gullett, 12–13; Bean, 164

CHAPTER 3 (pp. 27–32)
1 3 Mar., 1916, Blaksley, IWM
2 Wavell: *Pal.*, 36
3 Adderley, 49
4 3 Mar., 1915, Blaksley, IWM
5 MacMunn, 128
6 n.d., Souter to Maj.-Gen. Peyton, Fox, 90

7 Wavell: *Pal.*, 38
8 MacMunn, 10 , 65, 68, 101–45

CHAPTER 4 (pp. 33–36)
1 *Desert*, 16
2 Bean, 146
3 Anon. 'Operations of the Mounted Troops of the E.E.F.', *Cav. Jnl*, X, 1920 , 392; Darley, 33–8
4 McGrigor, I, 27
5 Badcock, 19, *et seq*; Blenkinsop, 154 *et seq.*
6 Badcock, 83–4

CHAPTER 5 (pp. 37–38)
1 Quoted in Wavell: *Pal.*, 92
2 Quoted in Gullett, 41
3 Murray, 193–5, 204
4 Murray, 189
5 Powles, 43

CHAPTER 6 (pp. 39–44)
1 Warrell: *Pal.*, 43–44
2 Blackwell, 55; Tallents, 47
3 MacMunn, 170; Kress, 505; Wavell: *Pal.*, 43; see also Cobham, 54, where 'a machine gun company of twelve guns' is mentioned; see also Murray, 26. The real number is impossible to ascertain
4 Cobham, 50
5 Blenkinsop, 164
6 The fullest accounts of the two actions are to be found in MacMunn, 161–9, Cobham, 46–56 and Fox, 92–5. Gullett, 81–8, is not very reliable
7 Gullett, 121; Cobham, 67

CHAPTER 7 (pp. 45–49)
1 BM Add. MS 52463; *Chauvel*, 69
2 *Chauvel*, 67
3 Badcock, 42
4 Olden, 94
5 Vernon, 106; Darley, 32
6 Kemp, 41
7 17 May, 1916, McGrigor, 150
8 Teichman, 58–9
9 Powles, 19

10 Brittain, 6; Gullett, 102; Powles, 24
11 Teichman, 58
12 Cobham, 72; Wilson, 65; Idriess, 91
13 Teichman, 88, 108; *Chauvel*, 73; Foster, 9; Blenkinsop, 139
14 Powles, 24; 15 May, 1916, Idriess, 91; Teichman, 108; Berrie, 103
15 Wilson, 65; Powles, 19; Blenkinsop, 139; Foster, 10; Gullett, 104

CHAPTER 8 (pp. 50–54)

1 *Desert*, 1; Idriess, 69
2 23 and 26 May, 1915, Blagg
3 Idriess, 193; 8 Aug., 1916, Berrie, 88
4 Wilson, 122–3
5 Idriess, 70 , 194, 313
6 Brittain, 10
7 Fox, 120; Teichman, 48, 107, 111; Idriess, 84; Cobham, 68; Gullett, 107
8 McGrigor, 312; Idriess, 81
9 Fox, 140; Teichman, 46–7, 107; Blackwell, 92; Gullett, 128
10 11 Jan, 1918, 20 Corps Order, Chetwode; Idriess, 186
11 Idriess, 186
12 Gullett, 105; Teichman, 127, 134, 138
13 Gullett, 105
14 Idriess, 69, 105; see also Darley, 42

CHAPTER 9 (pp. 55–73)

1 Kress, 506; Gullett, 126; *Chauvel*, 81; Wavell: *Pal.*, 47; Gullett, 119
2 Gullett, 99
3 Birdwood to Monro-Ferguson, 25 Feb., 16 Sep., 1915, 24 Mar., 1916; Hamilton to Monro-Ferguson, 6 Oct., 1915; Novar Papers, National Library of Australia, quoted in *Chauvel*, 54
4 Tpr Ingham in unidentified newspaper cutting, quoted in *Chauvel*, 86; 23 Sep., 1917, Col Rex Hall's diary, quoted in *Chauvel*, 122; *Chauvel*, 133–4; Allenby, 6/IX, 33, LHC, quoted in James, 121; Chauvel, 119
5 Chauvel to Hutton, 28 Jun., 1916, Hutton Papers, BM Add. MS 550089, quoted in *Chauvel* 97
6 Olden, 91
7 *Chauvel*, 81

8 Gullett, 354
9 Idriess, 126
10 Powles, 28
11 Gullett, 151
12 Berrie, 79
13 Gullett, 160
14 Gullett, 174; Idriess, 149; Teichman, 70
15 Darley, 46; Gullett, 167–9
16 Gullett, 171
17 MacMunn, 194
18 *Chauvel*, 83
19 Wavell: *Pal.*, 51
20 MacMunn, 201
21 *Chauvel*, 83
22 *Chauvel*, 81
23 Gullett, 191. The best short account of the battle is to be found in Wavell: *Pal.*, 47–51, closely followed by the more detailed one in *Chauvel*, 76–8.. MacMunn, 175–201, is slightly pedestrian but generally accurate, while Gullett, 140–194, gives the fullest account, though a little prejudiced in favour of the Australians. It recounts numbers of good details of an anecdotal character. Foster, 127–143, is particularly full on the details of the pursuit, especially on 9 August.

CHAPTER 10 (pp. 74–89)
1 Chetwode, 3; Wavell; *Pal.*, 59; Olden, 104; quoted in *Chauvel*, 96
2 WO telegram, 26174, 9 Dec. 1916, Murray, 130
3 WO telegram, 26624, 15 Dec., 1916, Murray, 131
4 WO telegrams, 27761 and 28297, 11 Jan. and 22 Jan. 1917, Murray, 131
5 Kress, 506
6 Wavell: *Pal.*, 71; Teichman, 92
7 Gullett, 199
8 Wilson, 86
9 MacMunn, 245; Gullett, 197–200; see also Foster, 143–6
10 MacMunn, 245–6; Gullett, 200–203; see also Foster, 146, where it is said that 'the enemy's camp was repeatedly bombed by our aeroplanes', which seems unlikely in view of the fog.
11 Quoted in Gullett, 209
12 Gullett, 215–16

13 Powles, 51
14 Powles, 52
15 Gullett, 219
16 Blackwell, 78
17 Darley, 61
18 Gullett, 221
19 Gullett, 224
20 Olden, 103
21 Gullett, 223–4
22 No. 1 envelope, Chetwode
23 Powles, 57
24 Powles, 54–5. The fullest account of Magdhaba is in Gullett, 214–228, but it is not very reliable; MacMunn, 251–258, is more so; Foster, 150–153 and 157, has a few interesting comments, while *Chauvel*, 87–89, gives the best brief general account of the battle; Powles, 50–56, is excellent for the New Zealand Mounted Brigade, while Darley (9th Regiment) and Olden (10th Regiment) are admirable for their regiments.
25 Murray, 104
26 Chetwode to Dobell, quoted in Gullett, 233
27 Foster, 157; McGrigor, 340
28 Foster, 158
29 Olden, 110
30 Powles, 76
31 Wilson: *Pal.*, 63
32 Blackwell, 82; Gullett, 356; Godrich, 56
33 Teichman, 104
34 Darley, 67
35 Gullett, 242; the fullest account of the Rafah raid is in Gullett, 229–343; MacMunn, 262–271, is shorter and more dependable; there are useful comments in Foster, 153–159; *Chauvel*, 90–95, is brief but excellent; Powles, 64–79, is especially good for the New Zealanders; the Australian and yeomanry regimental histories give a few interesting details and anecdotes. Godrich, 56, refers to the part played by the HAC Battery.

CHAPTER 11 (pp. 00–00)

1 Wavell: *Pal.*, 71; Kress, 506; Fletcher, 26; Gullett, 287; Cobham, 91
2 WO letter 01/45/151, 11 Jan., 1917, Murray, 131
3 See *Chauvel*, 97

4 Murray to Robertson, 5 Sep., 1916, BM Add. MS 52463, quoted in *Chauvel*, 96
5 Eastern Force Order, No. 33, 24, Mar., 1917, MacMunn, 413
6 MacMunn, 316; Wavell: *Pal.*, 76
7 Teichman, 120
8 Idriess, 246
9 Gullett, 273; MacMunn, 294
10 MacMunn, 297
11 MacMunn, 298
12 Wilson, 97
13 Idriess, 251–2
14 Gullett, 279
15 Gullet, 281
16 Powles, 92–3
17 Wavell: *Pal.*, 77
18 MacMunn, 321–2
19 MacMunn, 306
20 MacMunn, 310
21 MacMunn, 310
22 *Chauvel*, 104–5, 239; Browne, 229
23 Wavell: *Pal.*, 80
24 Gullett, 287; Darley, 80
25 Anon. *Experiences of a Regimental Medical Officer*, 15, AWM, quoted in *Chauvel*, 105
26 MacMunn, 332
27 *Chauvel*, 106
28 *Chauvel*, 106
29 Hogue, 122
30 Blenkinsop, 186
31 MacMunn, 348–9
32 For both Gaza battles, MacMu 276–350, is the most reliable, Gullett, 244–339, the fullest and most readable, *Chauvel*, 96–109, the best for the higher commands' thinking and Wavell: *Pal.*, 71–88, the most concise and accurate general account. Browne, 223–241, always dependable, adds some evidence not given elewhere. Powles, 844–106, is good on the New Zealanders' part.
33 Godwich

CHAPTER 12 (pp. 108–118)

1 Gullett, 339; Wilson: *Pal.*, 80–1; Chetwode, 1

2 27 Apr. 1917, Williams, Dr Orlo, MS Diary, 1914–18, IWM
3 Bostock, 74; Wilson, 108
4 Gullett, 360; Wilson: *Pal.*, 91; Fox, 152–3
5 Cobham, 109–10
6 Bostock, 75
7 For details, see MacMunn, 356 and Kress, 508
8 Fletcher, 40; Hogue, 209
9 Vernon, 127
10 *Chauvel*, 114
11 Teichman, 163
12 *Chauvel*, 100
13 Fox, 149
14 27 May 1917, Idriess, 298–9
15 Idriess, 290
16 Wilson, 104; Idriess, 31; McGrigor, 241
17 19 June 1917, Baker
18 Wilson, 118, 89
19 Vernon, 126
20 Vernon, 138
21 Darley, 71
22 Teichman, 165
23 Idriess, 361–2
24 Gullett, 344–5
25 Anon. *Experiences of a Regimental Officer*, 16, AWM, quoted in *Chauvel*, 114; McGrigor, 232
26 Kress, 508
27 Chetwode, 1
28 Browne, 243
29 Powles, 111; Idriess, 295
30 Chauvel to Gullett, Chauvel Papers, quoted in *Chauvel*, 116
31 Wilson, 120
32 Idriess, 318

CHAPTER 13 (pp. 119–124)
1 19 June, 1917, Baker; Meinertzhagen, 219; MacMunn, 372
2 Allenby's final despatch, quoted in Robertson *S&S*, II, 182–3
3 MacMunn, 368
4 Gullett, 357
5 Fletcher, 33
6 16 June, 1917, Teichman, 155; Allenby 6/VIII, 48, LHC, quoted in James, 115

7 Idriess, 306
8 Wavell, 197; 25 Aug., 1920, Allenby to Edmonds, Edmonds (unpublished memoir), II, 1, LHC, quoted in James, 116; Wilson: *Pal.*, 83; Allenby, 6/VIII, 43, LHC, quoted in James, 116
9 Darley, 95
10 Wilson: *Pal.*, 86; Powles, 101
11 Wasserstein, *The British in Palestine: the Mandatory Government and the Anti-Jewish Contest, 1917–1929*, 1978, 21
12 Allenby, 6/IX, 15, LHC, quoted in James, 115

CHAPTER 14 (pp. 125–135)
1 Idriess, 307; *Advance*, 1; Wavell, 208; Cobham, 112
2 Chetwode, 6
3 25 Jan, 1918, 4th Supplement, *London Gazette*
4 Gullett, 380
5 Barrow, 168
6 Wavell: *Pal.*, 103
7 Preston, 13–14
8 Bourne, 46
9 Teichman, 168
10 Kemp, 41–2
11 Fox, 153
12 Preston, 14
13 Meinertzhagen, 223
14 Meinertzhagen, 222–3, 283–6; Falls, 30–1
15 Falls, 30
16 Falls, 40
17 Gullett, 366

CHAPTER 15 (pp. 136–141)
1 Falls, 686; Wavell: *Pal.*, 116
2 Wavell: *Pal.*, 110
3 Falls, 35
4 Falls, 36
5 Falls, 40
6 Wavell: *Pal.*, 112
7 Osborne, I, 351
8 See Garsia, Lt-Col Clive *A Key to Victory: A Study in War Planning*, 1940, 211–16
9 Liddell Hart to Lloyd George, 25 June, 1934, LH Papers, LHC, 1/450/File 1

10 See Newell, 371, *et seq.* for a full discussion of the 'Gaza School's' plan

CHAPTER 16 (pp. 142–162)

1 Wavell, 208; Gullett, 392; Falls, 61; Wavell: *Pal.*, 125
2 Preston, 12
3 Osborne: I, 346
4 Wavell: *Pal.*, 117
5 *Chauvel*, 126
6 Wavell: *Pal.*, 120
7 Idriess, 321–2
8 Gullett, 391
9 Osborne: I, 352
10 Jones, fn, 37; Preston, 29
11 Idriess, 325; fns 29, 30, Jones, 37
12 Hill, 128; Jones, 31, 34; Gullett, 397–8
13 Gullett, 398
14 Gullett, 404
15 19 Nov., 1917, Von Kressenstein's report in Husni n.p. See also Pirie-Gordon, 2 and Anon 'The Campaign in Palestine from the Enemy's Side', RUSI Jnl, vol. 67, 1992, 510
16 Preston, 25
17 Sanders, Liman von *Five Years in Turkey*, Annapolis, 1927, 203
18 Husni, n.p.
19 Husni, n.p.
20 Jones, 29
21 Wavell: *Pal.*, 125–6. The battle of Beersheba is well described by Falls, 44–62; Gullett, 384–407 recounts numerous interesting anecdotes and is exhaustive; Preston, 23–32 gives a good general account; *Chauvel*, 126–129, too, presents a good concise picture, while, as usual, Wavell: *Pal.*, 115–126, is clear and authoritative. Some interesting details appear in Osborne: I, 332–58; Jones's detailed research adds to and alters many of the other accounts
22 Paterson, A.B. ('Banjo') *Happy Dispatches*, 1980, 84

CHAPTER 17 (pp. 163–166)

1 Wavell: *Pal.*, 126; Gullett, 424
2 'Report on 20th Corps Operations, 20 Oct. to 7 Nov., 1917', 8, Chetwode
3 Falls, 82

4 Falls, 84
5 Teichman, 180
6 'Report on 20th Corps Operations, 20 Oct. to 7 Nov., 1917', 9, Chetwode

CHAPTER 18 (pp. 167–172)

1 Falls, 111; Gullett, 444; Osborne: I, 358; *Chauvel*, 130; Wavell, 217
2 Allenby to Chauvel, Falls, 107
3 Wavell: *Pal.*, 138
4 Gullett, 433
5 Preston, 52
6 Falls, 117
7 Idriess, 337
8 Preston, 51

CHAPTER 19 (pp. 173–182)

1 Falls, 123; Alan-Williams, 7; Wavell: *Pal.*, 148
2 Quoted in Adderley, 139
3 Cobham, 129–30
4 Teichman, 184
5 Quoted in Falls, 121
6 Cobham, 131
7 Adderley, 126
8 Osborne; I, 363
9 Falls, 122; Alan-Williams, 5
10 Adderley, 126
11 Teichman, 184–5
12 Falls, 123

CHAPTER 20 (pp. 183–190)

1 8 Dec., 1917, *Chauvel*, 133; Preston, 65, 77; Gullett, 469
2 Gullett, 446
3 Gullett, 451
4 Gullett, 453
5 Falls, 129
6 Kress in *Sinai*, I, 51, quoted in Falls, 141
7 16 Dec., 1917, *Advance*, 6
8 Darley, 107
9 See Cobham, 140–44, for a good description of the action at Balin

10 16 Dec., 1917, *Advance*, 7

CHAPTER 21 (pp. 191–204)

1 Falls, 172; Osborne: I, 375
2 Yeodiv's official report, quoted in Osborne: I, 369
3 Falls, 168
4 Falls, 168
5 Yeodiv's official report, Osborne: I, 372
6 Preston, 82
7 Perkins
8 Falls, 169
9 Osborne: I, 372
10 Osborne: I, 373
11 Perkins
12 'in the words of the Turkish historian', quoted in Falls, 173
13 Preston, 89
14 Cobham, 145
15 Preston, 96–7

CHAPTER 22 (pp. 205–212)

1 Osborne, I, 376; Falls, 697; Hatton, 181–2; quoted in Gullett, 495–6
2 Smith, *Hist.*, quoted in Wavell: *Pal*, 157
3 Barrow, 169
4 16 Dec., 1917, *Advance*, 8
5 Inchbald, 99–101; Wavell, 224
6 Blenkinsop, 207, 213, 214
7 Blenkinsop, 207
8 Barrow, 178
9 Barrow, 183
10 Barrow, 187–8
11 Falls, 252
12 11 Dec., 1917, Wavell, 231
13 Wavell: *Pal.*, 172; the best accounts of the actions of 18 November to 9 December are to be found in Falls, 186–264 and in Wavell: *Pal.*, 156–167. For the part played by Yeodiv see Barrow, 169–188
14 Blenkinsop, 211–12
15 Blenkinsop, 219–20

CHAPTER 23 (pp. 213–228)

1 Wavell: *Pal.*, 176; Falls, 298; Falls, 420; Bostock, 144; Sep., 1918, *Ops: V Cav. Div.*
2 Joint Note of the Military Representatives, Supreme War Council, quoted in Falls, 297
3 Chetwode to Wavell, 28 Mar., 1939, Allenby Papers
4 *Advance*, 14, 15
5 Wavell: *Pal.*, 180
6 'To GOC, 60th Div.', G.Z.27/29, 16 Mar., 1918, Falls, 705–6
7 Quoted in Gullett, 583
8 Wavell: *Pal.*, 182; *Advance*, 20
9 Maunsell, 183
10 18 May, 1918, Dening
11 See Olden, 208 and Darley, 116
12 Wylly, 143; Osborne, II, 129, 134; Falls, 417; Teichman, 224; Whitworth, 119
13 Wavell: *Pal.*, 184
14 Gullett, 612; Falls, 374
15 Chetwode to Chauvel, 8 May, 1918, Chauvel Personal Record File, AWM, quoted in *Chauvel*, 152
16 Chetwode to Wavell, 28 Mar., 1939, Allenby Papers, quoted in *Chauvel*, 152
17 Chauvel in *Reveille*, 1 Jun., 1936, quoted in *Chauvel*, 151

CHAPTER 24 (pp. 229–241)

1 *Twentieth MG Sqn*, 96–98; Hudson, 218; Preston, 189
2 See Gullett, 639
3 Chauvel to Gullett in an 'Autobiography' (unpublished), pp. 113–14, quoted in *Chauvel*, 156. See Gullett, 603
4 Quoted in Preston, 177
5 *Twentieth MG Sqn*, 100; Berrie, 146
6 Wilson, 134; Wylly, 144
7 Preston, 180; Teichman, 246
8 Darley, 140; Vernon, 156; Gullett, 643
9 Wilson, 142; Gullett, 644; *Twentieth MG Sqn*, 104–5
10 GHQ figures, quoted in *Chauvel*, 241; Preston, 181; Falls, 424; Blenkinsop, 228; Wilson: *Pal.*, 108, 116
11 Wilson, 131; Berrie, 143; Olden, 237
12 Berrie, 143; Wilson, 138
13 Falls, 443
14 Badcock, 198

15 Falls, 434; *Chauvel*, 160; Falls, 438; excellent accounts of this action are given in Gullett, 660–73 and in Preston, 181–5
16 Wylly, 145–6
17 Wylly, 146
18 Falls, 436
19 *Advance*, 23
20 A detailed account of this action is in Falls, 434–7. See also Wylly, 145–7 and Tallents, 153–5

CHAPTER 25 (pp. 242–252)
1 *Advance*, 25, 27; Falls, 438
2 Massey: *Triumph*, 103; Falls, 431
3 Falls, 447–8
4 *Chauvel*, 162; Osborne, III: 355
5 Wavell: *Pal.*, 196
6 'Desert Mounted Corps Operation Order, No. 21, 12 Sep., 1918', Falls, 721
7 Wavell: *Pal.*, 199
8 Wavell, 224–5
9 Darley, 144; Barrow, 194; Falls, 458

CHAPTER 26 (pp. 253–275)
1 Gullett, 690–1; Quoted in Gullett, 687; Olden, 250; Hudson, 223, Massey: *Triumph*, 148
2 Massey: *Triumph*, 98
3 *Twentieth MG Sqn*, 110; Barrow, 191
4 Powles, 234
5 Osborne: II, 354; Maunsell, 211
6 Vernon, 165
7 Powles, 235
8 Preston, 197
9 Gullett, 689

CHAPTER 27 (pp. 258–267)
1 Wavell: *Pal.*, 203; Jeremiah, xii, 5
2 Barrow, 192
3 Barrow, 192–3
4 Quoted in Barrow, 195
5 Barrow, 195
6 Falls, 464–5; Barrow, 194–5; *Chauvel*, 163
7 Barrow, 193

8 The fullest account of the infantry's actions is to be found in Falls, 488–511

CHAPTER 28 (pp. 262–267)

1 Hudson, 222; Gullett, 692–3; Hudson, 238; Osborne: III, 35–6; Powles, 238
2 Osborne: III, 22
3 Osborne: III, 24; Gullett, 699
4 According to Osborne: III, 359. This seems rather too early
5 Fox, 260
6 Barrow, 195
7 Falls, 514
8 Preston, 205; Gullett, 695
9 Falls, 522–3
10 Falls, 523
11 Osborne: III, 25
12 A detailed account of the action at Nahr Falik appears in Weir, Brig.-Gen. G.A. 'Some Critics of Cavalry and the Palestine Campaign', *Cav. Jnl*, Oct., 1920, 535–7
13 Fox, 262–3
14 Cardew, 200–202

CHAPTER 29 (pp. 268–278)

1 Gardner, 184; Hudson, 228; Wavell, 273; Falls, 527; John, I, 46
2 Osborne: III, 26–7
3 Cardew, 203
4 Osborne: III, 27
5 Fox, 265
6 Falls, 526, 527
7 Fox, 266; Falls, 526
8 Wylly, 153; Poona, 393–4
9 Chauvel to Bean, 8 Oct., 1929, *Chauvel*, 169
10 Preston, 208
11 Barrow, 196
12 Barrow, 197
13 Whitworth, 134
14 Barrow, 197
15 Barrow, 198–9
16 Whitworth, 137
17 *Chauvel*, 168

CHAPTER 30 (pp. 279–294)

1 Gardner, 185; Preston, 209–10; Olden, 260
2 Whitworth, 138–40
3 Whitworth, 141–5
4 Letter quoted in Barrow, 202
5 Barrow, 203
6 Hudson, 242–3; Barrow, 204; Preston, 212
7 Watson, 396–7
8 Gullett, 697
9 Osborne: III, 36; Falls, 530
10 Osborne: III, 36; 36
11 Olden, 257
12 Olden, 257–9
13 Osborne: III, 38
14 Falls, 531
15 Osborne, III, 39
16 Gullett, 708
17 Olden, 259–60
18 Preston, 213–14
19 Preston, 219
20 Falls, 532
21 Gullett, 710; Osborne: III, 139
22 *Advance*, 30
23 Quoted in *Chauvel*, 171
24 *Advance*, 29
25 Wavell, 278

CHAPTER 31 (pp. 295–300)

1 Gullett, 712
2 Falls, 534
3 Preston, 227
4 Weir, 540
5 The fullest accounts of the Haifa battle are in Weir, 538–41 and Falls, 534–8

CHAPTER 32 (pp. 301–314)

1 Gullett, 728; Falls, 545–6; Falls, 542; Wavell: *Pal.*, 223; Osborne: III, 149; Gullett, 726; Home, 183
2 Barrow, 206
3 Osborne: III, 140
4 Maunsell, 228

5 Osborne: III, 141
6 Stonham, 193
7 Maunsell, 229
8 Osborne: III, 141
9 Cobham, 183
10 Falls, 542
11 Osborne: III, 141–2
12 Osborne: III, 142
13 Stonham, 194
14 Maunsell, 230
15 Quoted in Falls, 545
16 Grant's Report, quoted in Osborne: III, 145
17 Gullett, 732
18 Osborne: III, 146–7
19 Osborne: III, 147
20 Osborne: III, 147–8
21 Sanders, Liman von *Fünf Jahre in Turkei*, quoted in Wavell: *Pal.*, 220
22 Osborne: III, 154
23 Wavell: *Pal.*, 221–2
24 Gullett, 727
25 See Wilson, 145–8, for a detailed account of the Ziza situation, 28–30 September, 1918
26 Australian Army Archives, M.P. 367, 469/2/446 and Chetwode, both quoted in *Chauvel*, 173

CHAPTER 33 (pp. 315–331)

1 Falls, 723; *Chauvel*, 176; Preston, 247, 272; Wavell, 283; Falls, 596; Powles, 262
2 Wavell, 283
3 Watson, 398
4 Watson, 399
5 Whitworth, 157
6 Whitworth, 158
7 Whitworth, 158
8 Whitworth, 159
9 Whitworth, 160
10 Whitworth, 160
11 Whitworth, 161
12 Barrow, 208
13 Beside Whitworth's excellent first-hand account of the action

of Irbid (155–61), Watson, 398–401, gives a full explanation of the secondary part played by the Central India Horse

14 Watson, 401–2; a more detailed account of the action at El Remte is given in Falls, 581–2

15 Wavell: *Pal.*, 225

16 Olden, 269

17 Falls, 570–1

18 Gullett, 748

19 Osborne: III, 282

20 Powles, 244–5

21 Falls, 573

22 Darley, 156

23 [Anon] '34th Poona Horse: Narrative Account of Cavalry Engagements in 1918 in Palestine', *Cav. Jnl*, X, 1920, 396

24 Olden, 277–81

25 Chauvel to Director, AWM, 1 Jan., 1936, *Chauvel*, 182

26 Darley, 160–1

27 Blenkinsop, 242

28 *Advance*, 33

CHAPTER 34 (pp. 332–340)

1 Wavell, 287, 288; Liddell Hart: *Fog*, 101, 102

2 Report by Maj. C. Hercus, DADMS, Anzac Div., quoted in Powles, 259–62

3 Preston, 287

4 Falls, 611

5 Falls, 611–12

6 Falls, 613

7 Lambert, 315

8 Lambert, 316. See also Falls, 613–16

9 Lambert, 318–19

10 Falls, 616

11 Wavell, 289

12 23 Dec., 1931, Allenby to Murray, 79/48/3, IWM

INDEX

BRITISH ARMY FORMATIONS, BATTLES and REGIMENTS appear under these headings

Aaronsohn, Aaron, spy, 128
Abd el Kader, (1807?–1883), 328
air forces, British, achieve superiority at Gaza III, 134
air forces, British and German, active at Rafah, 88 – see also ground-to-air
air forces at Gaza III, 139, 150
Alan-Williams, Lt A.C., Warwickshire Yeomanry, at Huj, 177
Albright, Maj. M.C., Worcestershire Yeomanry, at Huj, 177
Alexander III, ('the Great'), 356–323BC, 340
Ali Bey Wakaby, 313
Ali Fuad Pasha, Maj.-Gen., 335
Allenby, F-M Sir Edmund, 1st Viscount (1864–1936), 8, 35, 104, 118, 184, 215, 220, 316, 332, 340; gets on well with Chauvel, 56; replaces Murray as C-i-C, 119; sacks staff officers in Cairo, 121; removes HQ to Rafah from Cairo, 121; character of, 121–4; adopts Chetwode's Gaza III plan, 126; decides October 1917 best for Gaza III, 129; his orders to Chauvel for post-Beersheba pursuit, 168; decides to take Jerusalem, 206; orders Jerusalem not to be fought over, 210; ordered to send nine yeomanry regiments to France, 220; decides to occupy Jordan Valley, 230; decides on 1918 operations, 244; basis of his Megiddo strategy, 246; asks Chauvel: 'What about Damascus?', 295; cables Australian Government after Megiddo, 314; confers with corps commanders, 26 Sep., 1918, 316; gives orders not to enter Damascus, 327
Antill, Maj.-Gen. John Macquarie (1866–1937), 58; at Romani, 62; withdraws 3 ALH Bde at Romani, 69
Apsley, Lt-Col Allen Algernon Bathurst, Lord (1895–1942), Gloucestershire Yeomanry, 325
Arab cavalry, 'all done up like wedding cakes', 96
Arab Legion, 12
Arab Revolt starts, 218
Armistice with Turkey signed, 339
Asquith, Herbert Henry, 1st Earl of Oxford and Asquith (1852–1928), succeeded by Lloyd George as PM, 75
Australian Comforts Fund, 81
Australian Flying Corps, inferiority in 1916, 59
Australian Flying Squadron, 1st, in post-Gaza III pursuit, 184
Australian light horsemen, types of, 4–15

Baker, Lt-Col Sir Randolf Littlehales, 4th Bt (1879–1959), Dorset Yeomanry, 119; at El Mughar, 197
Barrow, Gen. Sir George de Symons (1864–1959), 9, 127, 221; GOC, Yeomanry Mounted Division, 113; at Gaza III, 135; his 'Special Detachment' post-Beersheba, 168; in post-Gaza III pursuit, 184; at El Mughar, 194; in advance to Jerusalem, 206; pre-Sharon, 258; at Megiddo, 264; in advance to Damascus, 316
Bartlett, Lt J.S., 11th ALH Regiment, at Kh. Buteihah, 170
Basti Singh, Jem., 319

BATTLES (*including wars, campaigns, expeditions, actions, combats, engagements, skirmishes, sieges and unopposed entries*)
- Abu Shushe, 15 Nov., 1917, 201
- Abu Tullul, 14 July, 1918, 236
- Acre, 23 Sep., 1918, 296
- Agagia, 26 Feb., 1916, 30
- Aleppo, 25 OCT., 1918, 335
- Ameidat, 7 Nov., 1917, 170
- Amman, 24 Sep., 1918, 313
- Arras, 9 Apr., 1917, 119, 123
- Askar, 21 Sep., 1918, 305
- Asluj-Auja railway, expedition to destroy, May, 1917, 117
- Ayun Kara, 14 Nov., 1917, 202
- Balin, 12 Nov., 1917, 189
- Barada Gorge, 30 Sep., 1918, 326
- Balkan War, 1912, 21, 22
- Beersheba, 31 Oct., 1917, 149
- Beirut, 7 Oct., 1918, 334
- Bir el Abd, 8 Aug., 1916, 71
- Birket el Fuleh, 20 Sep., 1918, 282
- Cambrai, 20 Nov., 1916, 158, 210
- Caporetto, 24 Oct., 1917, 210
- Damascus, 1 Oct., 1918, 329
- Dueidar, 23 Apr., 1916, 43
- El Afule, 20 Sep., 1918, 285
- El Buggar, 27 Oct., 1917, 138
- El Hayira, 1 Oct., 1918, 327
- El Hinu, 14 July, 1918, 236

El Mezze, 30 Sep., 1918, 326
El Mughar, 13 Nov., 1917, 191
Es Salt raid, 11 Apr., 1918, 226
Es Salt, final capture of, 23 Sep., 1918, 313
Et Tine, 9 Nov., 1917, 188
Gallipoli, evacuation of, 1915, 39
Gaza I, 26 Mar., 1917, 97; plan for, 93
Gaza II, 17 Apr., 1917, 25, 105
Gaza III, 31 Oct., 1917, plan for, 125, 128; preparation for, 129; deception plans for, 133; dispositions of formations for, 135; Beersheba charge, 149; Gaza evacuated, 164
'Haifa Annexation Expedition', 22 Sep., 1918, 295
Halazin, 23 Jan., 1916, 28
Hareira, 6 Nov., 1917, 166, 168
Haritan, 26 Oct., 1918, 337
Homs, 16 Oct., 1918, 334
Huj, 8 Nov., 1917, 160, 173; village captured, 185, 198
Irbid, 26 Sep., 1918, 318
Issus, 333BC, 340
Jaffa, 16 Nov., 1917, 202
Jemmame, 8 Nov., 1917, 171
Jenin, 21 Sep., 1918, 290
Jericho, 21 Feb., 1918, 217
Jerusalem, battles for, 206; entered, 12 Dec., 1917, 210
Jifjaffa, 10 Apr., 1916, 33
Jisr el Damiye, 22 Sep., 1918, 312
Junction Station, 14 Nov., 1917, 200
Kadem, 28 Sep., 1918, 325
Katia, 23 Apr., 1916, 19, 41, 198
Kaukab, 30 Sep., 1918, 325
Kaukabia, 12 Nov., 1917
Khan Ayash, 2 Oct., 1918, 330
Khan Buteihah, 7 Nov., 1917, 170
Khuweifle, 164
Kimberley, 15 Feb., 1900, 129
Kiswe, 30 Sep., 1918, 325
Kuneitra, 28 Sep., 1918, 323
Magdhaba, 23 Dec., 1916, 20, 79, 86, 117, 341
Maghara, 13 Oct., 1916, 77
Makhadet Abu Naj, 22 Sep., 1918, 302
Masudi, 24 Sep., 1918, 306
Mazar, 17 Sep., 1916, 76
Megiddo, Sep., 1918, 258
Musmus Pass, Sep., 1918, 274
Nablus, 20 Sep., 1918, 289
Nahr Falik, 19 Sep., 1918, 266
Nazareth, 20 Sep., 1918, 269
Oghratina, 23 Apr., 1916, 40
Old Aqir, 13 Nov., 1917, 199
Rafah, 9 Jan., 1917, 20, 85
Romani, 3 Aug., 1916, 41, 54, 56, 58; 60
Sasa, 28 Sep., 1918, 328
Semakh, 25 Sep., 1918, 310
Sharon (Megiddo), Sep., 1918, 258
Sheria, 7 Nov., 1917, 168, 258
Surafend, 14 Nov., 1917, 200
trans-Jordan operations, 1st, Mar., 1918, 219
trans-Jordan operations, 2nd, Apr-May, 1918, 225
Tripoli, 13 Oct., 1918, 334
Tul Karm, 18 Sep., 1918, 263, 288
Wadi Hesi, 7 Nov., 1917, 171
Wadi Majid, 25 Dec., 1915, 28
Wadi Senab, 13 Dec., 1915, 28

Bengough, 2nd-Lt J.C., Gloucestershire Yeomanry, at Agagia, 32
Bessières, Marshal Jean Baptiste, Duc d'Istrie (1766–1813), 127
Betjeman, Sir John (1906–1984), 78
Birdwood of Anzac, F.M. William Riddell, 1st Baron (1865–1951), 56, 103
Blagg, 2nd-Lt., S., S. Notts Yeomanry, 50
Blaksley, 2nd-Lt., J.H., Dorset Yeomanry, 27; at Agagia, 30
Bols, Lt-Gen. Sir Louis Jean (1867–1930), 225
Bourchier, Lt-Col Hon. Murray William James (1881–1937), 4th ALH Regiment, 150
Bourne, Lt-Col G.H., 2nd ALH Regiment, 64
Bolton, Tpr, 4th ALH Regiment, 150; at Beersheba, 156
Bostock, Cpl Henry P., 10th ALH Regiment, 111, 213
Bridges, Brig.-Gen. William Throsby (1861–1915), Inspector-General of Commonwealth Forces, 14
bridging, pre-Megiddo, 251
Brierty, Lt A.R., 11th ALH Regiment, at Kh. Buteihah, 170

BRITISH ARMY FORMATIONS (*excluding regiments, batteries, etc.*)
XX Army Corps, 229; formed, 126; at Gaza III, 138; in advance to Jerusalem, 207; in advance to Jericho, 217; at Megiddo, 260
XXI Army Corps, 190, 200, 209, 229, 246; formed, 127; at Gaza III, 163; in post-Gaza III pursuit, 187; in advance to Jericho, 217; at Megiddo, 260
Bikanir Camel Corps, 2; at Romani, 70
Chaytor's Force, 312; at Amman, 314
Desert Column, formed, 78; at Rafah, 88; formations in, 92

Desert Mounted Corps, 126; formed, 112; at Gaza III, 145; in post-Gaza III pursuit, 184; re-organized, 220; occupies Jordan Valley, 230; at Megiddo, 258
Eastern Force, re-organized, 91; abolished, 126
Egyptian Camel Transport Corps, 35, 60
Egyptian Labour and Transport Corps, 46, 234
Imperial Camel Corps, 18, 36; at Maghara, 76; at Magdhaba, 79; at Rafah, 85; at Gaza I, 97; at Gaza II, 105; at Gaza III, 135; in post-Gaza III pursuit, 184
Anzac Mounted Division, 3, 14, 44, 45, 190; formed, 92; at Gaza I, 97; at Gaza II, 106; at Gaza III, 138; in advance to Jericho, 217; re-organized, 220; its work finished, 314
Australian Mounted Division, 3, 55, 58, 85, 190; at Gaza III, 138; in post-Gaza III pursuit, 188; enters Et Tine, 200; in advance to Jericho, 217; re-organized, 220; in trans-Jordan operations, 225; at Megiddo, 248; in advance to Damascus, 323
2nd Mounted Division, 3
3rd (Lahore) Division, at Megiddo, 261
4th Cavalry Division, 248; formed, 221; pre-Megiddo, 255; at Megiddo, 286; at Irbid, 321
5th Cavalry Division, 248; formed, 221; at Megiddo, 261
7th (Meerut) Division, 255; at Megiddo, 261
Hedjazian Cavalry, 327
Imperial Mounted Division, formed, 91; at Gaza I, 97; at Gaza II, 106
Royal Naval Armoured Car Division, 28
Yeomanry Mounted Division, 190; formed, 113; in post-Gaza III pursuit, 184; in advance to Jerusalem, 206
1st Australian Infantry Division, 3, 56
42nd Infantry Division, 45; at Romani, 69
52nd (Lowland) Division, 45, 93, 190; at Romani, 66; at El Mughar, 192; in advance to Jerusalem, 207
53rd (Welsh) Division, 45, 93; at Gaza I, 99; at Gaza III, 138
54th (East Anglian) Division, 93
60th Division, 168, 171, 175; in advance to Jericho, 217; in 2nd trans-Jordan operations, 225
74th (Yeomanry) Division, 91, 93, 114; at Gaza III, 166
75th Division, 200; st El Mughar, 192; in advance to Jerusalem, 206; at Megiddo, 261
1st Australian Light Horse Brigade, 45, 56; at Romani, 63; at Magdhaba, 79; at Wadi Hesi, 171; in post-Gaza III pursuit, 185
2nd ALH Bde, 43, 45, 76; at Romani, 63; at Gaza I, 95; at Gaza III, 145; at Wadi Hesi, 171; at Amman, 313
3rd ALH Bde, 45, 110; at Magdhaba, 81; at Rafah, 85; at Gaza I, 97; at Gaza III, 147; in post-Gaza III pursuit, 184; at Megiddo, 289; after Khan Ayash, 330
4th ALH Bde, 4, 63, 91; in Beersheba charge, 142; in 2nd trans-Jordan operations, 227
5th ALH Bde, 4, 5, 85, 221; at Gaza I, 101; at Gaza III, 153; in post-Gaza III pursuit, 184; at Megiddo, 288
6th ALH Bde, at Gaza I, 101; at Megiddo, 265
10th ALH Bde, 221; at Megiddo, 264
11th ALH Bde, 221; at Megiddo, 264
12th ALH Bde, 221; at Megiddo, 264, 265
13th ALH Bde, 221; at Megiddo, 265; at Nazareth, 269
14th ALH Bde, 221; at Megiddo, 266; at Haritan, 338
Camel Corps Brigade, 18, 219
Composite Mounted Brigade, in Western Desert, 28
Imperial Service Cavalry Brigade, 2; enters Gaza, 168; in post-Gaza III pursuit, 187
15th (Imperial Service) Brigade, 221
20th (Imperial Service) Brigade, 230
3rd Mounted Brigade, at Romani, 67, 76
5th Mounted Brigade, 202, 220; at Romani, 67; in post-Gaza III pursuit, 189; in advance to Jerusalem, 206
6th Mounted Brigade, at El Mughar, 198
7th Mounted Brigade, takes Ras al Nagh, 165; in post-Gaza III pursuit, 188
8th Mounted Brigade, 25, 191; at El Mughar, 195
New Zealand Mounted Rifles Brigade, 12, 200, 201, 202; at Romani, 65; at Magdhaba, 79; at Gaza I, 97; at Gaza III, 145; in post-Gaza III pursuit, 184; suffers from malaria, 333

Yeomanry Mounted Brigade, 2
5th (Yeomanry) Mounted Brigade, 85
7th (Yeomanry) Mounted Brigade, arrives from Salonika, 112; at Gaza III, 143
8th (Yeomanry) Mounted Brigade, arrives from Salonika, 112; at El Buggar, 138; at El Mughar, 191
22nd (Yeomanry) Mounted Brigade, 199; at Gaza I, 92

Brittain, Pte H.C., W. Somerset Yeomanry, 51
Brooke, Mrs Geoffrey, 343
Brooke, Maj-Gen. Geoffrey Francis Heremon (1884–1966), 343
Browne, Brig.-Gen. John Gilbert (1878–1968), 94; Chauvel's GS01, 57
Brusilov, Aleksei Alekseevich (1853–1926), 75
Buchan – see Tweedsmuir
Bulfin, Gen. Sir Edward Stanislaus (1862–1939), 284, 316; GOC, XXI Corps, 127; at Gaza III, 135; in advance to Jerusalem, 209; pre-Sharon, 258
Bullivant, Capt., Middlesex Yeomanry, 306
Bulteel, Maj. Sir John Croker (1890–1956), Bucks Yeomanry, at El Mughar, 196
Burton, Lt F.J., 4th ALH Regiment, *k.* at Beersheba, 156
Byng, F-M Julian, 1st Viscount (1862–1935), 123

camels, 212; types of, 19; desperate condition of near Jerusalem, 207
Cameron, Lt-Col Donald Charles (1879–1960), 12th ALH Regiment, 127; at Gaza III, 150
casualties at Megiddo, 331
Chatham, Maj. William, 5th ALH Regiment, at Gaza I, 97
Chauvel, Gen. Sir Henry George (1865–1945), 9, 14, 45, 55, 56, 57, 86, 222, 228; gets on Allenby's nerves, 57; his staff, 58; at Romani, 60; at Magdhaba 80; at Gaza I, 96; GOC, Descorps, 112; at Gaza III, 145; orders 4th ALH Bde to take Beersheba, 149; receives Allenby's orders for post-Beersheba pursuit, 168; agrees swords for 5 ALH Bde, 223; ready for 1918 operations, 244; at Megiddo, 273; replies; 'Rather!' to Allenby's question about Damascus, 295; orders 5 Div. to take Haifa and Acre, 295; his troubles at Damascus, 329, 330
Chaytor, Maj.-Gen. Sir Edward Walter Clervaux (1868–1939), 60, 249; at Romani, 62; at Magdhaba, 80; GOC Anzac Mounted Div., 112; at Gaza III, 143; elects mounted rifles role for Anzacdiv, 223; forms 'Chaytor's Force', 248; takes Jisr el Damiye crossing, 302
Cheape, Lt-Col Hugh Annesley Gray-(1878–1918), Warwickshire Yeomanry, at Katia, 42; at Huj, 175
Chetwode, F-M Sir Philip Walhouse, Bt, 1st Baron (1869–1950), 14, 116, 316; commands Desert Column, 77; at Magdhaba, 79; at Rafah, 85; at Gaza I, 96; GOC, Eastern Force, 112; his plan for Gaza III, 125, 128; reconnoitres pre-Gaza III, 133; in advance to Jerusalem, 207; tries to persuade Allenby not to undertake 2nd trans-Jordan operations, 228; at Megiddo, 249
Chisholm, Dame Alice (1856–1954), 11
Clarke, Brig.-Gen. Goland Vanhalt (1875–1944), commander, 14th ALH Bde, at Nazareth, 273
Cobham, John Cavendish Lyttelton, 9th Viscount (1881–1949), Worcestershire Yeomanry, at Huj, 176
Costello, Maj. E., 11th ALH Regiment, 310
Coster, Lt, Central India Horse, *k.*, 1918, 308
Cotter, Tpr A., 4th ALH Regiment, at Beersheba, 156
Coventry, Lt-Col Hon. Charles John (1867–1929), Worcestershire Yeomanry, at Katia, 42
Cox, Armourer Staff-Sgt A.J., 4th ALH Regiment, at Beersheba, 156
Cox Maj.-Gen. Charles Frederick (1863–1947), 58, 220; at Magdhaba, 80; in post-Gaza III pursuit, 185; at Abu Tullul, 236
Cripps, Lt-Col Hon. Frederick Heyworth (1885–1977), Bucks Yeomanry, at El Mughar, 194

Dallas, Maj.-Gen. Alister Grant (1866–1931), at Maghara, 77; at Gaza I, 96
Dammers, Capt. G.M., Dorset Yeomanry, at El Mughar, 197
Darley, Lt T.H., 9th ALH Regiment, 81, 84
Davison, Maj.-Gen. Douglas Stewart (1888–1929), 2nd Lancers, 275; at Irbid, 320
Dawney, Maj.-Gen. Guy Payan (1878–1952), 225; his and Chetwode's plan for Gaza III, 125
deception, methods of pre-Megiddo, 255; by dummy horses, 256
Dening, Maj.-Gen. Roland (1888–1978), 10, 222

dental service, 54
Deraa, Arab atrocities at, 322
Derby, Edward George Villiers Stanley, 17th Earl of (1865–1948), 120
Dickson, Lt A.F., Poona Horse, 240
Diplock, B.J., inventor of pedrail, 65
diseases in desert, 52; in Jordan Valley, 232
Djemal the Lesser, 249
Djemal Pasha, Akmed (1872?–1922) 33, 75, 85, 106; attacks Suez Canal, 1915, 25; comments on Murray's despatch after Gaza I, 104
Djevad Pasha, 312, 328; succeeds Kress i/c VIII Army, 248
Dobell, Lt-Gen. Sir Charles Macpherson (1869–1954), 75, 91; at Magdhaba, 84; at Gaza I, 96; replaced by Chetwode as GOC, Eastern Force, 112
Doig, 2nd-Lt P.W.K., 10th ALH Regiment, 290
donkeys, 212; first used in advance to Jerusalem, 207
Downes, Maj.-Gen. Rupert Major (1885–1945), 58
Dyal Singh, Risaldar, Central India Horse, 288

Edwards, 2nd-Lt J.W., Worcestershire Yeomanry, at Huj, 177
Eldridge, Tpr C., in Beersheba charge, 151
Enver (Bey) Pasha (1881?–1922), 27, 243
Essud Bey, Col, commander, 3 Cav. Div. at Beersheba, 159

Faisal I (1885–1933), 217, 218, 249, 312
Falkenhayn, Gen. Erich von (1861–1922), 220; C-i-C, Turkish forces, 136; launches counter-attack against Jerusalem, 211
Falls, Capt. Cyril Bentham (1888–1971), 139, 141
fanatis, 36
Fane, Maj.-Gen. Sir Vere Bonamy (1863–1924), commander 7 Meirut Division, 259; at Megiddo, 264
Ferid Bey, Col, at Beersheba, 159
Fetherstonhaugh, Maj. C.M., 12th ALH Regiment, at Beersheba, 157
Fevzi Pasha, Gen., at Gaza III, 138
Fitzgerald, Maj.-Gen., Gerald Michael (1889–1957), 331; i/c 5 Mounted Bde at Gaza III, 148; succeeded by Kelly, 177
Fletcher, Cpl A.W., Lincolnshire Yeomanry, 90, 114, 120
food and drink, all ranks, 50
Forrest, Pte, Gloucester Yeomanry, 267
Foster, Bt-Col William James (1881–1927), 113, 259, 275
Foulkes-Taylor, Lt Charles, 8th ALH Regiment, 328; in Es Salt raid, Apr., 1918, 226
French, F-M Sir John Denton Pinkstone, 1st Earl of Ypres (1852–1925), 37, 122
Fryer, Lt-Col F.A.B., i/c 22 Bde, 127, 199

Gell, Maj. P.F., Special Service Officer, Jodhpore Lancers, 237
George, L-Cpl T.B., 3rd MG Sqn, 291
Gobind Singh, Jemadar, VC, 2nd Lancers, 283
Godrich, Sgt E.V., Worcestershire Yeomanry, 88, 91, 107
Godwin, Lt-Gen. Sir Charles Alexander Campbell (1873–1951), 222; i/c 6 (Yeo.) Bde, 127; at El Mughar, 194; at Megiddo, 277
Goltz, Baron Kolmar von der, Pasha (1843–1916), 22
Gofton, Tpr Septimus Laws, 5th ALH Regiment, 98
Gordon, Maj. R.G.S., Dorset Yeomanry, at Mughar, 197
Gough, Gen. Sir Hubert de la Poer (1870–1963), 123
Gould, Maj. G., 2nd Lancers, 318
Grant, Brig.-Gen. William (1870–1939); i/c 4 ALH Bde, 127; at Gaza III (Beersheba charge), 148; at 2nd trans-Jordan operation, 227; criticized by Allenby, 227; at Jisr el Majame, 308
Granville, Lt-Col C.H., 1st ALH Regiment, at Abu Tullul, 236
Graves, Robert Ranke (1895–1985), 330
Gray-Cheape – see Cheape
Green, Brig.-Gen. Wilfrith Gerald Key (1872–1937), 286, 322
Gregory, Maj.-Gen. Charles Levinge (1870–1944), 221; teaches 5 ALH Bde use of sword, 223; near Beisan, 302
ground-to-air, air-to-ground contact methods, 132, 250

Haig, F-M Sir Douglas, 1st Earl (1861–1928), 123, 214
Hamilton, Gen Sir Ian Standish Monteith (1853–1947), 8, 56
Hammerstein-Gosmold, Lt-Col von, 331
Hankey, Sir Maurice Pascal Alers, 1st Baron (1877–1963), 123
Harbord, Brig.-Gen. Cyril Rodney (1873–1958), at El Tinu, 237; at Acre, 297; at Haritan, 335
Hatrick, Lt, Auckland Regiment, 147
Henley, Tpr A.E., 4th ALH Regiment, at Beersheba, 156
Hejaz railway, 230, 312, 316
Herbert, Capt. Endyr, Gloucestershire

Yeomanry, *k.* 1917, 189
Hercus, Maj. C., DADMS, Anzac Division, 315, 332
Hill, A.J., 45
Hill, Maj.-Gen. John (1866–1935), at El Mughar, 194
Hodgson, Maj.-Gen. Sir Henry West (1868–1930), 93, 113, 223; commander Imperial Mounted Div., 91; at Gaza I, 96; asks for swords for two Aus. bdes, 137; i/c Aus. Mounted Div. at Gaza III, 139; at Megiddo, 290; in advance to Damascus, 316; at Kaukab, 325
Holden, Col Hyla N., Jodhpore Lancers, *k.* 1918, 298, 337
Home, Brig.-Gen. Sir Archibald Fraser (1874–1953), 302
horses, without water, eighty-four hours, post-Gaza III, 188; endurance in post-Gaza III pursuit, 203; losses and condition of, 31 Oct. to 31 Dec, 1917, 211; condition of in Jordan Valley, 233; extraordinary powers of endurance of, throughout campaign, 341
Howard-Vyse, Maj.-Gen. Sir Richard Granville Hylton (1883–1962), Chauvel's COS, 57; at Musmus Pass, 274; sacked by Barrow, Megiddo, 277
Husein ibn-Ali (1856–1931), King of the Hejaz, 218
Husni, Col Hussein, 134
Hyman, Maj. E.M., 12th ALH Regiment, at Beersheba, 153, 157

Idriess, Tpr Ion L., 48 *et seq.*
influenza, incidence of, Oct., 1918, 333
Inglis, Lt, Gloucestershire Yeomanry, 272
Ismet Bey, Turkish commander at Beersheba, 158

Jackson, Capt., M.H., 29th Lancers, 303
Ja'far Pasha, Sennusite commander, at Agagia, 31
James, Lawrence, biographer of Allenby, 123
Jang Bahadur Singh, Ressaidar, 2nd Lancers, 281
Jones, Shoeing-Smith Thomas, 5th ALH Regiment, 98
Jordan Valley, horrors of summer in, 230–1

Kelly, Brig.-Gen. Philip James Vandeleur (*died* 1948), 202; at Huj, 177; at Megiddo, 268; sacked after Nazareth, 274
Kheri, Singh, Ress., Central India Horse, 319
King, Brig-Gen. Algernon D'Aguilar (1862–1945), 222; in 'Haifa Annexation Expedition', 295
King, Lt W.K., 29th Lancers, 305
Kirkpatrick, Sgt M., 2nd NZ MG Sqn, 263, 326
Kitchener of Khartoum, Horatio Herbert, 1st Earl (1850–1916), 3
Kress von Kressenstein, Col Baron, 25, 45, 75, 85, 101, 116, 159, 185, 188; at Katia and Oghratina, 41; at Romani, 60

Lafone, Maj. A.M., VC, Middlesex Yeomanry, 25
Lal Chand, 2nd Lancers, 283
Lambert, Maj. W.J., Special Service Officer, Mysore Lancers, at Haritan, 337
lance given to Poona and Deccan Horse, 223
Lawrence, Gen. Hon. Sir Herbert (1861–1943), at Romani, 60
Lawrence, Lt-Col Thomas Edward (1888–1935), 46, 57, 217, 218, 225, 249, 317, 330; takes Akaba, 112; and Arab atrocities at Deraa, 322; in Damascus, 329
Lawson, Lt-Col E.F., Middlesex Yeomanry, 286; at El Mughar, 197; at Masusi, 306
Lawson. Maj. J. ('Porky'), 4th ALH Regiment, at Beersheba, 153
leave, to Egypt from Jordan Valley, summer, 1917, 234, . . .
Le Marchant, Lt-Gen. John Gaspard (1766–1812), 171
Liddell Hart, Sir Basil Henry (1895–1970), 141
Liman von Sanders, Pasha, Gen. Otto (1855–1929), 22, 100, 122, 159, 220, 237, 243, 249; takes command of Turkish forces, 219; escapes capture at and returns to Nazareth, 272; retires to Baalbek to reorganize force, 329
Lloyd-Baker, Capt. M.G., Worcestershire Yeomanry, at Katia, 42
Lloyd George of Dwyfor, David, 1st Earl (1863–1945), 111, 119, 122, 141, 214; succeeds Asquith as PM, 75; announces capture of Jerusalem, 210
Loynes, Maj. J., 11th ALH Regiment, 310
Lucas, Maj. G.W.C., Poona Horse, 240
Ludendorff, Gen. Erich Friederich Wilhelm (1865–1937), 220
Lukin, Maj-Gen. Sir Henry Timson (1860–1925), 30
Lunt, Maj-Gen. James Doiran (*b.* 1913), 12
Lynden-Bell, Maj.-Gen. Sir Arthur Lynden (1867–1943), 112
Lyttelton, Maj. Hon. J.C. – see Cobham

Macandrew, Maj-Gen. Henry John Milnes (1866–1919), 237, 335; takes command 5 Cav. Div., 221; at Megiddo, 260; in advance to Damascus, 316
Macarthur-Onslow, Brig.-Gen. George Macleay (1875–1931), 221; at Megiddo, 263
McClellan, George Brinton (1826–1885), 287
McGrigor, Lt, Gloucestershire Yeomanry, 52, 113; at Rafah, 86
machine guns, types of, 16
Maitland-Woods, Chaplain, 115
malaria, incidence of Oct., 1918, 332
Mallet, Sir Louis du Pan (1864–1936), 2
Maude, Lt-Gen. Sir Frederick Stanley (1864–1917), 112
Maunsell, Capt., E.B., Jacob's Horse, 10
Maxwell, Gen. Sir John Grenfell ('Konky') (1859–1929), i/c Western Frontier Force, 27
Maygar, Lt-Col Leslie Cecil (1871–1917) 150
Mehtab Singh, Dafadar, Hodson's Horse, 266
Meinertzhagen, Col Richard (1878–1967), 78; his pre-Gaza III deceptions, 133
Meldrum, Brig.-Gen. William (1865–1964), at Romani, 65; i/c NZ Mounted Rifles Bde, 112; at Gaza III, 145; in post-Gaza III pursuit, 184
Meredith, Lt-Col B.P.G. 4th ALH Regiment, *k.* at Beersheba, 160
Meredith, Lt-Col John Baldwin; at Romani, 63; i/c 4 ALH Bde, 91; invalided home, 127
Mills, Maj., 18th Lancers, 270
Mitchell, Lt F.A., Gloucestershire Yeomanry, at Romani, 68
'Money's Detachment', 92
Money, Maj.-Gen. Sir Arthur Wigram (1866–1951), 92
Montgomery of Alamein, Sir Bernard Law, 1st Viscount (1887–1976), 122, 123
Morgan, Cpl, Gloucestershire Yeomanry, 111
Morgan, John Hunt (1825–1864), 287
Mueller, Lt, 9th ALH Regiment, at Huj, 180
Mukand Singh, R.M., 2nd Lancers, 280
Murray, Gen. Sir Archibald James (1860–1945), 8, 18, 35, 73, 75, 133, 340; forms EEF, 1916, 37; pressed to send units to France, 55; at Romani, 62; given go-ahead for 1917 campaign, 90; his despatch after Gaza I, 104; replaced by Allenby, 119
Mustafa Kemal Pasha (Kemal Ataturk) Gen. (1881–1938), 244, 335; at Haritan, 339

Nagington, Pte, 2nd Lancers, 284
Nand Singh, A.L.D., Central India Horse, 319
Nawab-Ali Khan, Jemadar, Hodson's Horse, 266
Newcombe, Col Stewart Francis (1875–1956); at Gaza, 140; at Dhahriye surrenders special detachment, 164
Newton, Lt-Col Frank Graham (1877–1962), at Gaza I, 97
Nur Ahmad, Risaldar, Hodson's Horse, 266

Olden, Lt A.C.N., 10th ALH Regiment, 74, 290, 328; at Magdhaba, 82
O'Leary, Tpr T., 4th ALH Regiment, 156
Onslow – see Macarthur-Onslow
Oppen, Col von, commander, Asia Corps, 303

Papen, Maj. Franz von (1879–1969), 206
Parsons, Lt-Col J.W., 11th ALH Regiment, 310
Patterson, Lt R.R.W., 3rd MG Sqn, 291
pay, Australian officers', 14
Pem Singh, Jemadar, Poona Horse, 239
Perkins, Lt C.H., Bucks Yeomanry, 194
Pertab Singhji, Gen. Sir (1845–1922), at El Hinu, 241
Picot (French Consul-General), 123
Plumer, F-M Herbert Charles Onslow, 1st Viscount (1857–1932), 123
Powles, Col Guy (1872–1951), 100; Chauvel's AA and QMG, 58; at Magdhaba, 80
Preston, Lt-Col Hon. Richard Martin Peter (1884–1965), 183, 204, 291
Primrose, Rt Hon. Neil Primrose (1882–1917), at El Mughar, 196; dies of wounds, 201

Ratcliffe, Pte J.C., Worcestershire Yeomanry, at Katia, 43
Rawlinson, F-M Henry Seymour, 1st Baron (1864–1925), 123
Refet Bey, i/c XXII Corps, at Old Aqir, 199

REGIMENTS (*and other units, e.g. independent squadrons, battalions, batteries*)

REGULAR CAVALRY
1st Dragoon Guards, 44
3rd Dragoon Guards, 2
9th Lancers, 44
21st Lancers, 44

YEOMANRY
Berkshire, at El Mughar, 194
Buckinghamshire, at Agagia, 30; at El Mughar, 194
City of London, 2, 62; at Maghara, 77
Composite Regiment of, 2nd, 28
Dorset, at Agagia, 30; at El Mughar, 194; at Megiddo, 264; horses of without water, 341
Duke of Lancaster's, 2; in post-Gaza III pursuit, 187
East Riding, 200
Gloucestershire, 52, 111, 114, 177, 206; at Katia and Oghratina, 41; at Romani, 62; at Rafah, 85; at Megiddo, 267
Hertfordshire, 2, 93; in post-Gaza III pursuit, 187
Lincolnshire, 90, 220; horses of without water for eighty-four hours, 202, 341
Middlesex, 25; at Afule, 285; near Beisan, 305; at Masudi, 306
Royal Glasgow, 93; in post-Gaza III pursuit, 187
Sherwood Rangers, 39; at El Hinu, 237; at Acre, 297
South Nottinghamshire, 12, 50
Warwickshire, at Katia and Oghratina, 41; at Romani, 62; at Huj, 173
Westminster, 112
West Somerset, 51
Worcestershire, 52, 107, 202, 249; at Katia and Oghratina, 41; at Romani, 62; at Rafah, 85; at Huj, 173; at Askar, 305; enters Beirut, 334

INDIAN CAVALRY
Central India Horse, at Megiddo, 264; at Irbid, 319
Deccan Horse, armed with lances, 223; at Afule, 285
Hodson's Horse, at Megiddo, 266
Hyderabad Lancers, 2; in post-Gaza III pursuit, 187; at Megiddo, 293
Imperial Service Cavalry, at El Hinu, 237
Jacob's Horse, at Megiddo, 264; near Beisan, 303
Jodhpore Lancers, 3; in post-Gaza III pursuit, 187; at El Hinu, 237; at Acre, 297; at Haritan, 336
2nd Lancers, at Megiddo, 264; at Irbid, 318
18th Lancers, at Megiddo, 267
19th Lancers, at Megiddo, 286
29th Lancers, near Beisan, 302
Mysore Lancers, in post-Gaza III pursuit, 187; at El Hinu, 237; at Acre, 296; at Haritan, 336
Patiala Lancers, 2
Poona Horse, armed with lances, 223; at El Hinu, 239

AUSTRALIAN CAVALRY
1st Australian Light Horse, at Ameidat, 170
2nd ALH, at Romani, 64; at Magdhaba, 80
3rd ALH, 25, 39; at Magdhaba, 80; at Ameidat, 170
4th ALH, 4, 91; at Beersheba, 150
5th ALH, 12, 24, 43, 76; at Romani, 65; at Wadi Hesi, 171
6th ALH, 66
7th ALH, 65; at Gaza I, 95; at Gaza III, 143; at Wadi Hesi, 171; at Huleikat, 185
8th ALH, at Jifjaffa, 33; at Romani, 71; at Rafah, 86; in Es Salt raid, 226, at Megiddo, 289
9th ALH, 33, 69; at Magdhaba, 81; at Gaza III, 150; at Huj, 180; at Barada Gorge, 327; in last charge of campaign, 330
10th ALH, 6, 7, 9, 25, 69, 111; at Gaza I, 102; at El Mughar, 192; in advance to Jerusalem, 206
11th ALH, 91; at Romani, 62; at Maghara, 77; at Gaza III, 153; at Khan Buteihah, 170
12th ALH, 77, 91; at Gaza III, 150; in post-Gaza III pursuit, 187
14th ALH, formed, 221; at Barada Gorge, 327
15th ALH, formed, 221; at Megiddo, 263
New South Wales Lancers, 58
Northern River Lancers, 5
Queensland Mounted Rifles, 5, 56
Royston's Horse, 59
Upper Clarence Light Horse, 5

NEW ZEALAND CAVALRY
Auckland, 12, 219; at Rafah, 86; at Gaza I, 99; at Gaza III, 146; in post-Gaza III pursuit, 184
Canterbury, 12; at Rafah, 86; at Gaza I, 99; at Gaza III, 146

ARTILLERY
Ayrshire Battery, Royal Horse Artillery (Territorial), 46, 70; at Magdhaba, 80
Berkshire Bty, RHA(T), at El Mughar, 197; at Megiddo, 264; at Irbid, 320
Essex Bty, RHA(T), 113; at Beersheba, 155; at Wadi Hesi, 171
Hampshire Bty, RHA(T); at Megiddo,

288; nr Beisan, 303
Hong Kong and Singapore Bty, 18; at Maghara, 77, 80; in advance to Jerusalem, 209
Honourable Artillery Company, 85, 114, 177; at Gaza III, 153; 'A' Bty loses guns, 227; 'B' Bty at Megiddo, 264
Inverness Bty, RHA(T), 46, 69; at Magdhaba, 80; at Rafah, 87
Leicester Bty, RHA(T), 46, 201; at Romani, 62
Nottinghamshire Bty, RHA(T), 28; at Gaza III, 153; loses guns, Apr., 1918, 227
Somerset Bty, RHA(T), 46; at Romani, 62; at Magdhaba, 80; at Gaza I, 99; at Gaza III, 146

ENGINEERS
15 Field Troop, Royal Engineers, at Haifa, 298

MACHINE-GUN UNITS
Cavalry Corps Motor Machine Gun Battery, in Western Desert, 30
2nd New Zealand MG Squadron, formed, 221
3rd MG Sqn, at Beersheba, 150; at Megiddo, 289
4th MG Sqn, at Beersheba, 150
17th MG Sqn, 209; at El Mughar, 195; at Megiddo, 264; at Irbid, 320
18th MG Company, at Megiddo, 286

ARMOURED AND LIGHT CAR UNITS
2nd Light Armoured Car Battery, 200; ordered to take Haifa, 295; at Haritan; 337
7th Light Car Patrol, at Rafah, 85; ordered to take Haifa, 295
11th Light Armoured Car Battery, at Musmus, 274
12th Light Armoured Car Battery, 200; ordered to take Haifa, 295; at Haritan, 337

MISCELLANEOUS
Regiment Mixte de Cavalerie, 221; at Megiddo, 263; at Kaukab, 325; at El Mezze, 326
West Indian Regiment, 230, 256

Reeves, Tpr, Staffordshire Yeomanry, 285
Richardson, Lt-Col K.D., 7th ALH Regiment, 301
Robertson, Lt-Gen. Sir Horace Clement Hugh (1894–1960), at Magdhaba, 80
Roberston, F-M Sir William Robert, 1st Bt (1860–1933), 8, 74, 92, 214
Robey, Capt. R.K., 12th ALH Regiment, at Beersheba, 157
Robinson, Maj. J.F.M., E. Riding Yeomanry, at Old Aqir, 199
Rome, Brig.-Gen. Claude Stuart (1875–1956), i/c 8 ALH Bde, 127
Rothschild, Maj. Evelyn de, Bucks Yeomanry, dies of wounds, 1917, 201
Rouse, Cpl, Somerset Battery, RHA(T), 99
Royal Air Force, pre-Megiddo, 252; Palestine Brigade's superiority at Megiddo, 254; at Sharon, 260
Royal Flying Corps, at Jifjaffa, 33; its inferiority in 1916, 59
Royal Military College, Duntroon, Australia, 14
Royal Navy lands stores, Nahr Sukereir, 188; moves Faisal's Arabs to Akaba, 218; two destroyers of shell coastal road, Sharon, 261
Royston, Brig.-Gen. John Robinson (1860–1942), 58; at Magdhaba, 82; at Rafah, 87; at Gaza I, 97; goes on leave to South Africa, 127
Rozkalla Salim, a Beduin, 26
Russell, Col Reginald Edmund Maghlin (1879–1950), 117
Ryrie, Maj.-Gen. Hon. Sir Granville de Laune (1865–1937), 9, 58; at Gaza I, 96; at Gaza III, 145

Said Abd el Kader, Emir, 328
Sanders, von – see Liman von Sanders
Sangster, Maj.-Gen. Patrick Barclay (1872–1951), 306
Sayyid Ahmed (the 'Grand Senussi'), 28
Scarlett, Gen. Sir James Yorke (1799–1871), 171
Scott, Tpr W., 4th ALH Regiment, 156
Senussites, the 18, 17, 91
Shaitan Singh, Risaldar, Jodhpore Lancers, 239
'Shea's Force', 218
Shea, Gen. Sir Stuart Mackenzie (1869–1966), 120; orders yeomanry to charge at Huj, 175, 219; at Megiddo, 260
Smith, Brig.-Gen. Clement Leslie, VC (*died* 1927), 18; i/c Mobile Column, Romani, 62
Smith, Tpr Vic, 4th ALH Regiment, at Beersheba, 155
Smuts, F-M Rt Hon. Jan Christian (1870–1950), 119, 215, 220
Snow, Lt G., 7th ALH Regiment, 99
Souter, Col Hugh Maurice Wellesley (1873–1941), Dorset Yeomanry, 30
Stansfield, Maj. William (1877–1946), 58
steel helmets, issued and later partially withdrawn, 224

Stuart, James Ewell Brown, *known as* Jeb (1833–1864), 287
swords given to two Australian brigades, 223

Tacitus, Corneliu (55?–*after* 117), 74
Tantia Topi (1819?–1859), 287
Taylor, F. Lindsay, Tpr, 4th ALH Regiment, at Beersheba, 154
Teichman, Capt. O., Medical Officer, Worcestershire Yeomanry, 88
Thakur Dalpat Singh, Lt-Col, Jodhpore Lancers, 299
Tiller, Maj. (German), 100
Timperley, Maj. L.C., 10th ALH Regiment, 184
Todd, Lt-Col T.J., 10th ALH Regiment, deals with massive diseases of Turks, Damascus, 334
transport of sick and wounded, 54
Trew, Brig. Gen. Edward Fynmore (1879–1935), 222

TURKISH ARMY
cavalry tactics of, 24
strength of, 21
IV Army, 245, 312; remnants of withdraw through Deraa, 316
VII Army, 206, 245; at Gaza III, 136; at Sharon, 260; escape route closed, Megiddo, 293
VIII Army, 185, 199, 245; at Sharon, 260; escape route closed, Megiddo, 293
Asia Corps, 303
II Corps, 312
XXII Corps, at Old Aqir, 199
3rd Cavalry Division, 159
26th Infantry Division, 171
27th Infantry Division, virtually destroyed at Gaza III, 161
54th Infantry Division, 171
7th Regiment, Caucasus Cavalry Brigade, 237
9th Regiment, Caucasus Cavalry Brigade, 237
11th Regiment, Caucasus Cavalry Brigade, 237
1st Cavalry Regiment, captured, 288
46th Infantry Regiment, standard captured at Khan Ayash, 330
146th Infantry Regiment, 331

Turner, Col Charles Edward (1876–1961), Gloucestershire Yeomanry, at Romani, 66
Turner, Lt R.H.F., 2nd Lancers, 280
Tweedsmuir, Sir John Buchan, 1st Baron (1875–1940), 21

Valintine, Capt. R., Warwickshire Yeomanry, at Huj, 177
Vaughan, Brig. Edward William Drummond (1894–1953), 280; at Irbid, 319

Wade, Abdul, 'camel king of Central Australia', 19
Waite, Lt Frederick Mitchell, 5th ALH Regiment, at Gaza I, 98
'Watson's Force', at Megiddo, 249
Wavell, F-M Archibald Percival, 1st Earl (1883–1950), 21, 23, 75, 78, 214, 250; takes Allenby's Gaza III report home, 187
Western Frontier Force, 28
Westminster, Hugh Richard Arthur Grosvenor, 2nd Duke of (1879–1953), 28
White, Lt R.U., Gloucestershire Yeomanry, 272
Whitworth, Brig. Dysart Edward (1890–1974), 281; at Irbid, 318
Wigan, Brig.-Gen. John Tyson (1877–1952), i/c 4 ALH Bde at Gaza III, 143; at Megiddo, 276
Wiggin, Brig.-Gen. Edgar Askin (1867–1939), at Katia and Oghratina, 41; at Huj, 176
William II (the Kaiser) (1859–1941), 210
Williams, Lt-Col Henry John (1870–1935), Worcestershire Yeomanry, at Huj, 176
Williams-Thomas, Lt-Col Frank S. (1879–1942), Worcestershire Yeomanry, at Katia and Oghratina, 41
Willisallen, Maj. T.L., 7th ALH Regiment, at Huleikat, 185
Wilson, Brig.-Gen. Lachlan Chisholm (1871–1947), introduces 'spear-point' water pumps, 48; i/c 3 ALH Bde, 127; at Gaza III, 148; at Es Salt raid, Apr., 1918, 226; at Megiddo, 289
Wilson, F-M Sir Henry Hughes, 1st Bt (1864–1922), 2, 214; urges Allenby to make cavalry raid on Aleppo, 332
Wilson, Lt R.H., Gloucestershire Yeomanry, 111
Wingfield-Digby, Maj. F.J.B., Dorset Yeomanry, at El Mughar, 197
wireless, used for first time to direct artillery from the air, 88
Wiseman, Cpl, Gloucestershire Yeomanry, 267
Woelworth, Capt., 277

Yorke, Brig.-Gen. Ralph Maximilian (1874–1951), at Katia, 42; at Romani, 68
Young, Capt. Hubert, 225
Ypres, Earl of – see French

Zalim Singh, Ressaisar, Poona Horse, 240